Health

OXFORD PHILOSOPHICAL CONCEPTS

OXFORD PHILOSOPHICAL CONCEPTS

Christia Mercer, Columbia University

Series Editor

PUBLISHED IN THE OXFORD PHILOSOPHICAL CONCEPTS SERIES

Efficient Causation
Edited by Tad Schmaltz

Sympathy
Edited by Eric Schliesser

The Faculties
Edited by Dominik Perler

Memory
Edited by Dmitri Nikulin

Moral Motivation
Edited by Iakovos Vasiliou

Eternity
Edited by Yitzhak Melamed

Self-Knowledge
Edited by Ursula Renz

Embodiment
Edited by Justin E. H. Smith

Dignity
Edited by Remy Debes

Animals
Edited by G. Fay Edwards and Peter Adamson

Pleasure
Edited by Lisa Shapiro

FORTHCOMING IN THE OXFORD PHILOSOPHICAL CONCEPTS SERIES

Health
Edited by Peter Adamson

Persons
Edited by Antonia LoLordo

Evil
Edited by Andrew Chignell

Space
Edited by Andrew Janiak

Teleology
Edited by Jeffrey K. McDonough

Love
Edited by Ryan Hanley

Human
Edited by Karolina Hubner

OXFORD PHILOSOPHICAL CONCEPTS

Health

A HISTORY

Edited by Peter Adamson

OXFORD
UNIVERSITY PRESS

OXFORD
UNIVERSITY PRESS

Oxford University Press is a department of the University of Oxford. It furthers the University's objective of excellence in research, scholarship, and education by publishing worldwide. Oxford is a registered trade mark of Oxford University Press in the UK and certain other countries.

Published in the United States of America by Oxford University Press
198 Madison Avenue, New York, NY 10016, United States of America.

Library of Congress Cataloging-in-Publication Data
Names: Adamson, Peter, 1972– editor.
Title: Health : a history / edited by Peter Adamson.
Description: New York : Oxford University Press, 2019. |
Series: Oxford philosophical concepts | Includes bibliographical references and index.
Identifiers: LCCN 2018016069 (print) | LCCN 2018022021 (ebook) |
ISBN 9780190921293 (online content) | ISBN 9780199916436 (updf) |
ISBN 9780190921286 (epub) | ISBN 9780199916443 (pbk. : alk. paper) |
ISBN 9780199916429 (cloth : alk. paper)
Subjects: LCSH: Medicine—Philosophy.
Classification: LCC R723 (ebook) | LCC R723.H394 2018 (print) | DDC610.1—dc23
LC record available at https://lccn.loc.gov/2018016069

1 3 5 7 9 8 6 4 2

Paperback printed by Sheridan Books, Inc., United States of America
Hardback printed by Bridgeport National Bindery, Inc., United States of America

Contents

Contributors

GLENN ADAMSON is the twin brother of Peter Adamson. He is also a curator, writer, and historian who works across the fields of design, craft, and contemporary art. Currently Senior Scholar at the Yale Center for British Art, Adamson has been Director of the Museum of Arts and Design, New York; Head of Research at the V&A; and Curator at the Chipstone Foundation in Milwaukee. His publications include *Fewer Better Things: The Importance of Objects Today* (2018); *Art in the Making* (2016, co-authored with Julia Bryan Wilson); *Invention of Craft* (2013); *Postmodernism: Style and Subversion* (2011); *The Craft Reader* (2010); and *Thinking Through Craft* (2007).

PETER ADAMSON is the twin brother of Glenn Adamson. He is also Professor of Late Ancient and Arabic Philosophy at the Ludwig Maximilians Universität in Munich. With G. Fay Edwards, he is the co-editor of another volume in the Oxford Philosophical Concepts Series, entitled *Animals: The History of a Concept* (2018), and the author of the book series *A History of Philosophy Without Any Gaps*, published by Oxford University Press. Two volumes collecting his papers on late ancient philosophy and philosophy in the Islamic world appeared recently with the Variorum Series published by Ashgate.

JAMES ALLEN is Professor of Philosophy at the University of Toronto and was formerly Professor of Philosophy at the University of Pittsburgh. He is the author of *Inference from Signs: Ancient Debates About the Nature of Evidence* (2001) and the co-editor of *Essays in Memory of Michael Frede* (special issue of *Oxford Studies in Ancient Philosophy*, 2011). He has published articles about ancient skepticism, Stoicism, Epicureanism, Plato, Aristotle and the relations between ancient medicine and philosophy spanning topics in ethics, logic, and natural philosophy.

TOM BROMAN is Emeritus Professor of History of Science and History of Medicine at the University of Wisconsin, Madison. He is the author of *The Transformation of German Academic Medicine, 1750–1820* (1996) and co-editor of *Science and Civil Society* (2002). He has authored articles that analyze the constitution of scientific and medical expertise, the early history of scientific journals, and the history of the public sphere in the eighteenth century. He is currently writing a comprehensive survey of science in the Enlightenment.

GUIDO GIGLIONI is Associate Professor of History of Philosophy at the University of Macerata, Italy. His research is focused on the interplay of life and imagination in the early modern period, on which he has written and edited several contributions. He has published two books, on Jan Baptiste van Helmont (2000) and Francis Bacon (2011).

ANITA GUERRINI is Horning Professor in the Humanities and Professor of History at Oregon State University. She has published widely on the history of animals, medicine, food, and the environment. Her books include *Experimenting with Humans and Animals: From Galen to Animal Rights* (2003) and *The Courtiers' Anatomists: Animals and Humans in Louis XIV's Paris* (2015). Current research projects concern skeletons as scientific and historical objects (for which she recently won a grant from the National Science Foundation) and the role of history in present-day ecological restoration. She blogs at anitaguerrini.com/anatomia-animalia.

JIM HOPKINS is Reader Emeritus in Philosophy at King's College and Visiting Professor in the Psychoanalysis Unit of the Research Department of Clinical and Health Psychology at University College London. His main work has been on psychoanalysis, consciousness, Wittgenstein, and interpretation.

LUDMILLA JORDANOVA is Professor of History and Visual Culture and Director of the Centre for Visual Arts and Culture at Durham University. Her books include *Lamarck* (1985), *Sexual Visions* (1989), *History in Practice* (2000, 2006, with the third edition near completion), *The Look of the Past* (2012), and *Physicians and their Images* (2018). Her research interests include portraiture and the cultures of science and medicine since 1600.

HELEN KING is Professor Emerita in Classical Studies, the Open University, and a Visiting Professor in Politics and International Studies, University of Warwick. She has published widely on gender and sexuality, myth, and ancient gynecology and obstetrics, and their reception into the early modern period and beyond. Her books include *The One-Sex Body on Trial: The Classical and Early Modern Evidence* (2013) and, co-edited with M. Horstmanshoff and C. Zittel, *Blood, Sweat and Tears: The Changing Concepts of Physiology from Antiquity into Early Modern Europe* (2012).

ELSELIJN KINGMA is Associate Professor in Philosophy at the University of Southampton and Socrates Professor in Philosophy and Technology in the Humanist Tradition at the University of Eindhoven in the Netherlands. She leads a major research project on the metaphysics of pregnancy and has published numerous articles on the philosophy of medicine and biology.

GIDEON MANNING is a Visiting Scholar at the Claremont Graduate University, having previously taught at the University of Pittsburgh, the College of William and Mary, and the California Institute of Technology. His recent publications include "Descartes and the Bologna Affair," "Descartes' Metaphysical Biology," and the edited volume *Professors, Physicians and Practices in the History of Medicine: Essays in Honor of Nancy Siraisi* (2017). He is co-editor with Lisa Jardine of the forthcoming volume *Testimonies: States of Body and States of Mind in the Early Modern Period*.

RICHARD SCOTT NOKES is an Associate Professor of Medieval Literature at Troy University and founder of Witan Publishing. He is co-editor of *Global Perspectives on Medieval English Literature, Language, and Culture* and *Conflict in Southern Writing* (2007), and is the academic editor and publisher of multiple volumes of medieval scholarship on subjects as varied as medieval combat, manuscript production, and linguistics.

PETER E. PORMANN is Professor of Classics and Graeco-Arabic Studies at the University of Manchester. Recent publications include special double issues *The Arabic Commentaries on the Hippocratic "Aphorisms"* co-edited with Kamran I. Karimullah (2017), and *Medical Traditions* (co-edited with Leigh Chipman and Miri Schefer-Mossensohn, 2017–2018); and edited books: *La construction de la médecine arabe médiévale* (with Pauline Koetschet, 2016); *Medicine and Philosophy in the Islamic World* (with Peter Adamson; 2017); *1001*

Cures: Contributions in Medicine and Healthcare from Muslim Civilisation (2017); and the *Cambridge Companion to Hippocrates* (2018).

MICHAEL STANLEY-BAKER is an Assistant Professor in the History Program at Nanyang Technological University, Singapore, and formerly a postdoctoral researcher at the Max Planck Institute for the History of Science, Berlin. With Vivienne Lo, he is the co-editor of the *Routledge Handbook of Chinese Medicine* (2017), and also co-editor with Pierce Salguero of a volume provisionally titled *Situating Religion and Medicine Across Asia*. He is project lead for the *Drugs Across Asia* digital platform, in collaboration with the Max Planck Institute, National Taiwan University and Dharma Drum Institute for Liberal Arts, and serves as the Vice-President of the International Association for the Study of Traditional Asian Medicine. He publishes on Daoism and Chinese medicine in early imperial China, on practice theory, close reading, and Digital Humanities.

Health

OXFORD PHILOSOPHICAL CONCEPTS

Introduction

Peter Adamson

A striking feature of the history of philosophy, stretching from the ancient period through medieval times and up to the modern period, is the fact that so many great philosophers had interests in medicine or were practicing doctors. Ancient and medieval examples include Empedocles, Aristotle, Sextus Empiricus, Philoponus, Avicenna, and Maimonides—not to forget the intertwining of medical and philosophical ideas in Indian Ayurveda and the Chinese traditions studied in this volume by Stanley-Baker.[1] As for the modern period, a few striking examples include Descartes, who commented to the Marquess

1 For Ayurveda see the texts gathered in D. Wujastyk, *The Roots of Āyurveda* (New Delhi: Penguin, 1998), and for connections to philosophy G. J. Larson, "Āyurveda and the Hindu Philosophical Systems," in T. P. Kasulis et al. (eds), *Self as Body in Asian Theory and Practice* (Albany: SUNY Press, 1993), 103–21; Dagmar Wujastyk, *Well-mannered Medicine: Medical Ethics and Etiquette in Classical Ayurveda* (Oxford: Oxford University Press, 2012); Dominik Wujastyk, "Medicine and *Dharma*," *Journal of Indian Philosophy* 32 (2004), 831–42; K. Zysk, *Asceticism and Healing in Ancient India: Medicine in the Buddhist Monastery* (New York: Oxford University Press, 1991).

of Newcastle, "the preservation of health has always been the principal end of my studies"; Locke, who served as a personal physician and worked closely with Sydenham; Boyle, who wrote about the methods of Galen; Berkeley, who worked on tar water; and Leibniz, who claimed repeatedly that he was a Hippocratic. Conversely, prominent figures who were primarily doctors had interests in philosophy. In the ancient and medieval periods one could name Hippocrates (or at least some authors of the Hippocratic Corpus), Galen, and Abū Bakr al-Rāzī, whose translated works made him a medical authority for readers of Latin under the name "Rhazes." For the modern period, examples would include English physicians like Harvey and Charleton, and early Cartesians like Regius and de la Forge. It is thus unsurprising that philosophical ideas often affected medical ones and vice versa.

At the heart of this fruitful interchange between medicine and philosophy is the concept of health. Needless to say, health is a medical concept, indeed, the concept that gives medicine both its definition and its raison d'être. Yet it is also a philosophical concept, and an intriguing one at that. Health is unusual in seeming to straddle the divide between descriptive, empirical concepts and normative, value-laden concepts. When your doctor gives you a clean bill of health, you probably think of her as just giving you an objective scientific description of the state of your body. Yet it is almost irresistible to think of health as a way we "ought" to be, that there is some standard the unhealthy person is failing to meet. After all nearly everyone wants to be healthy, and we go to great lengths to avoid or recover from illness and injury.

The interrelation between these two ways of thinking about health is something of a leitmotif in the papers collected here. We will see thinkers of different cultures offering a range of scientific accounts of what it means to be healthy, from the idea of circulating *qi* in ancient China, to the humoral theory of Galen and the long-lived tradition of Galenic medicine he inspired, to the new physiological and pyschotherapeutic theories that emerged in the nineteenth century.

At the same time, we will see how normative implications were routinely drawn from these descriptive theories, as when the balance of humors was compared to (and seen as a causal basis for) a good balance in the soul by Galen and his heirs in the Islamic world. As shown in the chapters by Giglioni and Manning, robust connections between "sound mind" and "sound body" were still fundamental to Renaissance philosophy and lived on as a "medicine of the mind" in early modernity.

The contrast between normative and descriptive accounts of health is however addressed most explicitly by Kingma in her concluding chapter. As she shows in her examination of modern-day attempts to define health, both approaches face serious objections. The descriptivist must meet the challenge of giving objective criteria for measuring health, but the criteria suggested by philosophers of medicine would seem vulnerable to counter-examples. For instance defining health as adherence to a statistical average would mean saying that high-performance athletes are unhealthy. The normative account of health has its own problems though, not least because of its implied prescriptivism. The doctors, scientists, and philosophers who determine what it means to be healthy would effectively be telling us how to live our lives. Kingma's paper opens with a notorious example, namely the fact that homosexuality was until recently considered a disease. Kingma also reminds us of debates over how we should conceive of people who have "disabilities." Is it really useful to conceive of, say, blindness as a lack of "health"?

Such difficulties notwithstanding, most of the figures covered in this volume would have taken it for granted that health is a concept with both a normative and a descriptive dimension. In this sense health is like nature, another scientific concept that has often been taken to have normative standing. Again homosexuality would be a case in point, given that its supposed sinfulness has often been linked to its purportedly being "unnatural." The kinship between health and nature is no coincidence. As we shall see, many thinkers identified the healthy state of the body or soul with a natural state. Another powerful intuition that encourages us to think of health as a value-laden concept emerges from the aforementioned fact that nearly everyone wants to be healthy. In this respect, health is unusual among life goals. It is easy to imagine someone preferring not to be wealthy or powerful. Nor is it difficult to think of people who have no wish to be virtuous, given that the tyrannical souls so vividly described by Plato still thrive on the contemporary political scene. By contrast, having a preference for illness over health would in itself probably be taken as a sign of mental illness. Beginning in the ancient tradition, the philosophical schools sought to explain the nearly undeniable goodness of health, with some schools finding the challenge easier than others. Aristotelians gave health as a key example of the way normativity applies to physical states and saw health as illustrating their teleological approach to nature. The Stoics agreed about the value of nature but saw human normativity as being grounded in virtue. So they had to devise a special category to handle health and a few other goals: these are "preferred indifferents," which are rational to pursue even though their absence does not preclude happiness.

Even though health is most obviously a concept applied to bodies, we already noted that in the history of philosophy it has often been applied to psychological states too. We still speak today of "mental illness" of course, but in the ancient, medieval, and early modern periods even ethical vice was seen as a sort of disease. Many authors have envisioned a kind of medicine that applies to souls rather than, or in addition to, bodies. One might be tempted to take this as a mere metaphor. On this reading, talk of "health" in the case of the soul would simply be a way of referring to the soul's best state. But authors also detected an intimate causal relationship between bodily health and psychological health, such that a temperate body is a necessary (even sufficient?) condition for a temperate character. Some thinkers also exploited the parallel between physical and psychological health to suggest that bodily treatments have analogues at the psychological level. Because of these powerful links between the health of body and the health of soul, the skilled doctor can manipulate both. In both ancient Chinese texts and in Galen, it is thus said that a good diet leads to physical well-being and wisdom or virtuous character.

Because the history of philosophical reflection on health is bound up with the development of medicine—and because, as noted at the outset, so many authors were both doctors and philosophers—the contributors to this volume include both historians of philosophy and historians of medicine. In examining the history of health, they touch on many philosophical questions raised by the discipline of medicine. To take just one example, one reason to discount medicine's claims to be a "science" in the full and proper sense was that effective medical treatment was seen as highly responsive to the particularities of each patient. A good doctor would tailor the cure to the victim of the illness, not just the illness itself, and the same for prescriptions of prophylactic regimen. Indeed many doctors prided themselves on their patient-sensitive expertise and emphasized the need for long practice to achieve it. But the same feature of medicine could be used to suggest that medicine is more like a knack than a proper science. A related notion, which again raises the question of how medicine relates to ethics, is that doctors had themselves to be virtuous in order to perform successful diagnosis and to effect a cure.

The concerns of ancient and medieval authors lived on into the early modern period, which is no wonder given that some of these authors (especially Hippocrates, Galen, and Avicenna) remained key sources into the seventeenth century and even later. As late as the nineteenth century, Galen, Hippocrates, and even Paul of Aegina were still printed, at least in part with medical ends in mind and not simply for antiquarian interest. And as Hopkins reminds us in his study of Freud in this volume, the Freudian approach to the mind is apt to remind us of the Platonic tripartite soul. Yet significant changes in philosophy did bring changes in medicine, and vice versa. As we move into the Renaissance and the early modern period, we find that the emerging naturalistic, even mechanistic, world picture had significant implications for medicine. To the extent that early modern thinkers moved away from a teleological or normative understanding of nature and toward a physical conception in which bodies simply interacted according to laws of nature, their conception of health was bound to change. Here we see authors grappling with the problem of defining health in the absence of a teleological account of the nature of the human.

Nowadays, when it would be a rare person who could claim specialism in both medicine and philosophy, the old philosophical issues regarding health remain with us. Just consider the movement toward emphasizing a holistic and patient-specific understanding of health or disputes about how health should be balanced with other human goods. Debates about the quantity of life as opposed to the quality of life cannot be resolved simply by an account of what health is. It is our hope that a better understanding of health as a concept in the history of philosophy will help shed some light on such issues. At the same time, our ambition has been is to survey attitudes toward health across an unprecedentedly broad chronological range, which we hope will be of interest to both historians of philosophy and of medicine, and, of course, to the general reader.[2]

2 My thanks to Gideon Manning for help with this introduction, to Hanif Amin for extensive assistance in getting the volume ready for press, and to Michael Lessman for preparing the index. I would also like to thank King's College London for hosting an initial workshop on the topic.

CHAPTER ONE

Health and Philosophy in Pre- and Early Imperial China

Michael Stanley-Baker

Freedom from disease, physiological signs of well-being, and avoidance of death appear as consistent themes in many writings from pre-imperial and imperial China, but there were no direct analogues to a conceptual category like the English term "health." These themes do not appear under a universally consistent term or bounded theoretical formulation. Instead they appear as common technical terms or practical foci across different literary genres. Chinese writers refer to material forces like breath (*qi* 氣), the force of life (*sheng* 生), and inner nature (*xing* 性), which are cultivated or nourished (*yang* 養) in the body, to lengthen one's life (*ming* 命). Engagement with health was as a quantitative object that could be stored in the body, as a processual goal rather than a conceptual object. While the term "nourishing life" (*yangsheng* 養生) only came to refer to a common, recognizable set of practices after the first century CE, the practices and goals themselves were understood to be coherent in sources as early as 400 BCE.

In this chapter I argue that three major periods in pre- and early imperial China produced dominant modes of writing on bodily ideals that have continued until the present day. These include philosophical works, self-cultivation manuals, canonical medical writings, transcendence writings, and formal Daoist texts. While different authors and communities adopted a variety of positions regarding the political, social, and moral values of robust bodily function, these negotiations took place through a roughly common and emergently coherent set of terms, concepts, and assumptions about the nature of the body, its functions, and relatedness to the cosmos. The three periods include two periods of political disunity—the Warring States 戰國 (475–221 BCE) and the early period of the Northern and Southern dynasties 南北朝 (220–589 CE)—and the first founding and expansion of the unified Chinese empire under a relatively stable, centralized, and highly bureaucratized government during the Qin-Han 秦漢 period (221 BCE–200 CE). The discourses and practices that emerged during this time set the tone for the rest of imperial Chinese history, which built on and grew out of these earlier currents.[1]

Qi in Warring States Philosophical Works

Although writings on health and philosophy differed at many points, there is one central topic that they shared in common—the material force called *qi*, and the best way to cultivate it. The Warring States period saw the most wide-ranging and varied debates on philosophy,

---

1 I would like to gratefully acknowledge financial support during the writing of this paper from the Max Planck Institute for the History of Science in Berlin, and the KFG for Multiple Secularities at Leipzig University. Thanks are also due for comments on earlier versions of this paper are due to Vivienne Lo, Pierce Salguero, and Dolly Yang. I found particularly useful for structuring this paper, L. Raphals, "Chinese Philosophy and Chinese Medicine," in E. N. Zalta (ed.), *The Stanford Encyclopedia of Philosophy*, 2015, http://plato.stanford.edu/entries/chinese-phil-medicine/#Rel. An in-depth study worth consulting on the transition of embodiment theory through the period covered is O. Tavor, "Embodying the Way: Bio-Spiritual Practices and Ritual Theories in Early and Medieval China" (PhD diss., University of Pennsylvania, 2012).

ethics, and statecraft, which was referred to as the Hundred Schools (*Baijia* 百家). During this time, the various fiefdoms and principalities of the Zhou 周 dynasty became consolidated into seven major states, each embroiled in heightened rivalry and factionalism in their bids for power.[2] Nobles of the various duchies and marquisates would host at their courts "masters" or "nobles" (*zi* 子) of warfare, economics, statecraft, as well as "technical masters" (*fangshi* 方士) of divination, medicine, and dietary culture. It is in these contexts that much of the philosophical and technical literature that survives was produced.[3] This culture is evidenced in a number of texts excavated from numerous sites in central China, mostly in Hubei, dating from the fourth to the second century BCE.[4] These texts are often mixed or appear in collections containing manuals of different types. Practices advocated in these texts include breathing exercises, diet, stretching, massage, sexual cultivation, numerological divination, drug therapy, and hemerology.

---

2 M. E. Lewis, "Warring States Political History," in M. Loewe and E. L. Shaughnessy (eds), *The Cambridge History of Ancient China: From the Origins of Civilization to 221 B.C.* (Cambridge: Cambridge University Press, 1999).

3 Variously translated as "recipe masters," "masters of esoterica," "masters of methods," *fangshi* have been the subject of biographical studies and broad histories of technology. K. J. DeWoskin, *Doctors, Diviners, and Magicians of Ancient China: Biographies of Fang-shih* (New York: Columbia University Press, 1983); G. Lloyd and N. Sivin, *The Way and the Word: Science and Medicine in Early China and Greece* (New Haven, CT: Yale University Press, 2002). Sivin argues that the pejorative connotations of *fangshi* as subaltern to elite literati meant that no actors self-identified with the moniker, and therefore it had no coherence or social force and did not refer to a discrete, identifiable, body of actors. N. Sivin, "Taoism and Science," *Medicine, Philosophy and Religion in Ancient China* (Aldershot: Variorum, 1995), 303–30. Also see M. E. Lewis, *Writing and Authority in Early China* (Albany: State University of New York Press, 1999), 53–98. The definitive studies on the technical arts (*fangshu* 方術) are Li Ling 李零, *Zhongguo fang shu xu kao* 中国方术续考 (Beijing: Zhonghua shuju, 2000); Li Ling 李零, *Zhongguo fang shu kao (xiuding ben)* 中国方术正考 (修訂本) (Beijing: Dongfang chubanshe, 2001).

4 D. Harper, "The Textual Form of Knowledge: Occult Miscellanies in Ancient and Medieval Chinese Manuscripts, Fourth Century B.C. to Tenth Century A.D," in F. Bretelle-Establet (ed.), *Looking at It from Asia: The Processes That Shaped the Sources of History of Science* (Dordrecht: Springer, 2010); D. Harper, *Early Chinese Medical Literature: The Mawangdui Medical Manuscripts* (London: Kegan Paul, 1998); V. Lo 羅維前 and Li Jianmin 李建民, "Manuscripts, Received Texts, and the Healing Arts," in M. Nylan and M. Loewe (ed.), *China's Early Empires: A Re-appraisal* (Cambridge: Cambridge University Press, 2010).

Many philosophical works that survived from this early period were catalogued by the libraries and histories of later dynastic empires as "Master [Literature]" (*zi* 子).[5] "Masters" was not an exclusively literary category, denoting discursive or argumentative style such as "philosophy" or "rhetoric." It was also a social category that referred to the literature of authoritative experts.[6]

Many works in the master literature refer to embodied practices and means to produce beneficial physiological responses or an ideal relation among the body, the self, and the cosmos. They refer to breathing and meditation exercises designed to cultivate *qi* 氣—the vital breath, a material considered to permeate and give substance and structure to all existence. Earlier on in the Warring States period, *qi* referred to weather and climatic conditions, *qi* bearing some similarities to the Greek *pneuma* and Sanskrit *prana* (wind, air, breath) as the lively fluid substance that "vitalizes the body, in particular as the breath, and which circulates outside us as the air." By the third century BCE, it became "adapted to cosmology as the universal fluid, active as Yang and passive as Yin, out of which all things condense and into which they dissolve."[7]

---

5 Dagmar Schafer writes of the *zi* category in late imperial China as "the main Chinese bibliographic category of matters where the Chinese intellectual world assembled exceptional thinkers who had a comprehensive, sometimes encyclopaedic approach to fields of knowledge . . . aimed at revealing essential truths about the relation between heaven, earth and men." D. Schäfer, *The Crafting of the 10,000 Things: Knowledge and Technology in Seventeenth-Century China* (Chicago: University of Chicago Press, 2011). For a brief overview of Chinese historical bibliography, see E. P. Wilkinson, *Chinese History: a New Manual* (Cambridge: Harvard University Asia Center, 2012), 936–40.

6 In the Han dynasties (206–220 CE), medical literature was gathered under a distinct section titled "Methods and Techniques" (*fangji* 方技), but by the Sui dynasty (581–618) it was included as a form of masters literature, along with mathematics, astronomy, military tactics, and others. *Han shu* 漢書, by Ban Gu 班固, Yan Shigu 顏師古, and Ban Zhao 班昭. (Beijing: Zhonghua shuju, 1962), 10/30.1776–78, and *Suishu* 隋書, by Wei Zheng 魏徵. *Xin jiao ben Suishu fu suoyin* 新校本隋書附索引, ed. 楊家駱 Yang Jialuo. (Taibei: Ding wen shuju, 1980), 34/29.1040–51.

7 A. C. Graham, *Disputers of the Tao: Philosophical Argument in Ancient China* (La Salle, IL: Open Court, 1989), 101. On the observation of *qi* in the human body in medical and *yangsheng* literature, and the production of a regularized theory of its movements, see V. Lo, "Tracking the Pain: *Jue* and the Formation of a Theory of Circulating *Qi* through the Channels," *Sudhoffs Archiv* 83/2 (1999), 191–211. A set of papers on how *qi* has been differently conceptualized and applied throughout

Qi came to be considered to have a direct relationship with consciousness and the mind, so that its cultivation was considered to develop not only physiological well-being but also perspicacity and moral force. Through embodied meditations on *qi*, elites participated in a higher moral and philosophical order, which Mark Csikszentmihalyi describes as a culture of "embodied virtue."[8] The Confucian moralist Mengzi 孟子 (372–289 BCE), for example, famously describes a direct relationship between his ability to function as a moral actor and his ability to cultivate "radiantly bright *qi*" (*haoran zhi qi* 浩然之氣), which gives him the resources to engage with the difficult ethical issues of his time.[9] The attainment of this radiance was considered not only a visible sign of the attainment of sagehood but also partook in the luminosity of heaven.

Inscription on Circulating *Qi* (*Xingqi ming* 行氣銘)

An early example of the practice of circulating *qi* within the body is the *Xingqi ming*, a short, thirty-six character inscription on the twelve sides of a jade knob dated to the Warring States period (453–221 BCE).[10] While the ornamental value of the knob, which probably

---

Chinese history, in medicine, philosophy, religious rituals, and self-cultivation, is Onozawa Seiichi 小野沢精一, Fukunaga Mitsuji 福永光司, and Yamanoi Yū 山井湧 (eds), *Ki no shisō: Chūgoku ni okeru shizenkan to ningenkan no tenkai* 気の思想: 中国における自然観と人間観の展開, (Tookyo: Tōkyō daigaku shuppankai, 1978). For a recent selection of papers on the topic, Sakade Yoshinobu, *Taoism, Medicine and Qi in China and Japan* (Osaka: Kansai University Press, 2007).

8 M. Csikszentmihalyi, *Material Virtue* (Leiden: Brill, 2004).

9 Csikszentmihalyi, *Material Virtue*, 150–57. Csikszentmihalyi argues cogently for the translation as "radiant" in contrast to the "flood-like *qi*" more commonly found in secondary literature, on the basis of many contemporary allusions to luminescence, the heavens, sun and stars, in ritual and other texts.

10 Earlier in the possession of the collector Li Mugong 李木公 from Hefei, it is now stored in the Tianjin Museum. In dating the piece to the middle sixth century, J. Needham and Ling Wang, in *Science and Civilisation in China*, vol. 2 (Cambridge: Cambridge University Press, 1956), 143, follow H. Wilhelm, "Eine Chou-Inschrift über Atemtechnik," *Monumenta Serica* 13 (1948). An argument dating it to 380 is Guo Moruo 郭沫若, "Gudai wenzi zhi bianzheng de fazhan 古代文字辯證的發展," *Kaogu xuebao* 考古學報 29 (1972). Chen Banghuai dates it to the late Warring States period. Chen Banghuai 陳邦懷, "Zhanguo xingqi ming kaoshi 战国行气铭考释," *Guwenzi yanjiu* 古文字研究 7 (1982), (cited in Li Ling 李零, *Fang shu zheng kao*, 344, n.3).

adorned the end of a staff, likely outweighed its role as a communicative device, it is one of the earliest testimonies to the high regard with which breath cultivation was held. The inscription describes breathing deep into the body, holding the breath there, and building it up until it rises upward again, and thereupon cycling it downward once again.

> To circulate the breath (*xing qi*), breathe deeply so there is great volume. When the volume is great, the breath will expand. When it expands, it will move downwards. When it has reached the lower level, fix it in place. When it is in place, hold it steady. Once it is steady, it will become like a sprouting plant. Once it sprouts, it will grow. As it grows, it will retrace its path. When retracing its path, it will reach the Heaven area. The Heaven impulse forces its way downward. Whoever acts accordingly will live, whoever acts contrariwise will die.[11]

Texts from the period argue for a relationship between moral cultivation and its physiological effects. Where the texts disagree, however, is on whether these meditations should or could be used for nourishing the body's well-being for its own sake. Some argue that the physiological benefits of *qi* cultivation are demonstrable, while others argue that there is a wide gap between the meditations they advocate and the nourishing life exercises promoted by technical experts.

Guanzi 管子

For example, the *Guanzi* 管子 places an emphasis on the combined physiological and spiritual benefits of *qi* cultivation. Although traditionally attributed to Guan Zhong 管仲 (d. 645 BCE), this work

11 There has been considerable debate over the orthography and the precise meaning of the text among the authors in the previous note. The rendering is from W. A. Rickett, *Guanzi: Political, Economic, and Philosophical Essays from Early China*: A *Study and Translation*, vol. 2 (Princeton, NJ: Princeton University Press, 1998), 19, with minor modifications.

is now thought to be most mainly composed in the fourth century CE.[12] It is predominantly concerned with government and the art of rulership, and it emphasizes the virtue of the ruler as a model for his followers and the use of ritual combined with a clear and strict legal code. It appears to have circulated within or been produced by specialists at Jixia 稷下, an academy of sorts in the city of Linzi 臨淄 (in modern-day Shandong province), a community that also hosted authors and editors of other important received works such as the *Zhuangzi*, *Mengzi*, and *Daodejing*.[13]

In addition to a primary focus on statecraft, the *Guanzi* contains the earliest received texts dealing with the cultivation of *qi*. Four chapters discuss metaphysical meditations on the Way (*Dao* 道), a cosmic guiding principle within all things. In contrast to the other contemporary philosophical works that survive, the *Guanzi* placed a strong emphasis on the physiological benefits of internal cultivation. The *Guanzi* is considered to be informed by the communities of technical experts on longevity, whose expertise in later centuries came to be referred to as "nourishing life" (*yangsheng*).

In addition to arguing that meditation grants insight into universal processes and increases the ability to avoid harm and excessive toil, one chapter titled "Inner Training" (*Neiye* 內業) advocates the positive benefits of abundant, flowing *qi* and vital essence (*jing* 精).[14]

> For those who preserve and naturally generate vital essence,
> On the outside a calmness will flourish.

---

12 Although the stable version of the text was compiled as late as 26 BCE by the bibliographer of the Han court, Liu Xiang 劉向 (77–6 BCE), the earliest sections of these chapters are generally considered to be as old as the oldest parts of the *Daodejing* or older. H. D. Roth, *Original Tao: Inward Training and the Foundations of Taoist Mysticism* (New York: Columbia University Press, 1999); Rickett, *Guanzi*; W. A. Rickett, "Kuan tzu," in M. Loewe (ed.), *Early Chinese Texts: A Bibliographical Guide*, Early China Special Monograph Series (Berkeley: Society for the Study of Early China, 1993).

13 Rickett, *Guanzi*, 16–17.

14 *Jing* was considered to be a vital fluid that circulated through and nourished the body and when expelled took the form of semen or menses.

Stored inside, we take it to be the well spring.
Floodlike, it harmonizes and equalizes
And we take it to be the fount of *qi*.
. . . The four limbs are firm . . . *Qi* freely circulates through the nine apertures. . . . [15]
When you have no delusions within,
Externally there will be no disasters,
Those who keep their minds unimpaired within,
Externally keep their bodies unimpaired.[16]

Meditation was considered not only to calm the mind but also to improve complexion, vision, and hearing, as well as strengthen the body. In addition, it was considered to provide great stamina, endurance, wisdom, and perspicacity.

If people can be aligned and tranquil,
Their skin will be supple and smooth,
Their ears and eyes will be acute and clear,
Their muscles will be supple and their bones will be strong.
They will then be able to hold up the Great Circle
[of the heavens]
And tread firmly over the Great Square [of the earth]
They will mirror things with great purity,
And will perceive things with great clarity.
Reverently be aware [of the Way] and do not waver,
And you will daily renew your inner power. [17]

---

15 That is, the two ears, two eyes, two nostrils, mouth, anus, and genitals.

16 Li Mian 李勉 (ed.), *Guan Zhong* 管仲: *Guanzi Jin zhu jin yi* 管子今註今譯. (Taibei: Taiwan shangwu yinshuguan, 2013), 49.777–78; Roth, *Original Tao*, 74; also see Rickett, *Guanzi*, 47–48.

17 *Guanzi Jin zhu jin yi*, 49.778; Roth, *Original Tao*, 76; Rickett, *Guanzi*, 48–49.

The relationship among self-control, calm mental focus, and good bodily function is reiterated throughout the text:

> When the four limbs are aligned
> And the blood and *qi* are tranquil;
> Unify your awareness, concentrate your mind,
> Then your eyes and ears will not be overstimulated
> And even the far-off will seem close at hand.[18]

The text even goes so far to suggest dietary advice, detailing the physiological effects of proper or improper diet. Moderation in eating and fasting was considered critical for the production of vital essence and knowledge, the quintessential materials of body and mind.

> Overfilling yourself with food will impair your *qi*,
> And cause your body to deteriorate.
> Overrestricting your consumption causes the bones to wither
> And the blood to congeal.
> The mean between overfilling and overrestricting:
> This is called "Harmonious Completion."
> It is where the vital essence lodges
> And knowledge is generated.[19]

In the end, breath control grants long life.

> Just let a balanced and aligned [breathing] fill your chest
> And it will swirl and blend within your mind,
> This confers longevity.[20]

18 *Guanzi Jin zhu jin yi*, 49.778; Roth, *Original Tao*, 82; Rickett, *Guanzi*, 51.
19 *Guanzi Jin zhu jin yi*, 49.779; Roth, *Original Tao*, 90, Rickett, *Guanzi*, 53.
20 *Guanzi Jin zhu jin yi*, 49.778; Rickett, *Guanzi*, 52.

Zhuangzi 莊子

The *Zhuangzi*, a roughly contemporaneous work, is traditionally attributed to the fourth-century BCE intellectual Zhuang Zhou 莊周, also an erstwhile visitor at Jixia. Textual scholars now agree it is a multiauthored work that later became identified as "Daoist," with textual layers dating from between the late fourth century and early second century BCE. Some of these, the "Inner Chapters" are thought to be original to Zhuang Zhou himself.[21] Contrary to the *Guanzi*, the authors of the *Zhuangzi* create much more distance between meditations for spiritual insight and exercises for physical cultivation.

The essays directly engage with contemporary philosophical positions, such as the utilitarianism of Mozi 墨子; the social conservatism of Confucius and Mencius or Mengzi 孟子; and the egoistic hedonism of Yang Zhu 楊朱.[22] The *Zhuangzi* authors adopted a skeptical relativist position, maintaining that philosophical argument

---

21 H. D. Roth, "Chuang tzu," in Loewe, *Early Chinese Texts*, 56–66.

The English-language term "Daoism" is one of the most misleading terms in the history of Chinese religions and philosophy. The twentieth century saw considerable debate about later "religious" Daoism being a degradation of a nobler, and earlier "philosophical" variety. With better research into imperial era Daoism, and recognition of the presence of much philosophical reflection in later works, and of meditative practice (as described below) in earlier literature, this distinction has largely been disregarded as the product of late imperial and early republican scholarly bias (ca. 1800–1949). I refer to these early works as "philosophical" simply to mark their inclusion in a set of texts about morality, statecraft, and cosmological order by scholarly traditions since the Warring States and Han dynasty, but not to set them against later so-called religious texts.

On the other hand, there is still ongoing debate about the origin of a self-conscious Daoist tradition, ranging from the Warring States, to ca. 180–150 CE or as late as the fifth century CE, and thus whether these early texts should rightly be considered part of those later traditions. While in this chapter I point out continuities and tensions between texts from multiple periods, no claims are made here about continuities or genealogies between these traditions and periods, or their role in identifying "Daoism" as a continuous tradition.

The seminal article on this problem is N. Sivin, "On the Word 'Taoist' as a Source of Perplexity: With Special Reference to the Relations of Science and Religion in Traditional China," *History of Religions* 17/3–4 (1978), 303–30. Also see R. Kirkland, "The Taoism of the Western Imagination and the Taoism of China: De-colonializing the Exotic Teachings of the East" (paper presented at University of Georgia, October 20, 1997).

22 For a survey of these positions, see various essays in V. H. Mair (ed.), *Experimental Essays on Chuang-tzu*, Asian Studies at Hawaii, no. 29 (Honolulu: University of Hawai'i Press, 1983).

is always positional with regard to the assumptions of the speaker, and that it is impossible to adjudicate a universal position on the basis of any one of them. The *Zhuangzi* is also naturalist, emphasizing that humans and their social ways are part of the larger order of nature rather than opposed to it. It draws on the same physiological and cosmological notions that inform the *Guanzi* and other texts of the time. Thus the cultivation of *qi* takes an important role, as can be seen in passages such as these:

> Man's life is a coming-together of breath. If it comes together, there is life; if it scatters, there is death.

> The True Man breathes with his heels; the mass of men breathe with their throats.[23]

Despite this reliance on *qi* practice for personal cultivation, the various hands in the text do not advocate *yangsheng* practices per se and deliberately distance themselves from them. For example a later school of "syncretist" authors derides practitioners who deliberately adopt practices to lengthen life, because these will fail to attain long life naturally, and use unnecessary and artificial extra effort.[24]

---

23 Guo Qingfan 郭慶藩 and Wang Xiaoyu 王孝魚(eds), Zhuang Zhou 莊周: *Zhuangzi jishi* 莊子集釋 (Beijing: Zhonghua shuju, 1995), 22.773 and 6.228. Translation from B. Watson, *The Complete Works of Zhuang Zi* (New York: Columbia University Press, 2013), 22.177 and 6.43.

24 Graham identifies chapters 12–16 as composed by the "Syncretist School," an eclectic group who likely compiled the text in 180. Guan Feng considers these to have been composed in the late fourth or early third century by the school of Song Xing (?360–?290 BCE) and Yin Wen (fourth century BCE), based at the Jixia academy in Linzi, the capital of Qi (present day Shandong). See A. C. Graham, "How much of *Chuang tzu* did Chuang Tzu write?" in *Studies in Chinese Philosophy and Philosophical Literature* (New York: State University of New York Press, 1986), 283–321; and Guan Feng 關鋒, "*Zhuangzi 'Wai, za pian' chu tan* 《莊子：外雜篇》初探," in Zhuangzi zhexue yanjiu bianji bu 莊子哲學研究編輯部 (ed.), *Zhuangzi zhexue taolun ji* 莊子哲學討論集, (Beijing: Zhonghua, 1962) cited in Roth, "Chuang tzu," in Loewe, *Early Chinese Texts*, 56–57.

> To pant (*chui* 吹), to puff (*xu* 噓), to hail (*hu* 呼), to sip (*xi* 吸), to spit out the old breath and draw in the new, practicing bear-hangings and bird-stretchings, longevity his only concern—these are favoured by the masters who practice guiding and pulling exercises (*daoyin*), the man who nourishes his body, who hopes to live to be as old as Ancestor Peng. To attain . . . long life without guiding and pulling . . . this is the Way of Heaven and Earth.
>
> The coming of life cannot be fended off; its departure cannot be stopped. How pitiful the men of the world, who think that simply nourishing the body is enough to preserve life![25]

These passages, by different hands within the work, criticize those who sought to artificially preserve life through breathing and stretching techniques and argue that following the Way should spontaneously align one with the forces of life without such effort. Deliberately nourishing the physical form of the body (*xing* 形) is *not* the same as nourishing life, and it will not achieve the desired goal.

In the chapter titled "Nourishing Life" (*Yangsheng*), thought to be original to Zhuang Zhou, he does not mention breathing exercises or physiological meditations but instead focuses on the technical, embodied skills of a butcher. Through the practice of cutting up oxen, Cook Ding learns deftness of hand and subtlety of spirit, discovering how to send forth his spirits to sense the gaps between the joints, until he is so skilled he never has to sharpen his blade. This, says his interlocutor Lord Wen Hui, is the secret to nourishing life. The Way in the *Zhuangzi* is often found in the practical non-theoretical skills of

25 *Zhuangzi jishi*, 19.630 and 15.535–37; Watson, *Zhuang Zi* 19.145 and 15.119.
On *daoyin* in later Daoist regimens, L. Kohn, "Daoyin among the Daoists: Physical Practice and Immortal Transformation in Highest Clarity," in V. Lo (ed.), *Perfect Bodies: Sports, Medicine and Immortality: Ancient and Modern* (London: British Museum, 2012).

peripheral figures: craftsmen, carvers, bug-catchers, butchers, and the like, but notably never physicians.[26]

The question in the *Zhuangzi* is not how to collect *qi* to live a long life. Nor is it the nature of life or life force (*sheng* 生) itself. Rather the text elucidates how to know and deal with life, as embodied, as something known through skilled practice and not easily rendered in language.

Daode jing 道德經

The *Daode jing*, often referred to by the name of its putative author Laozi 老子, was similarly compiled by multiple hands over time. Archaeological discoveries over the last forty years have unveiled buried editions in tombs that have changed scholarly understanding of the text as a fairly unitary and stable work to that of an emerging set of discourses, layered over time.[27] This work survives in various received editions, as well as copies recently excavated from Warring States and Han dynasty tombs in Guodian 郭店 of Jingmen, Hubei (ca. 300 BCE), in Mawangdui 馬王堆 in Changsha, Hunan (168 BCE), and in a set purchased in 2009 by Peking University (ca. 141–87 BCE).[28]

---

26 Raphals, "Philosophy and Medicine"; L. Raphals, "Craft Analogies in Chinese and Greek Argumentation," in E. Ziolkowski (ed.), *Literature, Religion, and East-West Comparison: Essays in Honor of Anthony C. Yu* (Wilmington: University of Delaware, 2005), 181–201.

27 Traditional dates for an original ur-text are unreliable, and the variation within the excavated and received editions, as well as citations in other contemporary works, indicates that while the text(s) circulated likely as early as the fifth century CE, they were only compiled into a recognized, named edition in the late third century. W. G. Boltz, "Lao tzu Tao te ching," in Loewe, *Early Chinese*.

28 On the Mawangdui edition, W. G. Boltz, "Textual Criticism and the Ma Wang tui Lao tzu, review of *Chinese Classics: Tao Te Ching, D. C. Lau*," *Harvard Journal of Asiatic Studies* 44/1 (1984), 185–224; R. G. Henricks, "On the Chapter Divisions in the Lao-tzu," *Bulletin of the School of Oriental and African Studies* 45 (1982), 501–24. On the Guodian version, S. Allan and C. Williams, *The Guodian Laozi: Proceedings of the International Conference, Dartmouth College, May 1998* (Society for the Study of Early China and Institute of East Asian Studies, University of California, 2000). On the Peking edition Han Wei 韓巍 (ed.), *Laozi—Beijing Daxue cang Xi Han zhushu* 老子—北京大學藏西漢竹書, vol. 2 (Shanghai: Shanghai guji chubanshe, 2012).

The *Daode jing*, which came to be venerated by later traditions as the quintessential Daoist text, calls for quiescence, humility, and dispensing with complex technology, values, and social hierarchy, arguing for a return to simplicity in a golden age of the past, simultaneously reversing the trends of both old age and of the corruption of society. It advocates politically for rulership that uses "non-action" (*wuwei* 無為), that is, spontaneous action that introduces no artificiality or contrivance, and aligns this with a meditative focus on "embracing the One." As part of this process it advocates an apophatic reduction of external stimuli and focus on interior qualities and breathing. While earlier versions of the text foregrounded strategies for dealing with political exigencies, later editions encapsulated these strategies into an overarching philosophy of following the Way.[29]

New passages that appear in the second-century BCE Mawangdui edition are not extant in the earlier Guodian edition and appear to introduce a stronger emphasis on embodied meditation than previous layers. Three of these mention the ideal of an "infant" (*ying'er* 嬰兒), an image that evokes spiritual and physiological return to the suppleness and vitality of a newborn, and mesh this idea with the return to the political and social simplicity that is the goal of earlier layers of the text.[30]

> *Stably securing and nourishing the whitesouls* and embracing the One, can you avoid separation? Focusing on *qi* until you produce suppleness, can you become like an infant?[31]

---

29 A. Kam-leung Chan, "Laozi," in Zalta, *The Stanford Encyclopedia of Philosophy*, 2014, http://plato.stanford.edu/archives/spr2014/entries/laozi/

30 These are chapters 10, 20, 28. I use the numbering and divisions in the received edition, based on the *Heshang gong zhangju* 河上公章句. On the Guodian and Mawangdui, see D. Murphy, "A Comparison of the Guodian and Mawangdui Laozi Texts" (MA thesis, University of Massachusetts, 2006), 10.

31 *Daode jing*, chap. 10. Translation amended from D. C. Lau, *Dao de Jing* (Penguin Books, 1963).

This image of the infant set the stage for the text's future adoption by specialists of cultivating vitality and immortality. The opening phrase used in this passage, *zai ying po* 載營魄, directly draws on technical language from earlier mediumistic traditions of the state of Chu 楚. The trance reverie "Far-off journey (*yuanyou* 遠遊)," found in the fourth-century BCE poetry collection *Songs of the South* (*Chu ci* 楚辭) uses the phrase to describe preparation for a visionary, out-of-body spirit journey through ethereal realms. David Hawkes's translation renders it as follows:

> I *restrained my restless spirit* and mounted the Empyrean;
> I clung to a floating cloud to ride aloft on.[32]

By the time the term appeared in the Han dynasty *Daode jing*, the phrase *zai ying po* came to refer not to practices of out-of-body travel, but to inward-looking, embodied cultivation of spiritual unity. By the Han dynasty the whitesouls (*po* 魄) were paired with the cloudsouls (*hun* 魂) and became the subject of a great deal of anxiety.[33] These souls respectively numbered seven and three, departed to the earth and to the heavens at the time of death, and were thought by medical writers to reside in the lungs and the liver. They were also the cause of dreams, the souls' impressions as they left the body at night and thus opened their owner to harm from ghosts and spirits,

---

32 The italics indicate Hawkes's translation of the same phrase. D. Hawkes, *Ch'u Tz'u, The Songs of the South: An Ancient Chinese Anthology* (New York: Beacon Press, 1962), 84; Bai Huawen 白化文 and Wang Yi 王逸 (eds), *Chuci buzhu* 楚辭補注: *Yuanyou* 遠遊 (Beijing: Zhonghua shuju, 1983), 5.168. I do not contend that his rendering was "wrong," but rather, that the insertion of the term into a Han edition of the *Daodejing*, and also into a new ontological nexus, probably adapts it to a new practice, which requires retranslation.

33 This pairing became widespread, so that the second century CE commentator, the Venerable on the Riverside (Heshang gong 河上公), found it natural to argue that the opening phrase obliquely implied the cloudsouls as well. *Daode zhen jing zhu* 道德真經註 DZ 682 1.4b. On this commentary, Alan Kam-leung Chan, *Two Visions of the Way: A Study of the Wang Pi and the Ho-Shang Kung Commentaries on the Lao-Tzu* (Albany: State University of New York Press, 1991).

DZ refers to index numbers in Schipper and Verellen (eds), *The Taoist Canon: a Historical Companion to the Daozang*, (Chicago: University of Chicago Press. 2004).

which manifested as disease and ill-fortune. Thus practitioners developed a broad array of visualization and ritual techniques to constrain them and keep them from leaving the body.

This shift in emphasis is clear in the Han *Daode jing* practice that consolidates and nourishes, and keeps things from splitting up and breaking apart—reflecting the anxieties of the then-recently unified state trying to maintain central control—and marking a clear contrast to the ecstatic, disembodied, heavenly journey by the wild shamans of the fringe state of Chu. In the passage from the *Daode jing* just cited above, the *po*-stabilizing practice is complimented by the *qi* practice which follows it, producing "suppleness" like that of an infant—at once a bodily sensation, a subjective attitude of compliance, and a political stance of non-engagement. These qualities converge in the image of the baby, who is politically inactive, and whose body possesses abundant vitality in the flush of rapid growth, highly vulnerable, requiring nourishing and security through seclusion. These qualities recursively evoke the practices of "nourishing" and "embracing the One" in the first line, and an emphasis on inwardness, passivity, and growth rather than outward journeying.

From the Eastern Han onward (25–220 CE), this *Daode jing* passage became the *locus classicus* for a host of Daoist, *yangsheng*, and immortality traditions that focused breathing and attention to the navel area, such as: "fetal breathing" (*taixi* 胎息); another practice called "clenching the fists" (*wogu*握固) where one clenches the thumbs, often while in a fetal position; the visualization and inner cultivation of a "ruddy infant (*chizi* 赤子)" within the body; or simply meditation on the breath in the lower abdomen or elixir field (*dantian* 丹田).[34]

---

34 Fetal breathing practices appear in biographies of first-century individuals and hosts of traditions are attested in Tang dynasty Daoist texts. C. Despeux, "Taixi," in F. Pregadio (ed.), *The Encyclopedia of Taoism* (London: Routledge, 2008). The *Laozi zhongjing* 老子中經 (Central Scripture of the *Laozi*) was highly influential on early body-god visualization practices, as well as containing the earliest attested reference to the elixir field in the abdomen. A. Iliouchine, "A Study of the Central Scripture of Laozi (*Laozi zhongjing*)" (MA thesis, McGill University, 2011); K. M. Schipper, "The Inner World of the Lao-tzu chung-ching," in Chun-chieh Huang and E. Zurcher

Although these developments began toward the end or after the Han dynasty, the appearance of the fetal passages in the Mawangdui *Daode jing* indicates that their roots lay in the expansion of interest in *yangsheng* in the early Han.

Han Dynasty Nourishing Life Literature and the Sciences

The broader sea-changes in *yangsheng* and medical culture that are reflected in the new physiological emphasis of the *Daode jing* took place at many levels of Han culture. Although *yangsheng* practices such as breathing, stretching, diet, drugs, and sexual hygiene existed long before the Han dynasty, they experienced a surge in popularity among Han elites and at court.[35] The cache of texts at Mawangdui is perhaps the best exemplar of this transition and provides an on-the-ground view into scientific knowledge in the Han.[36] This collection marks a pivotal stage in the development of medical theory in the first 200 years of the unified Chinese empire, just prior to the development of classical medicine in the late first century BCE. In the following section, I outline how recent scholarship has located the Mawangdui

(eds), *Time and Space in Chinese Culture* (Leiden: Brill, 1995); J. Lagerwey, "Deux écrits taoïstes anciens," *Cahiers d'Extrême-Asie* 14 (2004).

35 Harper, *Early Chinese Medical Literature* 30–31; Ngo Van Xuyet and Fan Ye, *Divination, magie et politique dans la Chine ancienne: Essai* (Paris: Presses universitaires de France, 1976).

36 Harper, *Early Chinese Medical Literature*. The facsimiles and transcripts of the original manuscripts are in MWD zhengli xiaozu (ed.), *Mawangdui Hanmu boshu* 馬王堆漢墓帛書 (Beijing: Wenwu chubanshe, 1985); Ma Jixing 馬繼興, *Mawangdui gu yishu kaoshi* 馬王堆古醫書考釋 (Changsha: Hunan kexue jishu chubanshe, 1992); Zhou Yimou 周一謀 and Xiao Zuotao 蕭佐桃, *Mawangdui yi shu kao zhu* 馬王堆醫書考注 (Tianjin: Tianjin kexue jishu chubanshe, 1988); Zhou Yimou 周一谋, *Mawangdui yixue wenhua* 马王堆医学文化, Di 1 ban. ed. (Shanghai: Wenhui chubanshe, 1994).

There are many other such finds that inform our knowledge in addition to the finds at Guodian and Mawangdui. Those with medical literature include Zhangjiashan 張家山 in Jingzhou, Hubei (196–186 BCE); Shuanggudui 雙古堆 at Fuyang, Anhui Province (165 BCE); Baoshan 包山 near Jingzhou, Hubei (323–292 BCE); Shuihudi 睡虎地 at Yunmeng, Anhui Province (217 BCE); Fangmatan 放馬灘 at Tianshui, Gansu (230-220 BCE) and Wuwei 武威 in Gansu (first century BCE).

literature within this broader emergence of medicine and scientific theory more generally.

The classical medical corpus of the late Western Han dynasty drew on and consolidated in new ways a set of cosmological theories—*yinyang* 陰陽, the five phases (*wuxing* 五行), and *qi*—to interpret the structure and internal workings of the body, disease formation, and ideal practices to keep the body attuned to the changes of the seasons.[37] These systems organized the complex multivariate phenomenal world into small sets of categorical groups of two (*yinyang*), three (heaven, earth, and man *tian di ren* 天地人), or five (five phases *wuxing* 五行). The homologies between objects of the same class were not just based on similar material properties but also because of their relational status in dynamic states of change. Thus, the same material when descending might be considered *yin* and when ascending *yang*.[38] Five phase theory holds that different phases of *qi* generate or control each other successively—so that wood generates fire, which generates earth, then in turn metal, water, and finally wood. In the controlling cycle, wood controls or constrains earth, which controls water, which controls fire, which controls metal, which controls wood.[39]

However, the use of *yinyang*, the five phases, and *qi* in complement with one another was not standard prior to the formation of classical medicine. Scholars in recent years have clarified the separate applications of these theories within excavated manuscripts and traced

37 P. U. Unschuld and H. Tessenow, *Huangdi Neijing Suwen: An Annotated Translation of Huang Di's Inner Classic—Basic Questions* (Berkeley: University of California Press, 2011); P. U. Unschuld, *Huangdi Neijing Suwen: Nature, Knowledge, Imagery in an Ancient Chinese Medical Text* (Berkeley: University of California Press, 2003).

38 Yang Weijie 楊維傑 (ed.), *Huangdi neijing suwen yijie* 黃帝內經素問譯解 (Taibei: Tailian guofeng chubanshe, 1984), chaps. 5, 6, 7.42–75. Translated in Unschuld and Tessenow, *Suwen* 5–7.95–154.

39 *Huangdi neijing suwen yijie*, 4.34–41, see translation in Unschuld and Tessenow, *Suwen* 4.83–94.

how they gradually became combined into a composite theory of nature that has been utilized since the first century BCE.[40] Prior to this convergence, these theories played critical roles in diverse technical domains, such as divination, astronomy, calendrics, cosmogenesis theory, and ritual practice and were adopted in critical arguments about the legitimacy of the state.[41]

Prior to the late third century CE, five phase theory was an omenological praxis used to divine the transmission of the heavenly mandate to different dynastic states. Although Zou Yan 鄒衍 (305–240 BCE) is credited with first developing and promoting five phase theory, none of his works survive. It was only with the *Spring and Autumn Annals of Mr. Lü* (*Lüshi chunqiu* 呂氏春秋), by the Qin 秦 chancellor Lü Buwei 呂不韋 (291–235 BCE), that *qi* became associated with five phase theory in discussions of political legitimacy and ritual theory. Organized as advice on political strategy and managing the state, the *Spring and Autumn Annals* was submitted circa 239 BCE, to Yin Zheng 嬴政 (260–120 BCE), before Yin became the first emperor of China. This complex encyclopedia contains many subjects on natural science, ranging across astronomy, ritual and music, agriculture, and natural and political philosophy. The following passage is the locus classicus most often cited for associating *qi* with the five phases.

---

40 An overarching survey of the development of technical theory during this period is Harper, "Warring States," in Loewe and Shaughnessy, *Cambridge History of Ancient China*. For an important corrective demonstrating how the appearance of the term for five phases (*wuxing*) in technical literature does not in each case denote the dynamic relational theory described above but only clearly emerges in late first-century BCE textual sources, M. Nylan, "*Yin-yang*, Five Phases, and *Qi*," in Nylan and Loewe, *China's Early Empires*. For a survey of all of the technical arts and thorough review of archaeological materials, see Li Ling 李零, *Zhongguo fangshu xu kao*; Li Ling 李零, *Fang shu zheng kao*.

41 On the overlaps between divination technology and pre-classical medicine, V. Lo, "The Han Period," in T. J. Hinrichs and L. L. Barnes (eds), *Chinese Medicine and Healing: An Illustrated History* (Cambridge: Belknap Press of Harvard University Press, 2013); Harper, *Early Chinese Medical Literature*.

> Whenever a true king is about to rise, Heaven invariably sends omens to the people below first. In the time of the Yellow Emperor, Heaven first made large earthworms and mole crickets appear. The Yellow Emperor said, "The *qi* of earth is getting strong and so he took yellow as his colour and earth as pattern for his activities. In the time of Yu, Heaven first made grass and evergreens appear. Yu said, "The *qi* of wood is getting strong," and so he took deep blue-green as his colour and wood as his pattern for his activities.[42]

The passage portrays the other three phases in similar form. Lü then describes the principle of stimulus-response (*ganying* 感應) by which materials in these corresponding categories were thought to interact.

> Things belonging to the same category naturally attract each other; things that share the same *qi* naturally join together; and musical notes that are close naturally resonate with one another. Strike the note *gong* on one instrument, and other strings tuned to the note *gong* will vibrate; strike the note *jue* and the other strings tuned to the note *jue* will vibrate. Water flowing across levelled earth will flow to the damp places; light evenly stacked firewood, and the fire will catch where it is driest.[43]

By creating a link between state authority, the five phases, and the natural processes of *qi*, Lü's work brought together state succession and principles for divination with the undergirding forces of the natural

---

42 Chen Qiyou 陳奇猷 (ed.), *Lü Buwei* 呂不韋: *Lüshi chunqiu jiao shi* 呂氏春秋校釋 (Shanghai: Xuelin chubanshe, 1984), 13/2.1; translation from Nylan, "*Yin-yang*, Five Phases, and *Qi*," 399–400. See also J. Knoblock and J. Riegel, *The Annals of Lü Buwei* (Stanford: Stanford University Press, 2000), 13.2.282–83.

43 *Lüshi chunqiu jiao shi*, 13/2.1., Knoblock and Riegel, *Annals*, 13.2.283–84. A parallel passage to this appears elsewhere in an extended discussion of resonance theory. *Lüshi chunqiu jiao shi*, 20/4.1. Knoblock and Riegel, *Annals*, 522. For a brief overview of varieties of stimulus-response theory, see Franklin Perkins, "Metaphysics in Chinese Philosophy," Zalta, *The Stanford Encyclopedia of Philosophy* 2015, http://plato.stanford.edu/archives/sum2015/entries/chinese-metaphysics/.

world, a move that went on to have far-reaching influence on technical thought and its relation to state power.[44] During the first two centuries of the Han dynasty, considerable debate arose concerning omenology when ritual specialists and political theorists sought to establish the emperor's divinity, while at the same time constraining his behavior and limiting his movements, encouraging him not to leave the court.[45]

However, Lü's intervention did not constitute the all-encompassing composite of cosmological correlative theory that was to become the legacy of Han thought—the interconnections between these theories took time to emerge. The medical works of the Mawangdui collection provide a close view of the development of *yangsheng* and medical culture at or before 168 BCE.[46] Although *yinyang* theory is applied in numerous places within the corpus, there is only one instance of five phase theory, indicating that the theories in the *Lü shi chunqiu* had not yet become widely incorporated.[47] The largest number of texts discuss how to diagnose or treat bodily channels (*mai* or *mo* 脈/脉), through which *qi* and blood circulated; also important are the *yangsheng* texts that discuss (or contain pictures of) techniques for augmenting, circulating, and storing *qi* in the body. Such techniques also included dietary and exercise regimens and sexual hygiene. Many techniques relied on a proprioceptive or tactile sense of the body's interior and heightened attention to the superficial conditions of the body: cold, heat, pain, redness.[48] Sexual hygiene was largely concerned not only

44 N. Sivin, "State, Cosmos, and Body in the Last Three Centuries B. C.," *Harvard Journal of Asiatic Studies* 55/1 (1995); Harper, "Philosophy and Occult Thought"; Lo, "The Han Period."

45 Aihe Wang, *Cosmology and Political Culture in Early China* (Cambridge: Cambridge University Press, 2006); Tavor, "Embodying the Way," 143–81.

46 The authoritative translation and exegesis is Harper, *Early Chinese Medical Literature*.

47 Harper, "Philosophy and Occult Thought," 866. This occurs in a text on fetal gestation and birth, *Taichan shu* 胎產書. *Mawangdui Hanmu boshu*, vol. 4, 376, and translation and commentary in Harper, *Early Chinese Medical Literature*, 379–81.

48 Lo, "Tracking the Pain." V. Lo, "The Influence of 'Yangsheng' Culture on Early Chinese Medicine" (PhD diss., School of Oriental and African Studies, 1998).

with the cultivation of *qi,* vital essence and spirits, but also with attaining a state of ecstatic and numinous insight (*shenming* 神明), which was associated with the attainment of sagehood.[49]

This literature on nourishing life had a strong structuring influence on the formation of classical medical literature in the next century, from which only the *Yellow Emperor's Inner Classic* (*Huangdi neijing* 黃帝內經) survives in the received tradition.[50] Compiled circa the first century BCE from older texts, such as those found in Mawangdui, the *Inner Classic* was formed at the same time that significant debates took place at court concerning correlative cosmology and ritual practice.[51] In this work, we find the most complete expression of five phase theory in the Chinese corpus. It integrates and correlates a host of phenomena under the notion of five phasal *qi*, including seasons, colors, flavors (of food and drugs), animals, weather patterns, as well as the internal organs, their related channels (*mai*), and acupuncture

---

49 V. Lo, "Spirit of Stone: Technical Considerations in the Treatment of the Jade Body," *Bulletin of the School of Oriental and African Studies* 65/1 (2002), 99–128; V. Lo, "Crossing the *Neiguan* 內關 'Inner Pass:' A *Nei/Wai* 內外 'Inner/Outer' Distinction in Early Chinese Medicine," *East Asian Science Technology and Medicine EASTM* 17 (2000), 15–65. Spirit (*shen* 神) had a double valence as both anthropomorphic spirits that could be embodied (in the living) or disembodied (i.e., the dead), as well as referring to the substance of cognition, located in the chest region, most often the heart (*xin* 心). *Shenming* could thus refer to the numinous spirits or to an expanded cognitive and ecstatic bodily state. The most thorough study of *shenming* in Warring States literature and scholarship is S. P. Szabó, "The Term Shenming: Its Meaning in the Ancient Chinese Thought and in a Recently Discovered Manuscript," *Acta Orientalia Academiae Scientiarum Hungaricae*, 56/2–4 (2003), 251–74.

50 The *Yellow Emperor's Inner Classic* survives in three recensions: the *Simple Questions* (*Suwen* 素問), the *Numinous Pivot* (*Lingshu* 靈樞), and the *Grand Basis* (*Taisu* 太素). These are all recombined passages from an originary work or works now lost. On textual filiation within these works, D. J. Keegan, "The Huang-ti nei-ching: The Structure of the Compilation, the Significance of the Structure" (PhD diss., University of California, 1988). On the dating and provenance of surviving editions, N. Sivin, "Huang ti nei ching 黃帝內經," in Loewe, *Early Chinese Texts*. On the influence of *yangsheng* culture, Lo 羅維前 and Li Jianmin 李建民, "Manuscripts, Received Texts, and the Healing Arts;" V. Lo, "The Influence of Nurturing Life Culture," in E. Hsu (ed.), *Innovation in Chinese Medicine* (Cambridge: Cambridge University Press, 2001).

51 For a comprehensive study of the *Yellow Emperor's Inner Classic*, see Unschuld and Tessenow, *Suwen*; Unschuld, *Nature, Knowledge, Imagery*. On debates surrounding the five phases in the late Western Han court, see Wang, *Cosmology and Political Culture in Early China*.

points along the channels.[52] The relationships between these phases structured notions of physiological operations, etiology, and curative practices such as acupuncture, diet, and seasonal habits and behaviors. Although the relationships between *qi, yinyang*, and the five phases are given their fullest expression in the *Yellow Emperor's Inner Classic*, they are taken up in a broad swath of scientific thought, ritual, and political theory. Under this scheme, none of these domains could ever be fully separate.[53]

These patterns of nature were never far off from the cosmo-political sphere. One of the most frequently used terms for curing is *zhi* 治, a politically resonant term meaning to control or govern. It was frequently contrasted with *luan* 亂, literally "chaos," which had overtones of rebellion and political uprisings that needed to be controlled in order to sustain the health of the state. These terms converged the curing of disease with the maintenance of political order, and frequent analogies were made between the state and the body, as in the chapter "Regulating the Spirit in Accordance with the Four Seasons":

> The sages did not treat those who were already ill, but treated those not yet ill,
> They did not put in order what was already chaotic, but put in order what was not yet in chaos.[54]

Homologies for governance abound within the text: the liver is described as a general, the heart as a ruler, the lung as a chancellor; *qi*,

52 E. Hsu identifies this composite as a "Body Ecologic," defined in relation to climatic factors. She distinguishes this from the earlier emphasis on moral cultivation tied to the cultivation of *qi*, which she identifies as the "Sentimental Body" of felt feelings, emotions, and values. E. Hsu 許小麗, "Outward Form (*xing* 形) and Inward *Qi* 氣: The "Sentimental Body" in Early Chinese Medicine," *Early China* 32 (2009).

53 Sivin, "State, Cosmos, and Body"; Raphals, "Philosophy and Medicine."

54 Amended from Unschuld and Tessenow, *Suwen* 2.56.

which circulates to the exterior defensive layer of the body, is referred to as a military camp and its supplies (*ying qi* 營氣).[55]

By the time the Han imperial "Bibliography of Arts and Letters" (*Yiwen zhi* 藝文志) was completed in the first century by the dynastic historian Ban Gu 班固 (32–92 CE), we find a clearer arrangement of the Chinese sciences.[56] As Nathan Sivin argues, there was no unified notion of science per se; rather the sciences were singular and individual.[57] This can be seen in the catalogue titles in the bibliography: "Arts and Calculations" (*shushu* 術數), which contained subsections on mathematics, divination, and "Methods and Techniques" (*fangji* 方技). The latter contains four subsections: "Medical Classics" (*yijing* 醫經); "Classic [drug] Recipes" (*jingfang* 經方); "Arts of the Bedchamber" (*fangzhong* 房中); and "Divine Transcendence" (*shenxian* 神仙).[58]

This catalogue, argues Miranda Brown, is the earliest recorded instance when the notion of medicine (*yi* 醫) was articulated as a unified, historical field of knowledge, and this framing was very influential on later medical writers, becoming a mainstay of medical orthodoxy.[59] Originally compiled by the court official and scholar Liu Xiang 劉向 (77–8 or 6 BCE) and edited by his son Liu Xin 劉歆 (50 BCE–23 CE), and later incorporated by Ban Gu into his work, the Lius' catalogue

---

55 *Huangdi neijing suwen yijie*, 8.76–80; Yang Weijie 楊維傑 (ed.), *Huangdi neijing lingshu yijie* 黃帝內經靈樞譯解 (Taibei: Tailian guofeng chubanshe, 1984), 16.186.

56 Ban Gu based his catalogue on work begun by his father Ban Biao 班彪 (3–54 CE) that assimilated the Seven Summaries (*Qilüe* 七略). A. F. P. Hulsewé, "Han shu 漢書," in Loewe, *Early Chinese Texts*.

57 N. Sivin, "Science and Medicine in Chinese History," in P. S. Ropp and T. H. Barrett (eds), *Heritage of China: Contemporary Perspectives on Chinese Civilization* (Berkeley: University of California Press, 1990); N. Sivin, "Why the Scientific Revolution Did Not Take Place in China—or Didn't It?" *Chinese Science* 5 (1982). See discussion in Raphals, "Philosophy and Medicine."

58 *Han shu*, 30.1776–81.

59 M. Brown, *The Art of Medicine in Early China: The Ancient and Medieval Origins of a Modern Archive* (Cambridge: Cambridge University Press, 2015), 95–96, 106ff and passim. Brown argues that references to an *yidao* 醫道 (Way of Doctors/Medicine), as in *Yellow Emperor's Inner Classic* chapter 75, were rare exceptions that proved the rule, and the majority of depictions of medicine as a body of thought emphasized variety and heterogeneity of the various schools, as in *Inner Classic*, 12. See *Huangdi neijing suwen yijie*, 12.104–07 and 75.670–72. For translations, see Unschuld and Tessenow, *Suwen* 12.211–18 and 75.645–50.

described medicine (*yi* 醫) as a collective, historical body of medical knowledge not limited to personal, lineage transmission but directly accessible through ancient texts. This framing of medicine as an historical body of knowledge reflected the intellectual politics of the Lius, who were sympathetic to the "old text" school, a group of academicians who argued that authoritative knowledge was better transmitted through early editions of classical texts. This position shaped the school's political outlook, which held the ruler as a rational actor who ought to be advised to take decisions based on logical argument. They were opposed to the "new text" school that argued orthodoxy was only transmitted through lineages of initiated academicians and supported a charismatic image of the ruler as a sage-king with divine powers.[60]

Alongside this framing of "Medical Classics" and "Classical Recipes" as historical categories, the general catalogue of "Methods and Techniques"—a title more suggestive of practical knowledge and skills than theoretical categories—exhibits tensions with regard to other kinds of knowledge. The two other subcategories, the "Arts of the Bedchamber" and "Divine Transcendence," receive no such historical framing and include texts on skills closely associated with the attainment, preservation, and enhancement of essence and *qi* for longevity, and for paranormal powers and immortality. Sexual cultivation was, as in Mawangdui, an important part of *yangsheng* practice, and this catalogue contained over 800 fascicles, or *juan*, dwarfing the other three, which contained roughly 200 each. Transcendence refers to the attempt to achieve supernaturally long life or immortality and miraculous bodily powers, and the editors were quite ambivalent about their status:[61]

> Transcendence is the perfection by which to guard one's inner nature and destiny, and thus wander, seeking, in the outer regions.

60 M. Nylan, "The 'Chin Wen/Ku Wen' Controversy in Han Times," *T'oung Pao* 80/1.3 (1994).

61 A major topic in the history of religions, it is impossible to describe even a representative bibliography here. A preliminary introduction is B. Penny, "Immortality and Transcendence," in L. Kohn (ed.), *Daoism Handbook* (Leiden: Brill, 2000). For a significant recent translation of transcendent biographies R. F. Campany, *To Live as Long as Heaven and Earth: A Translation*

> These gentlemen, free of cares and calm of mind, regard as equal the realms of life and death, and harbour no tremulous fears in their breast. However, there are those who make this their sole labour, and fill their writings with great lies and wondrous exotica to make them seem grander, these are not the teachings of the sage kings. Confucius said: "As for studying the occult and practicing wonders, later generations will record it, and I do not perform them."[62]

Liu Xiang argues here that while transcendents (*xianren* 仙人) are experts in preserving life and living at ease, there are also charlatans and fakers among their ranks, and thus unsuitable literature for the literate elite or for governing the state. They cite the authority of Confucius himself to create distance from this literature. This double-edged assessment makes it ambiguous whether it is acceptable literature—it is not so distant that it was excluded from the catalogue. While it is clear the Lius did not wholly embrace transcendent writings, it appears that some of it, probably that which was kept in the imperial library, was regarded as acceptable. Nevertheless, having been collected within the overall category of "Methods and Techniques," it is clear that these writings were thought to be part of a common body of related knowledge and thus related to medicine.

The roughly contemporaneous *Arrayed Traditions of Transcendents* (*Lie xian zhuan* 列仙傳), also shows transcendence and medicine to be related domains of activity.[63] This collection of biographies of immortals and wonder-workers included many of the health

---

and Study of Ge Hong's Traditions of Divine Transcendents (Berkeley: University of California Press, 2002). For a study of their social standing in medieval China, see R. F. Campany, *Making Transcendents: Ascetics and Social Memory in Early Medieval China* (Honolulu: University of Hawai'i Press, 2009).

62 *Han shu*, 30.1779–80. The passage cited from Confucius is from the *Doctrine of the Mean Zhongyong* 中庸 verse 11.

63 *Lie xian zhuan* 列仙傳 DZ 294, attributed to Liu Xiang. Because material in the text is verifiably from the second century CE, and the earliest attribution to Liu Xiang does not occur

practices witnessed in the Mawangdui collection, newly adopted into the regimes of transcendents. These included grain-fasting, sexual cultivation, breathing exercises, and stretching (*daoyin* 導引) for the attainment of supernatural longevity, immortality, and supernatural powers. It also describes transcendents as highly visible in curing and in the drug trade in local marketplaces.[64] While we cannot take the *Arrayed Traditions* as a veritable record concerning real events and real people, it certainly is a cultural record and attests to the imagination of what these practices were thought to achieve.[65] The catalogue was consistent with wider cultural notions that made no clear break between medicine and these broader collected health practices, which included those expressly used for religious cultivation.

The Religio-Medical Marketplace of the Six Dynasties

The decline of the Han dynasty took place amid the political, economic, and social upheaval of the second century CE, bringing famine, widespread epidemics, and an ideological vacuum—the destabilization of relations between heaven and humanity, individuals and their destiny.[66] The indiscriminate nature of disease, which strikes without regard to the moral worth of the individuals affected, gave rise to a

until the fourth century, this attribution and the early date are suspect. B. Penny, "Lie xian zhuan," in Pregadio, *The Encyclopedia of Taoism*. The seminal study of the *Arrayed Traditions* is M. Kaltenmark, *Le Lie-sien Tchouan: biographies légendaires des Immortels taoïstes de l'antiquité* (Pékin: Centre d'études sinologiques de Pékin, 1953). For a comparison of existing editions of the work, KUBO Teruyuki 久保 輝幸, "'Retsu sen den' no bōshitsu shita senden ni soku nitsuite 『列仙伝』の亡失した仙伝2則について," *Jinbun gaku ronshyū* 人文学論集 29 (2011).

64 As argued by Ogata Toru 大形 徹, "Rensen den ni miru doutokuteki sennin no houga 『列仙傳』にみる道徳的仙人の萌芽," *Jinbungaku ronshou* 人文学論集 33 (2015). Thanks to KUBO Teruyuki for this reference.

65 Campany, *Making transcendents*, 15–22.

66 Lin Fushi 林富士, "Donghan shiqi de jiyi yu zongjiao 東漢時期的疾疫與宗教," in *Zhongguo zhonggu shiqi de zongjiao yu yiliao* 中國中古時期的宗教與醫療 (Taibei: Lianjing chubanshe, 2008).

moral question about why bad things happen to good people. The theodicy of disease became a central question with which most religions of the time were concerned. In this climate, religious and technical actors of all stripes became involved in curing, caring, and embodied salvation.[67] Millennial cults such as the Celestial Masters (*Tianshi Dao* 天師) and the Great Peace (*Taiping* 太平) movement arose in the east and west of China respectively and became involved in fighting the Han government and attempting to set up sovereign theocratic states.[68] These movements identified the cause of disease in personal and familial sin, which were punished by the celestial bureaucracy, a divine simulacrum of the earthly state that resided in celestial realms and hidden caverns. The cure for such sin involved confessional rites and petitions to heavenly officials through rituals.[69] Transcendents such as those described in *Arrayed Traditions* and later works were well known for their skills in curing disease, as their search for longevity and immortality relied on many curative arts.[70] They sought to

---

67 On the religio-medical marketplace in medieval China, see C. P. Salguero, *Translating Buddhist Medicine in Medieval China* (Philadelphia: University of Pennsylvania Press, 2014); M. Stanley-Baker, "Daoists and Doctors: The Role of Medicine in Six Dynasties Shangqing Daoism" (PhD diss., University College London, 2013).

68 M. Loewe, "The Religious and Intellectual Background," in D. C. Twitchett and M. Loewe (eds), *The Cambridge History of China,* vol. 1: *The Ch'in and Han Empires, 221 B.C.–A.D. 220* (Cambridge: Cambridge University Press, 1986); P. Demiéville, "Philosophy and Religion from Han to Sui," in Twitchett and Loewe, *The Cambridge History of China*, vol. 1. On the Daoist formation of a theocratic state, see T. F. Kleeman, *Great Perfection: Religion and Ethnicity in a Chinese Millennial Kingdom* (Honolulu: University of Hawai`i Press, 1998), and most recently T. F. Kleeman, *Celestial Masters: History and Ritual in Early Daoist Communities* (Cambridge: Harvard University Asia Center, 2016).

69 B. Hendrischke, "Religious Ethics in the Taiping jing: The Seeking of Life," *Daoism: Religion, History and Society* 4 (2012); Li Jianmin 李建民, "They Shall Expel Demons: Etiology, the Medical Canon and the Transformation of Medical Techniques before the Tang," in J. Lagerwey and M. Kalinowski (eds), *Early Chinese Religion* (Leiden: Brill, 2009); M. Strickmann, *Chinese Magical Medicine* (Stanford: Stanford University Press, 2002); P. S. Nickerson, "The Great Petition for Sepulchral Plaints," in S. R. Bokenkamp (ed.), *Early Daoist Scriptures* (Berkeley: University of California Press, 1997).

70 Lin Fushi 林富士, "Zhongguo zaoqi daoshi de 'yizhe' xingxiang: yi Shenxian zhuan wei zhu de chubu tantao 中國早期道士的「醫者」形象：以《神仙傳》為主的初步探討," in *Zhongguo zhonggu shiqi de zongjiao yu yiliao* 中國中古時期的宗教與醫療 (Taibei: Lianjing, 2008); Lin Fushi 林富士, "Zhongguo zaoqi daoshi de yiliao huodong ji qi yiliao kaoshi: yi Han Wei Jin Nanbeichao shi de 'zhuanji' ziliao weizhu de chubu tantao 中國早期道士的醫療活動

escape the forces of destiny through alchemical drugs or arcane burial rituals.[71] In these writings, the ingestion of herbs was on a continuum with the ingestion of rare minerals, alchemical products, or the light of the sun, moon, and stars. Buddhists brought in foreign drugs, new disease treatments, and set up hospitals and hospices.[72] The notion of generating and spreading good karma, or merit, formed a broad ideological basis for extensive medical missionizing.[73] Health-related practices were widespread and straddled the (modern) line between normal good health and religious attainment.

Nourishing Life and Inner Nature

Within this broader context, no singular unified term emerged as a conceptual category directly parallel to "health." However, practices of the sort first evinced in Warring States *yangsheng* literature and the philosophical texts, such as breathing, bodily awareness, stretching, and diet, became adapted in late Han literature on transcendence.[74]

及其醫術考釋：以漢魏晉南北朝時期的「傳記」資料為主的初步探討," in *Zhongguo zhonggu shiqi de zongjiao yu yiliao* 中國中古時期的宗教與醫療 (Taibei: Lianjing, 2008).

71 On these methods, U.-A. Cedzich, "Corpse Deliverance, Substitute Bodies, Name Change, and Feigned Death: Aspects of Metamorphosis and Immortality in Early Medieval China," *Journal of Chinese Religons* 29 (2001); S. R. Bokenkamp, "Simple Twists of Fate: The Daoist Body and Its Ming," in C. Lupke (ed.), *The Magnitude of Ming: Command, Allotment, and Fate in Chinese Culture* (Honolulu: University of Hawai'i Press, 2005); R. F. Campany, "Living off the Books: Fifty Ways to Dodge *Ming* 命 in Early Medieval China," in Lupke, *The Magnitude of Ming*; M. Stanley-Baker, "Drugs, Destiny, and Disease in Medieval China: Situating Knowledge in Context," *Daoism: Religion, History and Society* 6 (2014).

72 Salguero, *Translating Buddhist Medicine in Medieval China*, and Liu Shufen 劉淑芬, "Tang, Song shiqi sengren, guojia, he yiliao de guanxi—Cong Yaofangdong dao huiminju 唐、宋時期僧人、國家和醫療的關係—從藥方洞到惠民局," in Li Jianmin 李建民(ed.), *Cong yiliao kan zhongguo shi* 從醫療看中國史 (Taipei: Lianjing, 2008).

73 C. P. Salguero, "Fields of Merit, Harvests of Health: Some Notes on the Role of Medical Karma in the Popularization of Buddhism in Early Medieval China," *Asian Philosophy* 23/4 (2013); Liu Shufen 劉淑芬, "Yaofangdong"; Zhang Ruixian 張瑞賢, Wang Jiakui 王家葵, and M. Stanley-Baker徐源, "The Earliest Stone Medical Inscription," in Lo, *Imagining Chinese Medicine*.

74 An early study of Daoist health practice is F. A. Kierman (trans.), *H. Maspero: Taoism and Chinese Religion* (Amherst: University of Massachusetts Press, 1981); On exercises and health cultivation after the Han, C. Despeux, "Gymnastics: The Ancient Tradition," in L. Kohn and Y. Sakade (eds), *Taoist Meditation and Longevity Techniques* (Ann Arbor: Center for Chinese Studies

Sects differed on how they understood the full potential of these practices, and this marked different types of practitioners and sectarian affiliation. The goals of achieving "long life" (*changsheng* 長生/*changshou* 長壽) and "immortality" (*busi* 不死) overlapped, since "long" life is an indefinite term.[75] By the fourth century CE, a clear hierarchy became quite widely established that distinguished between curing, *yangsheng*, and transcendence as three successive stages of progress on the spiritual path. This structure informed not only transcendent writings, like those of Ge Hong 葛洪 (?283–?343) and early Shangqing 上清 (Upper Clarity) Daoism, but also the materia medica, the *Shennong bencao jing* 神農本草經.[76] These created clear distinctions not just between the practices but also among their practitioners and the sects in which they were situated. The status of *yangsheng* practices was thus the focus of hot contestation—as different claims were made by competing sects about specific practices and what they could do. Thus the term *yangsheng* came to refer to a dynamic field of health practice consisting of many different actors and practices—practices that were also adopted into regimes of immortality and the attainment of divine powers.

While *yangsheng* literature in the main consisted of collections of techniques with little commentary, in the manner of recipe (*fang* 方) literature, increasingly over time the topic of "inner nature" (*xing* 性) came to the forefront of philosophical reflections. From the Han dynasty onward, the terms for "life" or "vitality" (*sheng* 生) and "inner nature" were treated as cognate terms and often interchangeable in

---

University of Michigan, 1989); U. Engelhardt, "Qi for Life: Longevity in the Tang," in Kohn and Sakade, *Taoist Meditation*.

75 M. Stanley-Baker, "Cultivating Body, Cultivating Self: A Critical Translation and History of the Tang Dynasty *Yangxing yanming lu* (Records of Cultivating Nature and Extending Life)" (MA thesis, Indiana University, Bloomington, 2006), 34–47.

76 On the place of drugs and other curative strategies within formalized cultivation regimes, Stanley-Baker, "Drugs, Destiny, and Disease."

variant editions of the same text.[77] Intellectuals in the Mystery School (*Xuanxue* 玄學) from the third to the fifth centuries came to engage in questions about inner nature and life force and their relationship to cognition, the self, morality, and individual agency.[78] In the seventh century, the medical prodigy Sun Simiao 孫思邈 (581–681? CE) argued that the cultivation of inner nature was more than just breathing exercises, rather it was rooted in ethical behavior.[79] Within the Daoist tradition of inner alchemy (*neidan* 內丹), which emerged in the tenth century, *xing* and *ming* came to refer to the two poles of inner consciousness and the material body, in which a complex of philosophical and personal and health were enmeshed. *Xing* referred to non-being, inner nature, humanity, innate nature, and original Buddha-nature, while *ming* referred to being, personal destiny, vital force, health, and the material world.[80]

Knowing Styles: On How and Not Why

Throughout this chapter I emphasized that while social and intellectual differences were made between different schools and disciplines, there were continuities of practice across these different domains. The practices of *yangsheng*—whether breath meditation, heightened bodily awareness, stretching, or diet—were adopted, adapted, and redefined within multiple communities, who thus situated them within or beyond

77 Witnessed by the commentarial traditions on the *Baopuzi neipian* 抱朴子內篇 in Wang Ming 王明 (ed.), *Ge Hong* 葛洪: *Baopuzi neipian jiaoshi* 抱朴子內篇校釋 (Taibei: Liren shuju, 1981). The *Shennong bencao* also refers to the middle level of cultivation as nourishing *xing*, not *sheng*.

78 See discussion of Ji Kang below.

79 Sun Simiao 孫思邈, *Beiji qianjin yaofang* 備急千金要方 (Taibei: Guoli zhongguo yiyao yanjiusuo, 1990), 28.476a–78b; S. Wilms, "Nurturing Life in Classical Chinese Medicine: Sun Simiao on Healing without Drugs, Transforming Bodies and Cultivating Life," *Journal of Chinese Medicine* 93 (2010).

80 F. Pregadio, "Destiny, Vital Force, or Existence? On the Meanings of *Ming* in Daoist Internal Alchemy and Its Relation to *Xing* or Human Nature," *Daoism: Religion, History & Society* 6 (2014); C. Despeux, *Immortelles de la Chine ancienne: Taoïsme et alchimie féminine* (Puiseaux: Pardès,

their own repertoires. The need to differentiate them arose precisely because of their similarity.[81] Historians have agreed that reference to epistemology was not commonly used to distinguish medicine or health practices from other kinds of knowledge, rather critiques were made of their character or genealogy.[82] Liu Xiang's critique of transcendents serves as an example: he differentiates the excessive sort, who told tall tales, from the noble type, who were above distinctions of death and life, and who knew the secrets of sustaining life. Liu did not foreground their knowing styles but their trustworthiness and their decorum.

While historians have noted the absence of epistemological argument, anthropologists observing similar habits in modern-day traditional Chinese medicine hospitals and clinics argue that a lack of focus on epistemology and metaphysics does not signify absence of reflection. Judith Farquhar argues that a deepened practical logic is situated in the bodies, institutions, materials, and methods by which Chinese doctors give treatment.[83] The knowing practice of medicine takes place within a temporal, processual domain that subordinates arguments about first causes to the exigencies of the clinic. Studied from this perspective, doctors' recourse to genealogies of practitioners is not merely social polemic but tacitly evokes inherited, collective experience. The metaphysics of the transformation of *yinyang* and the five phases are revealed not as static first principles that stand outside of time but as guiding orientations within the contingencies of the clinical encounter. Farquhar's emphasis on embodied practice and

1990), 223–27; I. Robinet with Chang Po-tuan, *Introduction à l'alchimie intérieure taoïste: de l'unité et de la multiplicité* (Paris: Editions du Cerf, 1995), 165–95.

81 On the construction of categories as an act of comparison and hierarchy-making, J. Z. Smith, "On Comparison," in *Drudgery Divine: On the Comparison of Early Christianities and the Religions of Late Antiquity* (Chicago: University of Chicago Press, 1990), 36–53.

82 Lloyd and Sivin, *The Way and the Word*, 205; Harper, *Early Chinese Medical Literature*, 156–83. Other scholars have also noted an absence of emphasis on analytical philosophy in traditional Chinese thought and the reframing of "philosophy." J. Kurtz, *The Discovery of Chinese Logic* (Leiden: Brill, 2011).

83 J. Farquhar, *Knowing Practice: The Clinical Encounter of Chinese Medicine* (Boulder, CO: Westview Press, 1994); J. Farquhar, "Problems of Knowledge in Contemporary Chinese

experience reveals classificatory strategies as strategic tools used in context, not as cosmological abstractions.

In this vein, modern Chinese doctors distinguish between the knowing practice of well-heeled clinicians and the measuring, standardizing, and objectifying the knowing style of western medical science. This is epitomized in the phrase "Knowing the why, but not the how" (*Zhi qi suo yi ran, er bu zhi qi ran* 知其所以然，而不知其然。), which contrasts arguments from western scientific medicine to the clinical know-how and textual traditions of China.[84]

This processual orientation toward knowledge as embodied and contingent is reflected in early texts like the *Zhuangzi* and expanded on in later commentaries, which anticipate the knowing styles of Chinese medicine today. The following passage from the chapter on "Nourishing Life" argues against the acquisition of knowledge for knowledge's sake:

> This life of ours, it has its limits but knowledge is limitless. You are already in danger if you chase the limitless by means of the limited. If you already know this, and still strive for knowledge, then you are in danger for certain![85]

Commentators in the medieval period argued that one should strive for virtuosity in nourishing inner nature (*xing*), and that this was an intuitive, skillful, and embodied praxis. Ji Kang 嵇康 (style name Shuye 叔夜, 223–262 CE) writes:[86]

> To have desire without thinking is a movement of one's inner nature (*xing*). [But] to first recognise something and then to produce

Medical Discourse," *Social Science & Medicine* 24/12 (1987). For a study of how the Daoist avoidance of excessively abstract, theoretical knowledge informs modern debates.

84 J. Farquhar and Wang Jun, "Knowing the Why but Not the How: A Dilemma in Contemporary Chinese Medicine," *Asian Medicine* 5/1 (2011).

85 *Zhuangzi jishi*, 2.3.115.

86 Style names were a formal term of address, usually given at coming-of-age rituals.

> a stimulus is a function of knowing (*zhi* 智). When inner nature moves, it is satisfied when it meets its object, there is no excess. But knowing follows a stimulus, and does not get exhausted; it is unable to stop. Thus the problems of the world are always to be found in knowing, not in the stirrings of *xing*.[87]

For Ji Kang, excessive fascination with the intellect and knowledge (both *zhi* 智) leads one to lose touch with the wisdom of inner nature (*xing*). Both are known through self-reflection. The intellect is portrayed as one step removed from the present moment—an initial cognitive recognition that only subsequently prompts activity. Inner nature moves spontaneously, without thought, and is intuitively and accurately responsive to the environment. This framing evokes the notion of enskilment, development of a fine-tuned sensitivity that reacts immediately, and is comparable to recent arguments for understanding knowledge as practice.[88] This is especially apparent when we consider that *Zhuangzi*'s heroes are craftsmen and skilled workers who, trained in seemingly lowly tasks, in fact portray a practical knowledge of the Dao as it moves through the world.

How one knows has a direct impact on health. Inner nature's desires are limited; they are framed by the possibility of the moment

87 The translation is informed by R. G. Henricks, *Philosophy and Argumentation in Third-Century China: The Essays of Hsi K'ang* (Princeton, NJ: Princeton University Press, 1983), 44–45. Henricks adds an alternate subject in places, a "one" that desires and seeks satisfaction. I choose to keep the subject as "intelligence," which has agency—it desires and seeks satisfaction. Note that Hsi Kang is an alternative spelling for Ji Kang. Not originally written as a commentary on the *Zhuangzi*, this passage was compiled by Zhang Zhan 張湛 (stylename Chudu 處度, fl. 370) along with other third- and fourth-century commentaries in his "Collected Essentials of Nourishing Life" (*Yangsheng yaoji* 養生要集), now lost. It survives in the redacted *Yangxing yanming lu* 養性延命錄 DZ 838 1.1.2b–3a. Stanley-Baker, "Cultivating Body, Cultivating Self."

88 On enskillment and practice as means to understand knowledge as embodied and performative, rather than abstract, subjective cognition, J. Lave and E. Wenger, *Situated Learning: Legitimate Peripheral Participation* (Cambridge: Cambridge University Press, 1991). On knowledge as ontological assemblage, in contrast to epistemological theory, as in B. Latour, "A Textbook Case Revisited. Knowledge as Mode of Existence," in E. J. Hackett, et al. (eds), *The Handbook of Science and Technology Studies* (Cambridge: MIT Press, 2007).

and satisfied when they achieve their object. But the search for knowledge, which is abstract, knows no bounds, and can exhaust the body. Whereas the spontaneous movement of inner nature nourishes health, attachment to artificial knowledge can destroy it.

Conclusion

The material terms *qi* and *sheng* were rarely subjected to epistemological inquiry—these terms were taken for granted, unlike *xing* and *ming*, which became hotly debated in both philosophy and inner alchemy. The majority of writings regarding *qi* and *sheng* were concerned with *pragmatic* issues of how to nurture them, whether and why the practices work, and with axiological or valuative frameworks within which to couch such practice regimes. Thus the object of inquiry was not epistemological or causal, that is, how to know what *qi* and *sheng* are, but rather how to cultivate them, and what it meant to do so. All of these questions implicitly or explicitly reflected on the status of experts in this knowledge, as we can see in early philosophers distinguishing themselves from those who merely cultivate the body, in the formation of medical theory, and in the bustle and jostle of the post-Han religio-medical market.

I have reviewed three main periods of writing on health, and how they have construed *qi* practices to nourish or protect life force (*sheng*), and thus lengthen lifespan (*ming*) and improve or transform destiny (also *ming*). After the Han dynasty, these deliberations also placed increased emphasis on the role of inner nature (*xing*) within this complex. Beginning with early philosophical writings that mention *qi* practices, different authors in the *Guanzi* and *Zhuangzi* sought to emphasize proximity or distance between physical and spiritual health. Different layers in the same texts, such as the *Daode jing*, evinced different periods of thinking about the body in relation to spiritual cultivation. A significant change in thinking about bodily cultivation and in how the body was conceived took place just prior to and during the Han dynasty, with the emergence of five phase thought and its eventual confluence with *yinyang* and *qi* theories. These configured bodily cultivation in much closer relation to conceptions of the natural world that reflected not just cosmological but also political powers. With the collapse of the Han dynasty and its ideological underpinnings, numerous new religions entered the scene, organizing healing, cultivation, and theodicy within different schemes.

In this regard, the early philosophers—who debated whether or not deliberate cultivation of the body was equivalent to cultivation of the mind and spirit—bear great similarity to the post-Han aspirants to transcendence—who evaluated whether *yangsheng* practices could lead to transcendence or whether they were merely useful for curing disease and achieving a normal lifespan. Both questioned whether practices of the interior body could lead to a transformed subjectivity, the status of a sage or superior being, while at the same time being anxious to distance such meditations from exercises that simply improved bodily function. These practices were not analyzed on the basis of their metaphysics, but on how to do them, who did them, and what they achieved. Knowledge about health in China was predominantly processual and embodied, rather than abstract; and it is better analyzed through theories of ontology and practice theory than through epistemological reflection.

CHAPTER TWO

Medical Conceptions of Health from Antiquity to the Renaissance

Peter E. Pormann

To counter the massive outbreak of foot-and-mouth disease in England in 2001, the government decided to resort to a massive cull. The strategy was not just to eliminate infected animals but also those that could potentially carry the disease. The massive pyres of burning lambs and sheep shocked the nation, and in letters and interviews, people expressed their revulsion at the "slaughter of innocent animals." Diseased animals were "guilty," whereas healthy ones were "innocent." A semantic shift occurred from bodily health to moral rectitude. The equation of bodily and moral integrity is an old one and goes back at least to the times of Homer, as illustrated by the example of Thersites from book 2 of the *Iliad*. In an assembly of the Achaeans, he was the only one who dared criticize Agamemnon for his greed and warmongering. Thersites is an antihero: he wants to go home rather than gain glory by taking Troy. "He was the most shameful man who came to Troy. He had crooked legs, a limp in one leg, bent shoulders slanting towards the chest, and moreover a pointed head, crowned with lank thin hair."[1] The shame resulted both from his bodily deformity and his moral depravity. In later times, the Greeks coined the term *kalokagathia* for the fact of being both "beautiful (*kalos*) and (*kai*) good (*agathos*)." This theme of linking moral and physical health is found not just in the Greek tradition but also in Latin and Arabic.

Other themes linked to the body that appear in Greek culture and then continue to be of importance in later traditions, both East and West, are those of purity and balance. In the Hippocratic Oath, for instance, the physician vows to keep his life and the art "in a pure and holy way (ἁγνῶς δὲ καὶ ὁσίως)."[2] Pythagoreans and other early Greek thinkers enjoined to abstain from eating beans (κυάμων ἀπέχεσθαι), in part, undoubtedly because they cause flatulence. The physician and philosopher Alcmaeon, who probably lived toward the end of the sixth century BCE, stressed the importance of balance in a famous fragment:[3]

> Alcmaeon says that what maintains health is the equality (*isonomia*) of the powers, of the moist and dry, cold and hot, bitter and sweet, and the other [opposites], whilst the monarchy of only one among them causes sickness, for the monarchy of the one of the two is destructive for the other. And sickness occurs, with regard to the agent, from excess of heat or cold; with regard to the [material] origin, from abundance or lack of nourishment; and with regard to place, blood, marrow, or the brain; it is also sometimes produced by external causes, certain kinds of water, the country, blows, dearth, and other causes similar to these, whilst health is the proportionate

1 Homer, *Iliad* 2.216–19.

2 H. von Staden, "'In a Pure and Holy Way': Personal and Professional Conduct in the Hippocratic Oath?" *Journal of the History of Medicine and Allied Sciences* 51 (1995), 404–37.

3 A. Laks and G. W. Most (ed. and trans.), *Early Greek Philosophy*, 9 vols. (Cambridge, MA: Harvard University Press, 2016), 5:762–65.

mixture of the qualities (τὴν δὲ ὑγιείαν τὴν σύμμετρον τῶν ποιῶν κρᾶσιν).

This fragment contains one of the earliest extant definitions of health: it is the proportionate mixture of the qualities. We find a political metaphor here as well: equality sustains health, whereas monarchy removes it. This balance of opposing qualities and humors will play a major role in the conception of health from antiquity to the Renaissance.

The need to keep the body in balance was even realized by philosophers who generally disdained the body. For instance, the Greek physician Galen of Pergamum (129–216) reports the following anecdote about the Cynic philosopher Diogenes of Sinope (fourth century BCE). Diogenes reportedly lived in a barrel, ate with the dogs (hence the name "Cynic," from Greek *kynes*, dogs), and despised all material comforts.[4] And yet, in order to get rid of an excess of sperm, he went to see a prostitute. The prostitute in question, however, was too slow, so that he "touched his penis with his hand and dismissed her, when she later entered, saying that his hand made him sing the wedding song already (τὴν χεῖρα φθάσαι τὸν ὑμέναιον ᾆσαι)."[5] Another philosophical school, that of the Stoics, had a similar approach. For them, bodily health belongs in the "indifferent (*adiaphoron*)" category: it does not matter whether one is healthy. And yet, they created a subcategory of the "preferable indifferent (*adiaphoron proêgmenon*)," under which health falls.[6]

Many of these ideas and principles found in ancient Greek thought about health and disease influenced later cultures. The purpose of this chapter is to highlight some of the developments in Greek medical

---

4 O. Overwien, *Die Sprüche des Kynikers Diogenes in der griechischen und arabischen Überlieferung* (Stuttgart: Franz Steiner, 2005).

5 Kühn, *Galen: Opera Omnia*, 8.419.

6 M. Forschner, *Die stoische Ethik* (Darmstadt: Wissenschaftliche Buchgesellschaft, 1995), 114–23.

literature, beginning with the Hippocratic Corpus, then moving to Galen's work, and the writing of late antiquity. I shall then trace ideas about health in the medieval Islamic world, offering a few vignettes on thinkers such as Abū Bakr al-Rāzī (d. ca. 925), Avicenna (d. 1037), and Ibn Buṭlān (d. ca. 1063). Likewise, two commentators on the Hippocratic *Aphorisms* will come under scrutiny, as they show how health could be defined in different manners. The Arabic medical tradition had a profound impact on the Latin Middle Ages, and I shall discuss some Latin translations of Arabic texts and see how they were adopted and adapted to suit the needs of a different cultural, religious, and linguistic context. I shall end my analysis with a short outlook on the Renaissance, when the Arabic heritage, although still dominant, came under ever increasing pressure. But first let us turn to the Hippocratic Corpus at the beginning of the Greek medical tradition.

Greece

Hippocratic Corpus

The Hippocratic Corpus is a collection of writings attributed to Hippocrates but which must, in reality, have been written by many different authors over a long period of time: the earliest date to the fifth century BCE, whereas the latest were perhaps written as late as the first or second century CE.[7] They also differ, sometimes radically, in their theoretical and practical approach. For instance, the treatise *Nature of Man* (fourth century BCE) sets out the doctrine of what is nowadays known as "humoral pathology." Health consists in the balance (*eukrasia*) of the four humors: blood (*haima*), phlegm (*phlegma*), yellow bile (*xantê cholê*), and black bile (*melaina cholê*). Each of the four humors has two primary (or "cardinal") qualities: blood is warm

7 For an up-to-date overview on the Hippocratic Corpus and the medical traditions contained in it, see the chapters by E. Craik and J. Jouanna in P. E. Pormann, *Cambridge Companion to Hippocrates* (Cambridge: Cambridge University Press, 2018).

and moist; phlegm is cold and moist; black bile is cold and dry; and yellow bile is warm and dry. Therefore, an excess of one of the humors also leads to an imbalance in the temperature and moisture of the body. Or, as the author of *On the Nature of Man* puts it:[8]

> Man's body contains in itself blood, phlegm, and yellow and black bile. These things make up the nature of his body, and through them, he suffers disease and enjoys health. He enjoys the greatest health, when these are in balance to each other in terms of mixture, power, and quantity, and when they are most mixed. He suffers disease, when there is too much or too little of one of them, or when it is separated in the body and not mixed with all of them.

Yet, another Hippocratic text, *On Ancient Medicine*, rejects as facile the notion that health can be reduced to a balance of warm and cold, and dry and moist. Rather, there are many more opposites that lead to health and disease:[9]

> For since they [sc. the first discovers of medicine] thought that it is not the dry or the wet or the hot or the cold or any other of these things that harms man—or that man has any need of them—but rather the strength of each thing and that which is more powerful than the human constitution, they regarded as harmful that which the human constitution was unable to overcome, and this they sought to remove. And the strongest of the sweet is the sweetest, of the bitter the bitterest, of the acid the most acidic, and of each one of all the things present, the extreme degree. For they saw that these things are also in the human being and cause it harm: for there is in

8 J. Jouanna, *Hippocratis De natura hominis*, Corpus Medicorum Graecorum 1:1.3 (Berlin: De Gruyter, 2002), 172, line 13, 174, line 3 [6.38–40 Littré].

9 M. J. Schiefsky, *Hippocrates: On Ancient Medicine*, Studies in Ancient Medicine, no. 28 (Leiden: Brill, 2005), 90–93 (translation slightly modified).

> the human being salty and bitter and sweet and acid and astringent and insipid and myriad other things having powers of all kinds in quantity and strength.

We therefore see that although the author of *On Ancient Medicine* rejected humoral pathology as developed in *On the Nature of Man*, he still embraced the notion of balance. Likewise, the author of *On Affections* opens his work with a statement that since "health is the greatest good (πλείστου ἄξιόν ἐστιν ἡ ὑγιεία)"[10] one needs to know about blood and phlegm, as these two cause diseases.

Therefore, health depends on the body being well balanced: an excess of a humor or a quality leads to disease. Yet, there are other factors beyond the mixture of the body that come into play as well. For instance, the Hippocratic author of *Airs, Waters, Places* investigates the effect of the environment on human health. Some locations, for instance, are more salubrious than others: if you live by a swamp or in a place where air is stale or water stagnant, this will have a negative effect. Bad air, so-called miasmas, causes diseases such as malaria (the Italian for "bad air"). Not only does the environment have an impact on health, but it also determines one's character traits. In the second half of *Airs, Waters, Places*, the author sets out an anthropology of different peoples according to where they live and contrasts Europe with Asia.

The Hippocratic Corpus remained highly influential throughout the centuries. Generations and generations of physicians read it, commented on it, and interpreted it in light of their own doctrines.[11] Likewise the historical figure of Hippocrates grew in stature, as more and more stories were told about him, most of them clearly

---

10 *Affections* 1, 6.208 ed. Littré. P. Potter (trans.), *Hippocrates*, vol. 5: *Affections. Diseases 1. Diseases 2* (Cambridge, MA: Harvard University Press,), 6–7, translates this as "health is of the utmost value to human beings."

11 This lively tradition has been traced in A. Anastassiou and D. Irmer, *Testimonien zum Corpus Hippocraticum* (Götingen: Vandenhoeck and Ruprecht, 1997–2012).

apocryphal.[12] One man did more than any other to perpetuate Hippocrates's fame and to ensure that his works continued to be read: Galen of Pergamum.

Galen

Whereas Hippocrates is nowadays the most famous Greek physician, Galen of Pergamum (ca. 129–216) was certainly the most influential.[13] He studied medicine and philosophy in Pergamum and Alexandria, and then headed to Rome, where he served as physician to the emperor Marcus Aurelius and his son Commodus. He reportedly wrote over 400 books in the course of his long life, and, today, his extant works make up a significant portion of the classical Greek literature that has come down to us. Toward the end of his life, he wrote not only a text titled *About His Own Opinions* but also *About His Own Books* and *About the Order of His Own Books*. He thus organized his works into a canon, and they certainly became canonical in later times. We can roughly divide Galen's oeuvre into three categories: works "for beginners" (τοῖς εἰσαγομένοις) with a mainly didactic purpose; monographs on individual topics such as simple and compound drugs, therapeutics, and so forth; and commentaries on Hippocratic works.

The first and key introductory text by Galen is *On the Sects for Beginners*. A short treatise on medical epistemology, it defines medicine as follows:[14]

> The aim of the art of medicine is health, and its end is the acquisition of health. Physicians ought to know by which means to bring

12 W. D. Smith, *The Hippocratic Tradition* (Ithaca, NY: Cornell University Press, 1979).

13 For an overview about Galen, see R. J. Hankinson (ed.), *The Cambridge Companion to Galen* (Cambridge: Cambridge University Press, 2008).

14 R. Walzer and M. Frede (trans.), *Galen: Three Treatises on the Nature of Science* (Indianapolis: Hackett, 1985), 3, with modification.

about health, when it is absent, and by which means to preserve it, when it is present.

Therefore, health and medicine are intrinsically linked in Galen's mind, much as the Hippocratic author of *On Affections* linked man's greatest good, health, to the "greatest benefit to mankind," medicine. According to Galen's definition, medicine has a twofold purpose: to retain health (prophylactic medicine) and to restore health (therapeutic medicine).

Galen was, above all else, a good Hippocratic, yet he shaped his Hippocrates in his own image. He achieved this not least by writing commentaries on some Hippocratic works to the exclusion of others. For instance, he authored an influential commentary on *Nature of Man* but considered *On Ancient Medicine* as spurious. Even where he regarded a text as genuine, he reinterpreted it in light of his own thinking. For instance, when commenting on the passage from *On the Nature of Man* quoted above, Galen paraphrases Hippocrates as saying that health ensues through the "symmetry (συμμετρία)" or "good mixture (εὐκρασία)" of the "elements (στοιχεῖα)."[15] None of these three technical terms, however, appears in the Hippocratic text; by using his own language, Galen simply overlays his ideas onto the Hippocratic original.

Therefore, a cornerstone of Galen's notion of health remains humoral pathology: the balance of the four humors (or "common elements," as Galen calls them here) takes center stage. We have already seen that the four humors are each linked to two of the four primary (or "cardinal") qualities, warm and cold, and dry and moist. In his work *On Mixtures*, Galen elaborated a typology of mixture, with one good mixture, resulting in health, and eight bad mixtures, resulting from an excess or a deficiency in one or two of the primary qualities.[16]

---

15 J. Mewaldt, *Galeni In Hippocratis De natura hominis commentaria* (Leipzig: Teubner, 1914), 32–33.

16 P. Singer (trans.), *Galen: Selected Works* (Oxford: Oxford University Press, 1997), 202–89.

In this way, health depends on a balance within the body, a balance of primary qualities linked to the four humors. There are, however, also external factors that influence health according to Galen, namely the so-called six non-naturals.[17] These six things that are not part of the nature of the individual are ambient air; food and drink; sleep and wakefulness; exercise and rest; excretion and retention (e.g., of urine, feces, and semen); and sadness and joy. Health is achieved by regulating these six non-naturals. For instance, excessive sex, leading to too much evacuation of semen, results in disease, whereas sex can be beneficial for certain conditions such as melancholy.

To preserve health means first and foremost to prescribe the right diet. Galen wrote a treatise *On the Powers of Foodstuff*, in which he arranges various food items according to their powers,[18] such as sweet or salty; beneficial or harmful to the stomach; favoring individual humors; digestible with ease or difficulty, and so on. In many cases, Galen provides recipes on how to prepare specific meals, saying that the physician "should be well versed in [the art of cookery]."[19]

Galen is also interested in other non-naturals. For example, he wrote a short and charming treatise *On Exercise with a Small Ball*, advocating physical activity. Toward the end of his life, he also penned an epistle *On the Avoidance of Grief*, in which he draws on Stoic and Epicurean philosophy to free the reader from distress; this text was recently (in 2005) rediscovered and has attracted much scholarly attention.[20]

According to Galen, the second main objective of medicine is the restoration of health when it is lost. In this area of therapeutics,

---

17 Galen sets out his theory in his *On the Medical Art*, chap. 23; see I. Johnston (ed. and trans.), *Galen: On the Constitution of the Art of Medicine. The Art of Medicine. A Method of Medicine to Glaucon* (Cambridge, MA: Harvard University Press, 2016), 246–49.

18 M. Grant, *Galen on Food and Diet* (London: Routledge, 2000), 11.

19 Kühn, *Galen: Opera Omnia*, 6.609.

20 See V. Nutton, "Avoiding Distress," in P. N. Singer (trans.), *Galen: Psychological Writings* (Cambridge: Cambridge University Press, 2014), 43–106.

Galen wrote numerous works focusing on a variety of themes. His *On Therapeutics for Glauco* in two books, for instance, is a beginner's guide to the subject, whereas his massive *On the Method of Healing* in seventeen books exhaustively discusses how to treat diseases. Restoring health of course involves manipulating the six non-naturals and, most importantly, adjusting the balance of the primary qualities and humors. One way is to employ simple and compound drugs. Galen wrote a sizeable work on the former, in which he develops a theory of degrees for different drug properties, such as primary (warm/cold; dry/moist), secondary (affecting the whole body), and tertiary (affecting a specific part of the body, e.g., diuretic). Galen insists that a competent physician ought to know these degrees:[21]

> The same applies to the drugs which have drying or moistening powers: one ought to know not only their general action, but also which is the first to depart from the balance and the mean between the opposite powers; and then which is the next one. Next, one ought to distinguish the third, fourth and fifth degree, if possible, by differentiating between them with clear definitions. Through such an accurate knowledge of their powers, we shall be able to use the simple drugs themselves with professional expertise [*technikôs*], and to compose compound drugs methodically [*kata methodon*], and, in addition, to use the drugs correctly [*orthôs*] that have already been composed.

The example of rue can illustrate how the different qualities and degrees combine in a simple drug:[22]

21 Kühn, *Galen: Opera Omnia*, 6.429; trans. C. Petit, quoted in P. E. Pormann, "The Formation of the Arabic Pharmacology: Between Tradition and Innovation," *Annals of Science* 68 (2011): 493–515, at 502.

22 Kühn, *Galen: Opera Omnia*, 7.100–101; trans. in Pormann, "The Formation of the Arabic Pharmacology," at 502, with modifications.

> Wild rue is even of the fourth degree of heating [drugs], and cultivated rue is the third [degree]. Not only does it taste sharp, but also bitter, and it therefore is able to cut and remove thick and viscous humours. Through the same power it is diuretic. It is also composed of small particles and removes flatulence [*aphysos*], so that it is fitting for flatulence, reduces and disperses the desire for sexual intercourse [*aphrodisia*], and dries well. For it belongs to the strongly drying drugs.

The actions of a simple drug can thus restore thc balance in the body, or one can resort to the use of compound drugs. Compound drugs are made according to sometimes elaborate recipes, and Galen wrote two seminal works on them, *On Compound Drugs According to Places* and *On Compound Drugs According to Genera*. The former is arranged according to the place in the body that is affected, such as the brain, the eye, the lungs, or the reproductive organs; the latter according to the type of compound drug.

Health of the body and health of the mind are interlinked for Galen. The psychic state of the patient is one of the six non-naturals, discussed above. Galen is also keen to point out "that the faculties of the soul follow the mixtures of the body," as a famous treatise by Galen is entitled. This interplay between mind and body is particularly visible in the area of mental health. There are a number of conditions affecting the brain such as phrenitis, lethargy, madness, and melancholy, all extensively discussed in his *On the Affected Parts*. Phrenitis, for instance, is characterized by the onset of delusions accompanied by severe fever (undoubtedly sometimes overlapping with what we call meningitis), whereas melancholy lacks fever. We shall return to the topic of melancholy in greater detail when discussing medieval medicine. For now, suffice it to say that Galen's system of health—how to preserve it and how to restore it—became dominant in late antique Alexandria and beyond.

Before turning to that topic, however, it is worth mentioning that in Roman imperial times, ideas of humoral pathology were by no means the only ones about health. One major challenge to Galen was Methodism—obviously the medical variety, not the Christian one due to the ministry of John Wesley.[23] Health, according to the Methodist school, ensued when the passageways of the body were moderately open. Disease, by contrast, occurred in three states: flux, when the passageways were too wide open; stricture, when they were too closed; and a mixed state, in which there is flux in some parts of the body and stricture in others. By the time of late antiquity in the Greek-speaking East, however, Methodism had given way to Galenism.

Late Antiquity

Late antique Alexandria was a bastion of Galenism.[24] So-called Iatrosophists, professors of medicine, taught the subject in its amphitheaters.[25] They resorted to a canon of books by Hippocrates and Galen that were core curriculum there. They include the Hippocratic *Aphorisms* and *Prognostic* and Galen's *On the Sects for Beginners, On the Elements According to Hippocrates*, and *On Therapeutics for Glauco*. Three genres of medical writing rose to prominence: abridgments, commentaries, and encyclopedias. The first two served didactic purposes: the commentaries were often verbatim notes taken during lectures (ἀπὸ φωνῆς); and the abridgments gave succinct summaries of these lectures. In both these genres, the principle of division (*dihairesis*) was a powerful mnemonic tool. For instance, in the

---

23 M. Tecusan, *The Fragments of the Methodists*, vol. 1: *Text and Translation* (Leiden: Brill, 2004).

24 O. Temkin, *Galenism: Rise and Decline of a Medical Philosophy* (Ithaca, NY: Cornell University Press, 1973) remains fundamental.

25 P. E. Pormann, "Medical Education in Late Antiquity: From Alexandria to Montpellier," in H. F. J. Horstmanshoff and C. R. van Tilburg (eds), *Hippocrates and Medical Education: Selected Papers Read at the XIIth International Hippocrates Colloquium, Universiteit Leiden, 24–26 August 2005* (Leiden: Brill, 2010), 419–41.

Alexandrian Summary of Galen's *On the Sects for Beginners* (extant only in Arabic, although most likely of Greek origin), the author sums up Galen's discussion of medicine as follows:[26]

> Soranus said in the context of defining medicine: "Medicine consists in knowing the matters of health and illness." Herophilus said: "Medicine consists in knowing the matters of health, that is, healthy bodies, the causes which preserve and effect health, and the signs indicating health; the matters of disease, that is, diseased bodies, the causes which effect disease, and the signs indicating disease; and the matters which are neither related to health nor disease, that is: the body, which is in this state [i.e. either health or disease]; the cause effecting this [either health or disease]; and the signs indicating it [i.e. either health or disease]." The causes comprise two groups: (1) health-related, and (2) disease-related. The health-related [sc. causes] comprise two sub-groups: (1a) [causes] preserving existing health, and (1b) [causes] restituting and bringing health after it [sc. health] has been lost. Disease-related [causes] also comprise two sub-groups: (2a) [causes] preserving existing disease, and (2b) [causes] creating disease which previously did not exist. The [subject of] causes of health which preserve existing health [i.e. (1a)] is called "regimen of healthy people"; it is effected through food and drink, venesection, exercise, and bathing. The [subject of causes] creating health which is non-existent [before, i.e. (1b)] is called "cure." Some of these causes expel from the body things which need expelling, like for instance venesection and purging through medication; others change the form as far as necessary,

26 O. Overwien (ed.), "Zur Funktion der *Summaria Alexandrinorum* und der *Tabulae Vindobonenses*," in U. Schmitzer (ed.), *Enzyklodädie der Philologie: Themen und Methoden der klassischen Philologie heute* (Göttingen: Ruprecht, 2013), 187–207, at 191–92; the end of the quotation has been supplied by collation of two manuscripts: Istanbul, Süleymaniye Kütüphanesi, MS Fatih 3538, reproduced by F. Sezgin, *Ǧawāmiʿ al-Iskandarānīyīn*, 2 vols. (Frankfurt: Maʿhad Taʾrīḫ al-ʿUlūm al-ʿArabīyah wa-l-Islāmīyah fī iṭār Ǧāmiʿat Frānkfūrt, 2001), 1:5; and London, British Library, MS Add. 23407, fol. 4a.

either from the outside, like for instance a bandage, or from the inside like drinking cold water.

We thus encounter Soranus, a methodist physician flourishing around 100 CE, and Herophilus, the great Greek anatomist of the third century BCE, who represents here the rationalist (or dogmatist) point of view. The latter defines medicine as knowledge of bodies, causes and signs, relating both to health and disease, with further subdivisions. We also find a similar division in the so-called *Viennese Tables*, containing medical branch diagrams. Here "healthy things (τὰ ὑγιεινά)" are divided into "things preserving an existing health; they are called healthy dietary measures" and "cures and remedies which restore non-existent health;" the former are further subdivided into "balanced food; drink; exercise; baths" and the latter into "phlebotomy; purging; drinking cold water; enemas; soft bandages."[27] This predilection for division also characterizes the development of medicine in the medieval Islamic world.

Medieval Islamic World

The late antique medical curriculum and medical practice in that period more generally had a profound impact on medicine in general and concepts of health in particular. In order to show some of the ways in which the Greek tradition persists and also is transformed, I shall offer four vignettes. First the oeuvre of the physician and philosopher Abū Bakr al-Rāzī offers insights into how he wanted to preserve health and how he innovated testing new ways of doing this. Second, Avicenna's definition of medicine and his debate about health being one of two states of the body illustrate a more fundamental (and theoretical) question. Third, the *Almanac of Health* by Ibn Buṭlān shows

---

27 Overwien, "Zur Funktion der *Summaria Alexandrinorum*," 197.

how the theoretical principals are put in to practice. Fourth, two commentators on the Hippocratic *Aphorisms* illustrate some of the exegetical strategies that help in the evolution of how health is conceived. To look at conceptions of health in the medieval Islamic tradition, let us first turn to an author who is arguably the greatest clinician of the medieval period and also an accomplished philosopher.

Abū Bakr al-Rāzī

Abū Bakr Muḥammad ibn Zakariyyāʾ al-Rāzī (d. 925) from the Persian city of Rayy wrote a large number of works on both philosophy and medicine, yet many of the former have not come down to us or only survive indirectly in the form of reports by usually hostile witnesses. On the medical side, however, we are far better served. We have a number of short treatises on a variety of topics, ranging from medical ethics and charlatans to whether one should consume mulberries after watermelons.[28] Al-Rāzī's massive medical encyclopedia, the *Book for al-Manṣūr* (*al-Kitāb al-Manṣūrī*), and his notes posthumously published as the *Comprehensive Book* (*al-Kitāb al-Ḥāwī*) also survive, as do two monographs on sexual intercourse and case notes published by his students after his death. In these works, al-Rāzī discusses health on numerous occasions. For the present purpose, I shall focus on his medical encyclopedia, the *Book for al-Manṣūr*, and notably on book 4, "On the Preservation of Health (*fī ḥifẓ al-ṣiḥḥa*)."[29] I shall also add a few remarks about al-Rāzī's interesting stance regarding medical epistemology, insofar as it concerns this topic.

Al-Rāzī's *Book for al-Manṣūr* continues a tradition of medical encyclopedias that dates to late antiquity. Oribasius (late fourth

---

28 P. E. Pormann and E. Selove, "Two New Texts on Medicine and Natural Philosophy by Abū Bakr Muḥammad ibn Zakarīyāʾ al-Rāzī," *Journal of the American Oriental Society* 137 (2017), 279–99.

29 The following references are to the edition by Ḥāzim al-Ṣiddīqī al-Bakrī, *Al-Kitāb al-Manṣūrī fī l-ṭibb* [*Book for al-Manṣūr on Medicine*] (Kuwait: Maʿhad al-Maḫṭūṭāt al-ʿarabīya, 1987).

century), Aëtius of Amida (sixth century), Aretaeus (sixth century), and Paul of Aegina (fl. seventh century) all wrote medical encyclopedias that also include long sections on the preservation of health. At the beginning of his *Book for al-Manṣūr*, al-Rāzī defines medicine as the science through which one can preserve present, and restore absent, health, echoing Galen's earlier definition. For al-Rāzī, medicine is an art that everybody ought to know because disease can strike at any time and physicians are not always in attendance.

In book (*maqāla*) 4 of his encyclopedia, al-Rāzī deals with the topic of how to preserve health. He largely covers in it various aspects of the six non-naturals. He discusses movement and rest (p. 203), sleep (204), food (204–6), drink (207), purging superfluities (208–9), location (209–10), chronic pain (210–13), worries and habits (213), countering damage caused by food (214–15) and drink (215–16), blood-letting (217), enemas (217–19), vomiting (220), sexual intercourse (220–21), bathing (221–22), care for the teeth (222–23), eyes (223–24) and ears (224), plague prevention (225–27), regimen according to seasons (227–28), and care for children and pregnant women, as well as obstetrics (229–34). He concludes this book with a page on how to test physicians (*fī miḥnat al-ṭabīb*).

In his discussion of wine, for instance, he has a paragraph on "how to replace wine (in prescriptions)." There, al-Rāzī gives the following advice:[30]

> Wine warms the stomach and the liver, dissolves flatulence, digests food, and has a diuretic and purging effect. Moreover, it makes one joyful and it entertains. This property cannot be replaced by any substitute to it.

30 Al-Ṣiddīqī al-Bakrī, 26.

Other properties can be partially replaced, although they, too, fall short of the original action. Al-Rāzī then provides a recipe for such a replacement for wine. Yet, he clearly thinks that there cannot be any real substitute.

In book 8 of his *Book for al-Manṣūr*, al-Rāzī lists various substances that harm one's health; they include venom of snakes and scorpions, bee and wasp stings, insect and spider bites, and so on. He also discusses the harmful effects on plants and medicinal substances such as arsenic, hellebore, hemlock and mercury. Al-Rāzī was not, however, content with merely quoting from past authorities; rather, he tested the damage that these substances cause in a variety of ways to health. For mercury, for instance, he gave a dose to an ape and observed its effects. Although the animal suffered pain, it merely secreted the mercury through the stool, and al-Rāzī therefore concluded that the harm to human health is not that great.[31] This animal experiment has been hailed by some as a precursor to modern medical testing.

Such Whiggish analysis is rather unhelpful,[32] yet it does raise the question of medical epistemology: how can we know that certain practices, therapies, or recipes contribute to the restoration of health? It is to this topic that al-Rāzī made quite significant contributions. His greatest claim to fame is undoubtedly his use of a control group in a medical experiment, which he describes as follows:[33]

> According to what I have seen by way of experience [*taǧriba*] and what I have seen in this book, regarding constant fevers: If [the patient] suffers from heaviness and pain in the head and neck lasting

31 Al-Ṣiddīqī al-Bakrī, 368; see A. Z. Iskandar, "Ar-Rāzī, the Clinical Physician (*Ar-Rāzī aṭ-Ṭabīb al-Iklīnī*)," in P. E. Pormann, (ed.), *Islamic Medical and Scientific Tradition*, Critical Concepts in Islamic Studies, 4 vols. (Routledge: London, 2011), 1: 207–53, at 225–26.

32 P. E. Pormann, "Medical Methodology and Hospital Practice: The Case of Tenth-century Baghdad," in P. Adamson (ed.), *In the Age of al-Farabi: Arabic Philosophy in the 4th/10th Century* (London: Warburg Institute), 95–118, at 111–12; reprinted in P. E. Pormann, *Islamic Medical and Scientific Tradition*, 2:179–206, at 193–94.

33 Pormann, "Medical Methodology and Hospital Practice," 109–10.

> for two, three, four, five days or more; and he avoids looking [directly] into the light, whilst tears flow; and he often yawns and stretches his body, having severe insomnia; and he perceives a feeling of extreme exhaustion, then the patient will progress to brain fever. He becomes daring like a drunkard, not paying any attention to food or drink, until the crisis supervenes. . . . So when you see these symptoms, resort to bloodletting. For I once saved one group [of patients] by [bloodletting], whilst I intentionally left another group, so as to remove the doubt from my opinion through this. Consequently all of these [latter] contracted brain fever.

The health of one group of patients is here preserved through bloodletting, whilst the control group contracted the condition. In this way, al-Rāzī innovated in the area of medical epistemology in order to preserve and restore health. Although this medical encyclopedia by al-Rāzī, and especially the ninth book, had great success both East and West, its popularity is eclipsed by Avicenna's *Canon of Medicine*.

Avicenna

Avicenna (d. 1037) is arguably the most influential medical writer of all times or at least of the medieval period.[34] His *Canon of Medicine*, a medical encyclopedia in five books, spawned a massive tradition of commentaries and abridgments, not just in Arabic but also in Persian, Latin, and Hebrew, among other languages. At the very beginning of this work, Avicenna defines medicine in terms that are very much reminiscent of Galen at the beginning of his *On the Sects for Beginners*, quoted above:[35]

---

34 See: P. Adamson (ed.), *Interpreting Avicenna* (Cambridge: Cambridge University Press, 2013).

35 Ibn Sīnā, *Kitāb al-Qānūn fī l-ṭibb,* 3 vols. (Būlāq, 1877), 1:3, lines 13–14.

> Medicine is the science through which one knows the states of the human body insofar as they are healthy or unhealthy, in order to preserve health when it is present, and to restore it when it is absent.

Avicenna continues to divide medicine into theory and practice, stating that he is concerned with theoretical knowledge and practical knowledge, but not actual practice. Then he offers an interesting argument against a tripartite division of human health:[36]

> An opponent cannot claim that the states of the human body are three: health [*ṣiḥḥa*], disease [*maraḍ*], and a third state that is neither health nor disease, saying "you [sc. Avicenna] have fallen short by only dividing into two [sc. into health and disease]." If such an opponent were to think, he would notice that neither of the following two things are necessary: neither the tripartite division, nor our abandoning it. For if such a tripartite division were necessary, we would say that absence of health [*al-zawāl ʿan al-ṣiḥḥa*] includes [both] disease [*maraḍ*] and this third state which they made not to fall under the definition of health. For health is a natural disposition [*malaka*] or a state [*ḥāla*] through which the actions proceed from a subject in a sound way [*salāmatan*]. It [this third state] does not have an equivalent definition [*muqābil hāḏā l-ḥadd*], unless they define health as they desire and attach to it [sc. the definition] conditions that they do not require. Physicians have no issue with this and do not argue about such things, nor does such a discussion provide them or anybody else with any benefit. The truth of the matter is more appropriate for the fundamental principles [*uṣūl*] of another art, namely fundamental principles of the art of logic; it should be studied under that heading.

36 Ibn Sīnā, *Kitāb al-Qānūn fī l-ṭibb*, 1:3,3–4.

In other words, Avicenna maintains the dichotomy of health and disease, already found in Galen, and argues against the position that there is an intermediary, third state, which, incidentally, he calls *taṯlīṯ*—a term denoting not just the division into three but also the Christian Trinity. At the end, Avicenna dismisses debates about whether the states of the body are two or three to the realm of logic; having no practical applications, physicians don't really care.

This discussion of where a particular question sits in the hierarchy of knowledge is of concern to Avicenna. He argued that medicine is a corollary branch of knowledge (or "science" in the sense of Latin *scientia*, Arabic *ʿilm*), akin to agriculture.[37] Questions of a higher ontological order, for instance about the makeup of matter, or even the existence of the four humors and the interplay of the four primary qualities, fall outside the purview of medicine; rather, they belong in the realm of physics (*ṭabīʿīyāt*). We shall return to Avicenna and his definition of medicine and health later, but now, let us turn to a highly practical work that aims at preserving and restoring health.

Ibn Buṭlān's Almanac of Health

> Health is rectified through the six [non-natural] causes which everyone who wants to have lasting health needs to balance and employ. [They are] first, improving the air reaching his heart; second, measuring food and drink; third balancing movement and rest; fourth preventing oneself from sinking into sleep or wakefulness; fifth measuring the expulsion and retention of superfluities; and six moderating one's joy, anger, fear, and despondency. This is how to have them in balance and through it, these six [non-natural] causes preserve health, yet when they depart from [this moderate state], they cause disease.

37 D. Gutas, "Medical Theory and Scientific Method in the Age of Avicenna," in D. C. Reisman (ed.), *Before and After Avicenna: Proceedings of the First Conference of the Avicenna Study Group* (Leiden: Brill, 2003), 145–62; reprinted in Pormann, *Islamic Medical and Scientific Tradition*, 1:33–47.

Thus begins Ibn Buṭlān's *Almanac of Health*, and we find here again the six non-naturals encountered earlier.[38] His *Almanac* is, in fact, a collection of tables with fourteen columns specifying "nature," "degree," "best type," "usefulness," "harmfulness," and so on until the last column, containing a brief narrative about "choice (*iḫtiyār*)," which is a catch-all category allowing Ibn Buṭlān to add his own comments. He lists 280 substances and activities coming under the heading of non-natural causes: numbers 1–210 are largely varieties of food, ranging from figs (no. 1) and raisins (no. 2) to rice (no. 36), chickpeas (no. 57), asparagus (no. 71), and different types of meat (nos. 92–5, 97–100), honey (no. 170), and different types of wine (nos. 190–94). Other things include music (nos. 211–13); joy, fearfulness, and anger (nos. 214–16); vomiting (no. 219); sleep (no. 221), sleep companion (no. 222), nightly entertainment (no. 223), and wakefulness (no. 224); sexual intercourse (no. 227) and sperm (no. 228); different types of exercise (nos. 232–38); different types of baths (nos. 239–44); seasons (nos. 271–74); locations (nos. 275–78); plague-infested air (no. 279), and the theriac (no. 280). For instance, Ibn Buṭlān describes sexual intercourse (*ǧimāʿ*) as follows:

> 1. number: 227; 2. name: "sexual intercourse;" 3. nature: uniting two partners to project seed; 4. degree: none; 5. good variety: when the destination of the sperm has been chosen; 6. usefulness: preservation of the species; 7. harmfulness: to people having cold and dry testicles; 8. removing its harm: through drugs generating sperm; 9. effect: none; 10. mixture: warm and moist; 11. age: adolescence and youth; 12. time: spring after having been cleansed from menstrual blood; 13. location: moderate; 14. people's opinions: H[ippocrates]; G[alen]; Ru[fus of Ephesus]; 15. choice: for procreation, nature

38 H. Elkhadem, *Le Taqwīm al-ṣiḥḥa (Tacuini sanitatis) d'Ibn Buṭlān: Un traité médical du XIe siècle* (Leuven: Peeters, 1990); all subsequent references are to this edition.

> has made a base pleasure leading to a noble goal in the universe. Through [this base pleasure], some people are seen to be jumping like brute beasts, especially when bitten and tickled, when the object [of the sexual encounter] is seen or thought about in a dream or whilst awake; what brings energy and happiness is prepared for him. The active [partner in the sexual intercourse] should be neither replete [with food] nor hungry, lest a blockage or dryness occur. The time should be balanced, the air not plague-like, and the location not conducive to disease.

In the margin, Ibn Buṭlān provides further astrological guidance:

> To produce a boy, one chooses male-producing signs [of the zodiac], the best of which are Libra and Sagittarius; for girls, Pisces and Virgo; Taurus is not bad either.

This arrangement in columns with marginal astrological notes proved extremely popular in both East and West. It made it easy to look up an activity or a substance to see how it fit one's own regimen and disposition. It was a practical solution to the problem of preserving one's health. Before looking at how Ibn Buṭlān and others influenced the Latin tradition, we will first briefly consider some more theoretical debates about health in Arabic commentaries on the Hippocratic *Aphorisms*.

Hippocratic Commentaries

Medical debate in Arabic flourished in particular in a genre that modern readers do not always associate with innovation, namely that of the commentary. In fact, there is a rich Arabic tradition on the Hippocratic *Aphorisms*, ranging from the ninth century to the fifteenth century.[39] The commentator who wrote by far the longest and

39 P. E. Pormann and N. P. Joosse, "Commentaries on the Hippocratic *Aphorisms* in the Arabic

most thorough work on the *Aphorisms* was the Christian physician Ibn al-Quff (d. 1286). He offers particularly interesting discussions of health as a concept. In the commentary on aphorism 2.19, he argues that health is beneficial not just in temporal but also in spiritual terms, saying:[40]

> We said that health is nobler than disease based on two reasons. First, health is achieved through balance while diseases occur due to a deviation from balance. Balance is nobler than its opposite. Second, through the presence of health, we can achieve happiness in this life and the next. Disease, however, prevents this. What helps achieve these two forms of happiness is nobler than that which prevents them.

Christians (just like Muslims) believed in the resurrection of the body, and therefore bodily health is important even for the afterlife; this, at least, appears to be al-Quff's point here. In his commentary on another aphorism, 2.1, he states that the parts of the body reach perfection (their ultimate aim or *entelecheia*) when they are healthy (*ṣaḥīḥa*), and continues:

> Health is brought about by two factors, first balanced mixture, and second a well-preserved structure. Therefore, whatever opposes this state of perfection constitutes pain. What opposes the condition of mixtures is what is called "noxious mixture" and [what opposes] the composite structure is the dissolution of continuity. Therefore, both [these things] are painful.

Tradition: The Example of Melancholy," in P. E. Pormann (ed.), *Epidemics in Context: Greek Commentaries on Hippocrates in the Arabic Tradition* (Berlin: De Gruyter, 2012), 211–249; and P. E. Pormann and K. Karimullah, "The Arabic Commentaries on the Hippocratic *Aphorisms*: Introduction," *Oriens* 45/1–2 (2017), 1–52.

40 The corpus of Arabic commentaries on the *Aphorisms* is now available at Manchester's institutional repository: www.research.manchester.ac.uk; the quotations from Ibn al-Quff's commentary on book 2 are available here: dx.doi.org/10.3927/52131995.

We have already talked about an imbalanced mixture as a cause for disease; here Ibn al-Quff adds the "dissolution of continuity (*tafarruq al-ittiṣāl*)" as an impediment to health. By this, he means blows, cuts, and so forth that clearly have a negative effect. Furthermore, Ibn al-Quff makes an interesting point about what degree of health is achievable. We cannot have absolutely perfect health, where all parts of our body are totally in sync. He calls this kind of health "imagined" (*ṣiḥḥa mutawahhama*). He continues:

> Such a type of health does not exist in the world outside us, owing to the fact that causes that change the body in this state surround it and influence it. This [type of] health is the one against which others are measured and which is used by physicians as a guide to whatever deviates from it. Then there is [health] that [really exists]; it has two types: that which is close to [the ideal type of health], and that which is far away from it. The former is called "best structure" and "balanced health," whereas the latter is, for instance, the health of someone suffering from a fever or cold, a boy or old man, or a convalescent person. Thus, health has a range within which it varies. The closer it is to imagined health the better; and the further the worse; and what lies between is mediocre health. Now that you know this, we can say that [health] which is close to the imagined health can be preserved by using that which is similar [to it].

In this rich quotation, Ibn al-Quff's predilection for division is visible. In each of the three texts quoted above, health is subject to a twofold division: temporal and spiritual; mixture and structure; perfect and imagined, and imperfect and really existing. More interestingly, Ibn al-Quff insists that health is an ideal that one can try to get close to but never reach. In this, he resembles other physicians and philosophers such as al-Rāzī and ʿAbd al-Laṭīf al-Baġdādī who insist that medicine is an approximate science in which one can never obtain absolute certainty, just as it is impossible to obtain absolute health.[41]

Latin Middle Ages

In the Latin West, Methodism remained popular in late antiquity and the early Middle Ages. Things, however, changed dramatically with the arrival of Arabic medical texts in Latin translation. The three focal points of what one might call the Arabo-Latin translation movement were southern Italy, Spain, and the Crusader States, especially the city of Antioch. With this translation movement, notions of health also traveled across the linguistic barriers.

Constantine the African

The first major translator was Constantine the African (d. before 1099), working mostly in southern Italy.[42] He translated numerous works, especially originating in his native North Africa. For instance, he rendered the medical encyclopedia *Provisions for the Traveller and Sustenance for the Sedentary* (*Zād al-Musāfir wa-Qūt al-Ḥāḍir*) into Latin, often in a paraphrastic way. There also is a Greek translation of this work—entitled Ἐφόδια τοῦ ἀποδημοῦντος—by a Constantine the Protosecretary of Rhegion, who should not be confused with Constantine the African. The *Provisions for the Traveller* was written by Ibn al-Ǧazzār, a Tunisian physician who died in 980 CE. Constantine the African translated two other texts without acknowledging authorship. The first is a monograph on melancholy by the North African

41 P. E. Pormann, "Qualifying and Quantifying Medical Uncertainty in 10th-century Baghdad: Abu Bakr al-Razi," *Journal of the Royal Society of Medicine* 106 (2013), 370–72; N. P. Joosse and P. E. Pormann, "Archery, Mathematics, and Conceptualising Inaccuracies in Medicine in 13th Century Iraq and Syria," *Journal of the Royal Society of Medicine* 101 (2008), 425–27.

42 C. S. F. Burnett and D. Jacquart, *Constantine the African and ʿAlī ibn al-ʿAbbās al-Maǧūsī: the Pantegni and Related Texts* (Leiden: Brill, 1994).

physician Isḥāq ibn ʿImrān (d. 907),[43] and the second on sexual intercourse by an unknown author.[44] Both illustrate the continuity of ideas from the Greek to the Arabic and Latin traditions.

Melancholy is a disease caused by an excess of black bile (Greek *melaina cholê*).[45] It is characterized by delusion, groundless fear, and despondency, all occurring without the presence of any fever. Therefore, in the case of melancholy, health appears to be impaired because of a natural cause, the excess of a humor, and one would imagine that it can be cured by restoring this humoral balance. In fact, the picture is much more complicated. In the Greek, Arabic, and Latin traditions, we find different types of melancholy, innate and acquired, and among the acquired type various subtypes, such as general, hypochondriac, and encephalic.[46] The treatment, likewise, did not just encompass medication but also diet, entertainment, and lifestyle more generally. For his own *On Melancholy* (*De melancholia*), Constantine the African largely drew on Isḥāq ibn ʿImrān's work.[47] Yet, Constantine adds a section on therapy, in which he recommends the moderate consumption of wine and sexual intercourse.

Sexual intercourse is also the topic of Constantine's *On Sexual Intercourse* (*De coitu*). This work is, again, an unacknowledged translation of an Arabic original, although the source has not been identified. The treatise is divided into 17 chapters: 1–12 deal with theory and 13–17 with practical advice. Chapters 8–10, for instance, deal with the

---

43 A. Omrani [ʿĀdil ʿUmrānī], *Maqāla fī l-mālīḫūliyā* (*Traité de la mélancolie*) (Carthage: Académie tunisienne des Sciences, des Lettres et des Arts Beït al-Hikma, 2009).

44 Translated in F. Wallis, *Medieval Medicine: A Reader* (Toronto: University of Toronto Press, 2010), 511–23.

45 There is an overabundance of literature on melancholy; however, the classical study remains R. Klibansky, E. Panofsky and F. Saxl, *Saturn and Melancholy: Studies in the History of Natural Philosophy, Religion, and Art* (London: Nelson, 1964).

46 See P. E. Pormann (ed.), *Rufus of Ephesus on Melancholy* (Tübingen: Mohr Siebeck, 2008).

47 K. Garbers (ed.), *Isḥāq ibn ʿImrān: Maqāla fī l-mālīḫuliyā, Abhandlung über die Melancholie, und Constantini Africani libri duo de melancholia: Vergleichende kritische arabisch-lateinische Parallelausg* (Hamburg: Buske, 1977).

benefits of sexual intercourse and when the best time to have it is. Constantine begins his discussion as follows:[48]

> In their books, the ancients said that the things that preserve health are exercise, baths, food, drink, sleep, and sexual intercourse. We should say how sexual intercourse is beneficial and when it ought to be carried out, and how it benefits or harms, and what happens to those who do it frequently.... But there is a suitable time for intercourse: when the body is in a tempered state with respect to all external influences, that is, it is neither full of food nor totally empty, nor cold nor warm nor dry nor moist, but tempered... intercourse is better before sleep than after sleep, because when one falls asleep, one rests from exertion.

We thus have here a list of six items, modified from the six non-naturals discussed above, and again, the idea of balance dominates. At the end of his *On Sexual Intercourse*, Constantine provides recipes for various aphrodisiacs, and it is for this reason that he is called "the cursed monk" in the "Merchant's Tale," one of Chaucer's *Canterbury Tales* (v. 1810). Be that as it may, the Latin translations by Constantine had a profound impact on notions of health, as they displaced Methodism in favor of humoral pathology. Even greater, however, was the influence of Avicenna's *Canon of Medicine*.

The Latin Avicenna

It was through the Latin translation by Gerard of Cremona (1114–87) that Avicenna's *Canon* would have a lasting influence on western medicine. It formed a major part in the curriculum of the nascent universities of Europe, and, during the Renaissance alone, it was

48 Translated in Wallis, *Medieval Medicine*, 516–17.

printed more than sixty times in various editions.[49] Avicenna's definition of medicine and health, quoted above, also featured prominently in the Latin tradition. In Latin it runs:

> *Dico quod medicina est scientia qua humani corporis dispositiones noscuntur ex parte qua sanatur uel ab ea remouetur ut habita sanitas conseruetur.*

This definition is quoted verbatim, for instance, by the great Spanish physician Arnald of Villanova (d. 1311),[50] and, as part of the *Generalities*, became core curriculum in the Italian universities from the High Middle Ages onward.[51] Teaching the *Canon* also inspired a vast commentary literature, where, again, we find notions of health transmitted and discussed in the new Latin context. One of the most famous commentators on Avicenna in Latin was Gentile Da Foligno (d. 1348), who devoted most of his life to writing an enormous commentary on Avicenna's *Canon*.[52] Whereas Avicenna's aims are largely theoretical—as we saw, he refrains from discussing actual practice—other more practical texts also had a great impact on the Latin tradition, including the fertile regimen of health literature.

Tacuinum Sanitatis (Almanac of Health)

In the late thirteenth century, we see another wave of Arabo-Latin translations, notably in Norman Sicily. The Jewish translator and physician Farağ ibn Sālim, known as Farragut, translated Abū Bakr al-Rāzī's *Comprehensive Book* (*al-Kitāb al-Ḥāwī*) into Latin at the behest

49 N. G. Siraisi, *Avicenna in Renaissance Italy: The Canon and Medical Teaching in Italian Universities after 1500* (Princeton, NJ: Princeton University Press, 1987; repr. 2014).

50 Arnoldus de Villa Nova, *Hec sunt opera Arnaldi de Villanova que in hoc volumine continentur* (Lyon: Fradin, 1504), sig. a_1^r.

51 Siraisi, *Avicenna in Renaissance Italy*, 1250–500, 160.

52 R. K. French, *Canonical Medicine: Gentile da Foligno and Scholasticism* (Leiden: Brill, 2001).

of Charles I of Anjou, king of Sicily (d. 1285); he also translated the *Almanac of Bodily Health* (*Taqwīm al-abdān*) by Ibn Ǧazla.[53] These translations date back to 1279 and 1280, respectively. One manuscript of the Latin translation of Ibn Buṭlān's *Almanac of Health* attributes it to the same Farağ ibn Sālim, again at the behest of Charles I of Anjou; another claims it was carried out on the order of Manfred, king of Naples and Sicily from 1258 to 1266.[54]

Be that as it may, arranging information in tables relating to regimen and preserving health proved very popular in the European tradition. The Latin version of Ibn Buṭlān's *Almanac* was first printed in 1531 in Strasburg, and then reprinted there in 1533. Vernacular versions also appeared, such as the German *Chessboard Tables of Health* (*Schachtafelen der Gesuntheyt*), which appeared in the same year. The Latin version is charmingly illustrated at the bottom of each page with pictures of the foodstuff or activities in question; plate 1 below purging and constipation; sexual intercourse and sperm; cleansing; drunkenness; and *foca*, a transliteration of the Arabic *fawqāʿ*, a sort of hangover drink. Ibn Buṭlān's *Almanac of Health* marked the beginning of a fertile genre of Latin texts on regimen.[55] It is just one of the many Arabic works in Latin translation that remained popular during the Renaissance, when the Arabic heritage remained highly relevant but also became fiercely contested.

Renaissance Medicine

We have seen that Avicenna's *Canon* continued to dominate medical discourse during the Renaissance and remained highly popular, not just in its Latin translation but also in its original Arabic. In fact, it is

53 M. Steinschneider, *Die hebraeischen Uebersetzungen des Mittelalters und die Juden als Dolmetscher*, (Berlin: Kommissionsverl. des Bibliographischen Bureaus, 1893), sec. 582, 974–75.

54 Elkhadem, *Le Taqwīm al-ṣiḥḥa*, 43.

55 M. Nicoud, *Les régimes de santé au Moyen Âge: Naissance et diffusion d'une écriture médicale en Italie et en France (XIIIe–XVe siècle)*, 2 vols. (Rome: École française de Rome, 2007).

one of the earliest books printed in Europe with Arabic type (Rome, 1593). We find an interest in Avicenna's definition of medicine and of health in the oldest Arabic manuscript now in the possession of the Royal College of Physicians, London, MS Tritton 12.[56] There on the first page of the text (see plate 2), a Renaissance hand adds the Latin translation to the Arabic. Yet, perhaps the same western hand, or, more likely another one, added in Arabic a definition of health (*taʿrīf al-ṣiḥḥa*), which runs as follows:

> Health is a bodily state from which all functions proceed at all times in a sound [*salīma*] fashion, without there being any immediate disposition for it to cease.

This shows that interest in the Arabic legacy continued. Yet there were also significant challenges to this Arabic heritage, which some physicians wanted to expunge from their own medical tradition. As time went on, fewer and fewer Arabic texts in Latin translation formed part of the medical curriculum, and by the nineteenth century, a lot of this Arabic heritage had been forgotten—or rather expunged for a variety of reasons.[57]

Conclusions

This brief survey has shown the common threads that run through notions of health from classical Greece to Renaissance Italy and beyond. We have often only briefly touched on the main aspects of health: balance of humors and qualities; the interaction between

---

56 P. E. Pormann, *Mirror of Health: Medical Science during the Golden Age of Islam* (London: Royal College of Physicians, 2013), 20–24.

57 P. E. Pormann, "The Dispute between the Philarabic and Philhellenic Physicians and the Forgotten Heritage of Arabic Medicine," in Pormann, *Islamic Medical and Scientific Tradition*, 2:283–316.

body and soul (as illustrated in the six non-naturals); and the shift of physical and moral health, which can even have effects on one's afterlife. This picture hides the many different approaches to health and medicine that existed over the centuries. I could only briefly allude to the differences, for instance, between works included in the Hippocratic Corpus (*On Ancient Medicine* versus *On the Nature of Man*); between the various schools of medicine, for instance those active in the Roman Empire (Methodism versus Galenic medicine); or among Renaissance physicians, for instance, between those favoring Arabic medical ideas and those advocating a return to the Greek and Roman sources.

We have also seen that health has philosophical implications. In the area of medical epistemology, for instance, al-Rāzī innovated in interesting ways, while drawing on debates that began in antiquity. Much more could have been said about the topic of "spiritual medicine," of philosophy as a means of keeping both body and soul healthy, just as Ibn al-Quff and others realized that health is a necessary condition for moral rectitude—*mens sana in corpore sano*.[58] The focus here was largely on written works, on medical texts with both theoretical and practical aims, and, again, a lot more could be said about the actual practice on the ground. To what extent, for example, did the ideas of Hippocrates, Galen, al-Rāzī, Avicenna, or Ibn al-Buṭlān really affect the fates and fortunes of individual patients along the centuries? Undoubtedly they did have a significant practical impact, although it is also clear that many treatments, recipes, and prescriptions were probably never or hardly ever employed.

"Health is the greatest good (πλεἱστου ἄξιόν ἐστιν ἡ ὑγιεἱα)," as the Hippocratic author of *Affections* 1 put it.[59] Medicine aims at preserving and restoring health. This is the most fundamental division of medicine: into prophylactics and therapy, more fundamental perhaps

58 For more on this topic see the contribution of Peter Adamson in the present volume.

59 See note 10 this chapter.

than even the division into theory and practice. The Hippocratic and Galenic framework of humoral pathology defined notions of health and disease over the centuries. And yet, we find subtle differences and new developments across cultures and creeds. This, perhaps, is the greatest lesson that this investigation about health can teach us: a shared common discourse, whether in Greek, Arabic, or Latin, united physicians and philosophers in their quest to understand and improve this "greatest good."

CHAPTER THREE

The Soul's Virtue and the Health of the Body in Ancient Philosophy

James Allen

Already in the fifth century BCE, medicine (*iatrikê*) enjoyed a high reputation among philosophers as an art dedicated to a manifestly valuable goal, health, and which, in the hands of the best practitioners, satisfied the highest scientific standards. It became commonplace for philosophers to hold up medicine as a model for their own discipline in two ways. They compared the function of philosophy to the curative or therapeutic function of medicine, and they viewed the kind of knowledge achieved in, or aspired to by, medicine as a standard against which to measure the knowledge to be achieved in their own discipline.

The aim of this essay is to explore—selectively—the two ways in which medicine served philosophy as a model together with some of their implications and the problems they raise. The point of departure for my first set of reflections is the analogy to the therapeutic *function* of medicine, the way in which it imparts health and cures disease.

I shall be especially concerned to discover whether the *content* of the knowledge that belongs to philosophy corresponds or fails to correspond to that belonging to the art of medicine in ways suggested by the analogy. The point of departure of the second is the way in which the standards satisfied by medical knowledge were held up as an example for philosophy to follow. I shall be especially interested in certain reflections in and about the discipline of medicine regarding the *character* of the knowledge needed by medicine to discharge its function and their implications for philosophy conceived as a counterpart to medicine, occupied with the care of the soul.

Medicine and Philosophy as Philosophy

The analogy between the functions of the two disciplines belonged to a family of comparisons. To the object of medicine's "care" (a translation of the Greek term, *therapeia,* often but not exclusively used with medical treatment), the body, corresponds the object of philosophy's care, namely the soul. The end of medicine is health; that of philosophy, virtue (*aretê*), regarded as the health of the soul (or the happiness that virtue produces).[1] The Roman Stoic Seneca (first century CE) even held that human beings owe the original acquisition of the concept of virtue to an analogy with bodily health and strength.[2] Opposed to health in the body are diseases; their counterparts in the soul are unruly passions and false judgments about good and evil, which it is the task of philosophy to cure or expel.

"While medicine cures the sicknesses of the body, wisdom (*sophia*) rids the soul of its affections (*pathê*)." This observation, which is due to Democritus, the fifth-century BCE atomist and ethical thinker, is

---

1 See J. Pigeaud, *La maladie de l'âme* (Paris: Les Belles Lettres, 1989).

2 *Letter* 120, 5; translation in B. Inwood (trans.), *Seneca: Selected Philosophical Letters* (Oxford: Oxford University Press, 2007).

an early but representative example of the philosophers' attitude.[3] And this sentiment was to be echoed repeatedly in the centuries to follow, constituting common ground between philosophers who agreed about little else. In the *Charmides*, often regarded as one of Plato's early dialogues meant to depict the historical Socrates roughly as he was, Socrates seizes the opportunity afforded by the young Charmides's headache to cite the view of a Thracian physician, who holds that the body should properly be treated as part of the whole consisting of soul and body and turns the discussion to the soul and one of its virtues, temperance (156d). He proceeds to describe the dialectical examination to which he then subjects Charmides as "doctoring" or "physic" (158e).

In Plato's *Sophist*, usually regarded as a later dialogue incorporating new and distinctively Platonic ideas, we find an account of a "noble form of sophistry," which resembles nothing so much as Socratic dialectic (226aff.). Like medicine, to which it is explicitly compared, it discharges a purgative function, freeing the soul of errors by refutation as medicine cures the body of disease by therapy. In Plato's *Gorgias*, Socrates elaborates an especially detailed analogy, to which we shall return, between, on the one hand, the arts that care for the body (medicine and gymnastics), and on the other, their counterparts in the care of the soul, which he treats as a highly philosophical form of politics, if not philosophy itself. Justice, which is a part of politics in this scheme, he calls "medicine for wickedness" (*iatrikê ponêrias*, 478d, meaning by "medicine" the discipline, not the medications it applies).

Epicurus, who because of his hedonism and rejection of divine providence was in other ways something of an odd man out in ancient philosophy, is in perfect accord with his rivals on this point.

---

3 H. Diels and W. Kranz (eds and trans.), *Die Fragmente der Vorsokratiker, Grieschisch und Deutsch*, 3 vols. (Berlin: Weidmann, 1951–1952), B 31; A. Laks and G. W. Most, (eds and trans.), *Early Greek Philosophy*, 9 vols. (Cambridge, MA; London: Harvard University Press, 2016) vol. 7, pt. 2, D 235. Unless otherwise noted the translations are my own.

> The *logos* of that philosophy is vain by which no affection (*pathos*) of a human being is treated. For just as there is no benefit to medicine if it does not heal the sicknesses (*nosos*) of bodies, so too there is none to philosophy unless it expels the affections of the soul.[4]

Chrysippus, the third head of the Stoa (third century BCE)—whose views about providence and pleasure could not have been more different from those of Epicurus—embraced the same system of analogies. "It is not the case," he maintained, "that that there is an art called medicine, which is occupied with the diseased body, and not an art occupied with the diseased soul." He went on to speak of the "physician of the soul," of the "affections" (*pathê*) treated by both types of physician and to develop an analogy between the therapies (his word) employed by each.[5]

Philo of Larissa (late second to early first century BCE), the last head of Plato's Academy and a so-called Academic skeptic, worked out a detailed and systematic comparison between medicine and philosophy, and their concerns, methods, and goals.[6] His student Cicero (106–43 BCE) made extensive use of the analogy between philosophy and medicine, and the maladies of the soul and the diseases of the body the treatment of which is their object, above all in the *Tusculan Disputations*, which tackles some of the most pressing issues faced by human beings, such as the fear of death, pain, mental distress, other dangerous emotions, and the sufficiency of virtue for happiness.[7]

---

4 H. Usener (ed.), *Epicurea* (Leipzig: Teubner, 1887), frag. 221.

5 P. H. De Lacy (ed. and trans.), *Galen: On the Doctrines of Hippocrates and Plato*, 3 vols. (Berlin: Akademie Verlag, 1978–84) vol. 1, 298, 27–300, 12.

6 Text, translation and discussion in C. Brittain, *Philo of Larissa: The Last of the Academic Sceptics* (Oxford: Oxford University Press, 2001), 278–90. More in M. Schofield, "Academic Therapy: Philo of Larissa and Cicero's Project in the Tusculans," in G. Clark and T. Rajak (eds), *Philosophy and Power in the Graeco-Roman World: Essays in Honour of Miriam Griffin* (Oxford: Oxford University Press, 2002), 91–109.

7 *Tusculan Disputations* 2.43, 3.6, 4.23, 58; text and translation in J. E. King (trans.), *Cicero: Tusculan Disputations* (Cambridge MA; London: Harvard University Press, 1927).

Sextus Empiricus, who was active in the second century CE and is the best known to us as one of the Pyrrhonian skeptics and a member of the empirical school of physicians, maintained that the enormous mass of skeptical arguments collected by his school over the course of its existence were comparable to the doctor's therapies and like them an expression of philanthropy, meant to free mankind of mental distress analogous to bodily disease.[8]

The main analogies gave rise to others. Antisthenes, a fifth-century BCE follower of Socrates who inspired the Cynics, defended his severity toward his students by comparing it with physicians' treatment of their patients, and he pointed to the fact that physicians spend time in the company of the ill without contracting their illnesses to explain his, and more generally the philosopher's, willingness to converse with the wicked.[9] Epictetus, the second-century CE Stoic philosopher, compared the philosopher's school to the doctor's surgery, and he observed that, like the patients who leave the latter still in pain from their treatment, the philosopher's pupils should expect their studies to be tough going.[10]

Health is the object of medicine both in the sense of its end or aim and in the sense of the subject of the knowledge or understanding by which it is constituted—together with, and in systematic relation to, the human body, the unhealthy affections to which it is liable and the measures by which they can be counteracted. The ancient philosophical authorities cited above insist that the function of philosophy is likewise therapeutic. Epicurus uses the verb *therapeuein*. Sextus speaks

8 *Outlines of Pyrrhonism* (hereafter *PH*) 3.280–81, translation in J. Annas and J. Barnes (trans.), *Sextus Empiricus: Outlines of Scepticism* (Cambridge: Cambridge University Press, 2000). See A.-J. Voelke, "Soigner par le Logos: La Therapeutique de Sextus Empiricus," in A.-J. Voelke (ed.), *Le Scepticisme antique: perspectives historiques et systématiques* (Geneva: Cahiers de la revue de Théologie et Philosophie), 1990.

9 Diogenes Laertius 6.4, 6; text and translation in R. D. Hicks (trans.), *Diogenes Laertius: Lives of Eminent Philosophers*, 2 vols. (Cambridge, MA: Harvard University Press, 1925).

10 *Discourses*, 3.23. 30; text and translation in W. A. Oldfather (trans.), *Epictetus: Discourses* (Cambridge, MA; London: Harvard University Press, 1925–1928).

of "curing" (*iasthai*), the verb from which the terms for doctor or physician (*iatros*) and the art of medicine (*iatrikê)* are derived. As disease is to the body, so "affection," *pathos*, meaning passion, emotion or psychic disturbance, is to the soul. Though Democritus and Epicurus use the term *nosos* for bodily diseases and *pathos* for affections of the soul in the above passages, as they were well aware, the term *pathos* can also be used of bodily disease (whence our term "pathology"). Chrysippus chose to emphasize the affinity between medicine and philosophy by using *pathos* for both bodily disease and psychic affection, and he spoke of the diseased body and the diseased soul using the verb corresponding to *nosos*. Cicero entertained the idea of translating *pathos*, meaning affection of the soul, as *morbus* (disease), before opting for *perturbatio* (disturbance).[11] Unstated but implied is an analogy between medical therapy—bleeding, dieting, cauterization, cold baths, and other measures in the ancient physician's therapeutic repertoire—and philosophical therapy—argument, instruction, and explanation.

Democritus and Epicurus, especially the latter, use the medical analogy to emphasize the instrumental value of the knowledge or wisdom pursued by philosophy. There is, then, a possible tension between philosophy, on a certain lofty elevated or exalted conception of the discipline, and philosophy conceived along therapeutic lines. Notoriously, Epicurus has no interest in knowledge for its own sake. Part of the burden of his observation cited above is to reject the idea that there might be such a thing as philosophical knowledge that is of value in its own right.

It seems fair to ask, is the knowledge that medicine needs in order to discharge its therapeutic function of a relatively low, ordinary, or garden variety, as regards its content, its character, or both? As we shall see, on some conceptions of the discipline the answer is yes. If this is so, might not the same be true of philosophy mutatis mutandis? At

11 *De finibus* 3.35; translation in J. Annas (ed.) R. Woolf (trans.), *Cicero: On Moral Ends* (Cambridge: Cambridge University Press, 2001).

least in the case of Epicurus, the answer is no. The knowledge philosophy needs to perform its therapeutic function is an exalted business, embracing a complete grasp of the fundamental atomic nature of the universe along with the capacity to explain and specify the causes of all the major natural phenomena, which include but are not confined to life and the soul. Nothing less will serve as a means to the end in view, namely to dispel false opinions about the gods, death, and the afterlife that are the chief impediments to happiness. These opinions are the affections of the soul that need to be cured or expelled. Still, the reasons for Epicurus' commitment to natural philosophy may appear peculiar to his own distinctive outlook and, from the perspective of an outsider, accidental, so that the question may be worth raising again. We shall come back to it.

For the time being, let us accept that, at least for many ancient philosophers, to secure the good of the beings with whose care they were charged, both medicine and philosophy had to command knowledge or understanding of a high and serious order, satisfying the strictest standards. This is certainly true of Plato's version of the analogy, which is set out in most detail in the *Gorgias*. The presentation there makes explicit much that was left implicit elsewhere, while adding some new and idiosyncratic elements of Plato's own. The context is a scathing examination of rhetoric's pretensions to the status of an art (*technê*). In order to explain his view that rhetoric is not a true art at all, but is rather an imposter that professes knowledge it does not have and flatters instead of caring for the object in its charge, Socrates erects the imposing system of correspondences that I mentioned.

Two pursuits are counterparts by standing in the same relation to different objects or domains. Thus rhetoric's place is that of a counterpart (*antistrophos*) to "cookery," because rhetoric and cookery are false likenesses or counterfeits of the true arts of justice and medicine respectively. Justice belongs to politics, which cares for the soul, while medicine is one of the arts that care for the body. They are counterparts to each other by discharging a corrective function, as contrasted with the regulative function of legislation and gymnastics, whose tasks are to set the soul and the body in good order in the first place. Medicine and justice possess genuine knowledge of the body and soul, the restoration of whose good conditions, health and virtue respectively, are their objects. By contrast rhetoric and cookery aim not at the good, but at pleasures of soul and body. They lack genuine understanding and rely instead on mere experience (*empeiria*) (463b, 499e–501b). As we shall see, the disparaging attitude toward "experience" here is highly tendentious, both in the context of the dialogue and that of ancient epistemology more generally.

For the present, especially worth noting is how, in the scheme laid out by Socrates, an inferior goal, pleasure, is paired with an inferior form of knowledge, experience—or rather, to do justice to the view taken by Socrates in the dialogue, a cognitive condition inferior to knowledge. The analogy between philosophy and medicine appears to reach its high watermark here. Objects of concern, the soul and the body, loftiness of aims, virtue and health, and the demand for knowledge of the highest order in both arts are all in harmony, as I put it above. Caution may be called for, because the dialogue speaks not of "philosophy" but of "politics" and its two species, "legislation" and "justice." On any account, politics as conceived by Socrates in the dialogue is exceedingly philosophical business, but is it identical to philosophy or a part of it? Evidence that it is comes from Socrates's assertion, late in the dialogue, that he is one of few present day Athenians, perhaps the only one, to practice the true art of politics, apparently by doing what he is depicted doing here and in other dialogues, namely philosophizing (521d ff.). That is what Callicles, the third and most formidable of the interlocutors Socrates faces in the dialogue, calls it in his word to the wise, warning Socrates to desist from *philosophizing* for his own good (484c–486d).

There is, however, an issue concerning the content of philosophical knowledge raised by the use of the medical or therapeutic model. In the *Gorgias* there gradually emerges a picture illustrated by appeal to the example of other arts, above all medicine, beginning with Socrates's efforts to elicit an answer to the question "what is the (alleged) art of rhetoric *about*?" (449c). Each art gives rise to characteristic activities or performances by the artist (450bc). It is productive of some good, for instance health in the case of medicine (451e–452d). As we have observed, each has an object of concern, such as the body. It must, if it has a corrective function like medicine, also know about the defective conditions to which the object of its concern is liable, for instance diseases, and know how to remedy them (449e–450a). Of central importance in the knowledge that belongs to a true art is the form or order that constitutes the good condition of its object of concern (506de).

Alongside his principal, therapeutic function, the physician, like the practitioners of other arts, has a second, complementary function: he makes or produces other physicians by instructing them in his art (449b, 455c, 458a). He does this by transmitting the knowledge that enables him to discharge his primary, therapeutic function. On this model, we would expect the philosopher or the philosophical politician above all to have, and to convey to his pupils, knowledge of the form or order that is the soul's good condition. If however the soul is, in whole or in part, mind or reason, its good or virtuous condition will be, in whole or in part, knowledge, and the soul will be treated or cared for by instruction. The primary, therapeutic function of philosophy may, then, not be distinct from its didactic function. This raises an inevitable question: When the philosopher teaches and cares for the soul by imparting knowledge, what is this knowledge *about*? It was easier to answer the analogous question about the content of the knowledge conveyed in medical instruction. In that case, it is easy to draw a clean distinction between the knowledge that belongs to the trained physician's mind and the condition of his patients' bodies, about which it is knowledge. With philosophy, by contrast, we are threatened with a regress of knowledge that is about knowledge.

A version of this puzzle occupies a prominent place in Plato's *Euthydemus*, which contains two sustained interventions by Socrates. The purpose of the first is *protreptic*, or exhortatory. In collaboration with his principal interlocutor, Clinias, Socrates argues that human beings who would be happy by securing what is good for themselves must pursue knowledge or wisdom, that is, practice philosophy. In the second, he takes up a question left unanswered at the end of the first intervention: what is the knowledge that will make us happy (288d ff)? It appears that the knowledge they seek is the art of politics, which they agree is the same as the kingly art. It is further agreed that each art furnishes a product (*ergon*) in the sphere of that which it governs; medicine and its product, health in the body, are Socrates's first example (291e). The product of the kingly art must, it appears, be knowledge, but not just any kind of knowledge. Apparently it is the knowledge in which it consists itself, and this is the knowledge by which one makes other human beings good (292d). This one will do by imparting the knowledge of how to make others able to impart the knowledge . . . and so on apparently without end. Approached in this way, the knowledge that is our soul's virtuous condition, which is analogous to the body's health, threatens to elude our grasp.

In the *Republic* Plato has Socrates complain about those (possibly including the real life Socrates himself) who hold that the good is knowledge or wisdom, namely knowledge of the good (505bc). And in another dialogue, the *Clitophon* (which may not be by Plato), after paying sincere homage to Socrates's mastery of protreptic and his ability to inspire ardor for virtue in his auditors, the title character complains that Socrates fails to say anything about the knowledge at whose acquisition their new-found zeal for virtue should be directed. This, he maintains, puts Socrates at a disadvantage compared with the professors of other arts, notable among them medicine.

I conclude that the analogies between medicine and philosophy and between bodily health and virtue will take us only so far. Useful and illuminating though the likeness between philosophical instruction and medical therapy may be, to the extent that the condition of the soul that is analogous to the health of the body is knowledge or wisdom, "knowledge of the soul and its good condition" will not serve to specify the content of philosophical wisdom and philosophical teaching as "knowledge of the body and the health of the same" serves to specify that of the medical art and medical instruction. As we have seen, in the case of Democritus and Epicurus, both of whom subscribed to the family of analogies with which we began, philosophical wisdom must range far beyond this. The same is true in different ways of Plato, the Stoics, and others.

The Character of Medicine and Philosophy

So far I have pursued some of the implications for the *content* of philosophy of the first of two affinities philosophers recognized between their discipline and medicine, that based on their shared therapeutic function. I now turn to the second, the way in which the nature and character of medical knowledge served as a model for the character of philosophical wisdom. As we have seen, in Plato's *Gorgias*, to the correspondences between the object of concern and ends pursued by medicine and justice, Socrates joined another between the kinds of knowledge they employ. It embraces genuine understanding of the natures of the matters that fall under its care, on which its choice of measures is based; by contrast, the imposter arts, rhetoric and cookery, rely on mere experience, or as Socrates puts it, the memory of what is accustomed to happen divorced from an understanding of the causes of what happens (499e–501a).

Plato's *Phaedrus*, which again takes up questions about the possibility of a rhetorical art, contains a famous appeal to the method of Hippocrates (271c). If we are guided by Hippocrates, it is agreed, we cannot hope to understand the soul or the body satisfactorily without knowledge of "the whole." It is a matter of scholarly controversy whether the whole in question is that of the body or of the universe, and also whether this clue can throw light on the Hippocratic problem, namely which, if any, of the heterogeneous collection of medical works ascribed to Hippocrates in antiquity were actually written by him. There can however be no mistaking Plato's respect for medical knowledge at its best.[12]

Aristotle, whose father was a physician, turns to medicine to illustrate the grasp of causes and principles—the mark of real knowledge, which sets it apart from mere experience—to illuminate the nature of wisdom or first philosophy, the subject we have come to call metaphysics (*Metaphysics* 981a7–20). What is more, he holds not only that there is an analogy between the knowledge of medicine and of philosophy but also that the content of medical knowledge and natural philosophy overlap. Where natural philosophy leaves off, medicine begins, and natural philosophers will conclude their studies by tackling medicine, while physicians, at least of the more serious and better sort, will begin with natural philosophy.[13]

Nevertheless one might wonder, does a physician really need knowledge of such an exalted order to discharge his therapeutic responsibilities? As we saw, in the *Gorgias* superior knowledge was paired with a superior end. There could be no true knowledge or science of pleasure, because genuine knowledge of the nature of the body or soul has to be, in the first instance, knowledge of its good, not something as superficial as pleasure (though it might well include knowledge of pleasure as a consequence). A pursuit concerned simply with pleasure would have to rely on an inferior form of cognition, mere experience. In the *Laws*, however, Plato allows that experience can serve as the basis of medical treatment directed toward health, the same end as medicine founded on real knowledge of the nature of the body (720a–e; cf. 857c–d). He pictures slaves or servants of doctors, called

12 J. Mansfeld, "Plato and the Method of Hippocrates," *Greek, Roman and Byzantine Studies* 2 (1980), 341–62. Cf. King, *Cicero: Tusculan Disputations* 3:7.

13 *On Sense and Sense-Objects* 463a19–b1; *On Youth and Old Age* 481b21–30.

"doctors" themselves in a manner of speaking. By observing their masters at work, they will have acquired the ability to treat many kinds of patients for many kinds of illness. Their patients will, to be sure, typically be drawn from the lower ranks of society, and these so-called doctors will lack the ability to converse with them about the nature of the body, the diseases from which it suffers, and the like, in the way true doctors can with their patients, who will be drawn from more elevated and better educated strata of society. Still, Plato seems to concede that they will be to a high, if not perhaps the highest, degree effective. The fact that it may be possible to achieve the end of medicine without possessing knowledge satisfying the highest standards gains added importance if such knowledge, apart from being unnecessary, is also unattainable, either in present conditions or ever. Patients need effective treatment when they are ill and cannot afford to wait until the art of medicine has satisfied all the expectations that philosophers and some of its philosophically minded practitioners have entertained.

Physicians and medical writers also had much to say about their discipline and its relation to philosophy. Paradoxically they were never more philosophical than when defending the autonomy of medicine from encroachments by the discipline of philosophy—for it is then that they tackled fundamental questions about the scope and nature of medical and other forms of knowledge. These defenses did not have to insist that medical knowledge should meet lower or different standards than those held up by the philosophers whose views we have been exploring. Indeed, on some plausible interpretations, the author of the Hippocratic treatise *On Ancient Medicine* thinks that the kind of hypothesis-based inquiry, as he styles it, pursued by natural philosophers and the physicians influenced by them is not merely out of place in the medical realm but also incapable of achieving genuine knowledge in any sphere, unlike medicine itself, whose discoveries put it ahead of all other disciplines.[14]

14 M. Schiefsky (trans.), *Hippocrates: On Ancient Medicine* (Leiden: Brill, 2005).

By contrast, other medical authors were willing to set a lower bar for their discipline. Diocles of Carystus, probably active in the fourth century BCE and one of the most eminent physicians of the age, is an especially important case in point. In a fragment about the properties of foodstuffs from a book on dietetics, he sets out several objections to the ways in which other physicians have tackled the subject.[15] Diocles objects in particular to those who think it is always necessary to specify the underlying cause why a foodstuff is, for instance, a diuretic or a laxative. They fail to realize that effective practice often does not require this and, further, that many existing things are in a certain way like principles by nature without satisfying the expectations of his opponents. He goes on to say that some physicians—meaning presumably those in the grip of this picture—tend to appeal to causes that are unknown, disputed, and implausible, and he maintains that, instead of following them or those who demand that a cause be specified in every case, it is better to put one's faith in what has been discovered through long experience. He takes this position despite acknowledging that in some cases the cause can, indeed, be discovered and despite thinking that inquiry into causes is worth pursuing.

On Diocles's view, then, it seems that there is a place, even a permanent place, in the art of medicine for precepts that, though in a certain way like principles because they serve the doctor as a basis for diagnostic and therapeutic reasoning, are not true principles of the kind demanded by philosophers like Plato and Aristotle. Such precepts would not comprehend fundamental natures or causes. Some of these quasi-principles will be empirical generalizations, based on the observation and memory of what is accustomed to happen, for which they are able to provide no explanation. In other words, Diocles countenances including as part of the art of medicine truths of the

15 P. J. van der Eijk, *Diocles of Carystus* (Leiden: Brill, 2000), frag.176. On Diocles and Herophilus, see M. Frede, "An Anti-Aristotelian Point of Method in Three Rationalist Doctors," in K. Ierodiokonou and B. Morison (eds), *Episteme etc. Essays in Honour of Jonathan Barnes* (Oxford: Oxford University Press, 2011), 115–37.

kind many philosophers had excluded from it, and which they had contrasted with principles of the kind that are fit objects for genuine knowledge and understanding. For Diocles the therapeutic function of medicine can be, indeed must be, discharged by relying on a mixture of precepts of different types.

Herophilus, a physician and medical theorist of the late fourth and third centuries BCE whose eminence was at least equal to Diocles's, theorized freely about natures and causes and thought it appropriate to appeal to them in medical reasoning, but he also raised doubts about the truth of the causal explanations that he employed, even perhaps the possibility of discovering true causal explanation at all.[16] A student of his, Philinus of Kos, active in the mid-third century BCE, is credited with founding the school of medical empiricism, which counted among its members many figures like Sextus Empiricus who were also members of the philosophical school of Pyrrhonian skeptics. Citing the enormous mass of conflicting theories produced by those who claimed to understand the nature of the body and the causes that produce health and disease, and the seemingly unending disputes to which they had given rise, they too argued that it was difficult and perhaps impossible to obtain the knowledge that Plato and those who agreed with him regarded as indispensable to an art of medicine. And they tried to show that, in any event, such knowledge was unnecessary, arguing for the possibility of an art of medicine dispensing with insight into the nature of the body and based solely on experience.

Far from requiring deep reflection and profound insight, a medical theorem—their term—might owe its discovery to a chance observation and its status to confirmation by subsequent observation, while the underlying causes because of which it obtained remained completely unknown. Galen (second century CE) illustrated the empiricists' ideas with a story about the chance discovery of a cure for

16 On him see H. von Staden, *Herophilus: The Art of Medicine in Early Alexandria* (Cambridge: Cambridge University Press, 1989).

elephantiasis. A sufferer who unwittingly consumed water in which a dead snake had decayed recovered from the illness afterward; later attempts to duplicate the result intentionally also met with success, and the resulting therapeutic theorem, unexpected and inexplicable as it was, could be accepted as a perfectly sound basis for treating cases of elephantiasis.[17]

We have come a long way from the picture with which we started, according to which philosophy and medicine are set over soul and body, whose excellent conditions, virtue and health, can be obtained only by means of the most profound knowledge of the relevant domain. If we are swayed by the lines of argument that moved the empiricists, and in a different way Diocles and Herophilus, we will be persuaded that the art of medicine can and perhaps must do without a full understanding of the nature of its defining object of concern, the body and its healthy condition. Can medicine so conceived and the health of the body it pursues, divorced from any deeper understanding, serve any longer as models for philosophy and the excellent condition of the soul that is its object?

For many philosophers the answer must obviously be no, but I shall conclude by mentioning some for whom the answer was yes, though with an important qualification. These were the Pyrrhonian skeptics, some of whom, as I mentioned, were also members of the empiricist school of medicine, most notably Sextus, whose works on Pyrrhonism survived to become our main source for this form of skepticism. Their attachment to the therapeutic ideal is plain from the passage I cited above, comparing the Pyrrhonists use of arguments to the doctor's therapies. It repays closer attention.[18]

---

17 Galen, "Subfiguratio Empirica," in Deichgräber (ed.), *Die griechische Empirikerschule: Sammlung der Fragmente und Darstellung der Lehre* (Berlin: Weidmann, 1965), chap. 10, 75ff. Translation in R. Walzer and M. Frede (trans.), *Galen: Three Treatises on Science* (Indianapolis: Hackett, 1985).

18 *PH* 3.280–81, trans. Annas and Barnes lightly modified. On this topic see also J. Allen, "Pyrrhonism and Medicine," in R. Bett (ed.), *The Cambridge Companion to Ancient Scepticism* (Cambridge: Cambridge University Press, 2010), 232–34.

> Sceptics are philanthropic and wish to cure by argument the conceit and rashness of the Dogmatists. Just as doctors for bodily affections have remedies which differ in potency, and apply severe remedies to patients who are severely afflicted and milder remedies to those mildly afflicted, so the sceptics propound arguments that differ in strength—they employ weighty arguments, capable of vigorously rebutting the dogmatic affections of conceit, against those who are distressed by severe rashness, and milder arguments against those who are afflicted by a conceit that is superficial and easily cured and which can be rebutted by a milder degree of plausibility.

It is noteworthy that, in the hands of the Pyrrhonists, the analogy to medical treatment extends to the individualized treatment of patients, which was such an important part of the Greek medical ideal. This is true despite the fact that the analogue on the side of philosophy is an odd one: the use of arguments of different levels of cogency adjusted to the spiritual sufferer's particular needs, not to persuade those to whom they are addressed, but rather to help them suspend judgment by balancing them against opposing arguments.

Analogous to the diseases that the physician aims to cure is the target at which Pyrrhonian philanthropy directs its fire, namely rashness or dogmatism. The analogue to bodily health that the Pyrrhonists aim to produce is tranquility or freedom from mental distress (*ataraxia*), which they sometimes identify as the end or goal of life.[19] The puzzlement we are likely to experience upon learning that a school of self-professed skeptics could have identified an "end of life" may be diminished if we keep in mind that they seem to have regarded the desirability of *ataraxia* as a given, akin to a desire for a healthy body, an assumption no one would question, rather than the deliverance of a deep insight into human nature or a contentious theory. That this happy condition was promoted by

---

19 *PH* 1.25, 215.

suspension of judgment for the sake of which the Pyrrhonists made use of therapeutic arguments of different strengths, was likewise not a conclusion derived from insight into human nature, but they maintained, a chance discovery.

> What is said about the painter Apelles belongs also to the sceptic. They say that he was painting a horse and wanted to represent in his picture the lather on the horse's mouth; but he was so unsuccessful that gave up, took the sponge on which he had been wiping off the colours from his brush, and flung it at the picture. And when it hit the picture, it produced a likeness of the horse's lather. Now the sceptics were hoping to achieve tranquility by resolving the anomalies in what appears and is thought, but being unable to do this they suspended judgment. When the suspended judgment, however, tranquility followed as it were by chance as a shadow follows a body.[20]

The medical empiricists did not have a monopoly on the idea of chance discovery confirmed by subsequent experience, but the resemblance between the two schools on this point is striking, even more so if we attend to the way they were willing to base a philosophy or way of life, on the one hand, and a system of medicine, on the other hand, on something as insubstantial and unsatisfying—from the point of view of other philosophers and physicians—as chance observations and experience entirely divorced from any deeper understanding of causes.

A significant part of Pyrrhonian skepticism as presented by Sextus corresponds nicely to the therapeutic method of medicine. Drawing on the work of his Pyrrhonian predecessors, Sextus describes in considerable detail techniques or modes for the production of arguments that can be opposed to those put forward by the school's dogmatic

---

20 *PH* 1.27–9; trans. Annas and Barnes, modified.

opponents, the better to bring about suspension of judgment and the tranquility attendant upon it. And he devotes substantial parts of his oeuvre to the exposition and skeptical examination of competing dogmatic views on a full range of philosophical issues from all three standard parts of philosophy—ethics, physics, and logic—as well as in the disciplines of rhetoric, grammar, astronomy, and mathematics. Among many other things, he tackles questions about the nature and possibility of truth, knowledge, and proof; about the nature and existence of the gods; and about the nature and reality of number and motion and change.

The Pyrrhonians are called "skeptics" from the term *skepsis,* meaning inquiry. They contrast their way of philosophizing with that of dogmatists, on the one hand, who, they maintain, have called off the search for truth prematurely because they think they have discovered and grasped it, and the Academics, on the other, whom they accuse (unfairly) of having given up the search for the opposite reason, because they are convinced that knowledge is unobtainable. By contrast they, the skeptics, continue inquiring.

The Pyrrhonists were no less attached to the therapeutic model of philosophy than any other school of ancient philosophy. But if their affinity with medical empiricism led them to think that tranquility, the health of the soul, could be achieved without discovering the fundamental truths about the world and human beings' place in it, which their dogmatic rivals regarded as an indispensable means to the same end, the suspension of judgment that they, to their surprise, find so congenial is not the result of disengagement or withdrawal from philosophical activity. On the contrary, it is produced and sustained by continual, energetic engagement in the business of philosophy. As a result, the condition desirable for human beings according to Pyrrhonism resembles that envisaged by many of their dogmatic opponents. To be sure, people realizing the Pyrrhonian ideal will not have *knowledge* in the way required by dogmatic philosophy, but they will *know* about, in the sense of being familiar with, or well versed in, all the major philosophical questions and the answers the philosophers have proposed, and they will be occupied with thinking and arguing about them, even if they never achieve (and are perfectly content not to achieve) the settled condition of knowing where the truth lies. The life of psychic health for the skeptics, no less than for many of the dogmatists, will be the life of philosophical theory.

Reflection

PHRONTIS: THE PATIENT MEETS THE TEXT

Helen King

Perhaps the best-known parts of the Hippocratic Corpus are the case histories collected in the seven treatises known as the *Epidemics*.[1] Like the other medical texts from the Corpus, which date from the fourth century BCE onward, these are mostly concerned with disease, but it is not difficult to read between the lines to identify what is thought to constitute "health."

The *Epidemics* bring into sharp focus one of the challenges of using historical materials on health and disease; finding ways to avoid making assumptions based on our experience of medicine. For us, and for generations of earlier readers of these texts, the naming of the individual patients, the account of the symptoms—often following a day-by-day format—and the strong authorial presence of the writer make the most obvious label for them that of "case histories." Yet these accounts appear to have been written more as "notes to self" than as records for others. Institutionally, within ancient societies, who can have been intended as the audience? There is a longstanding historical myth that the case histories of *Epidemics* 1 and 3 were written by Hippocrates himself,

1 For the Greek edition see E. Littré (ed.), *Oeuvres complètes d'Hippocrates,* 10 vols. (Amsterdam: Adolf M. Hakkert, 1973–1982).

with the *Aphorisms* coming later as a distillation composed for teaching purposes of the conclusions he had drawn inductively from the case histories. But as Volker Langholf showed, the reverse is more likely, with the *Epidemics* assuming a nosology already established by other "Hippocratic" treatises, including the theory of "critical days."[2] This suggests that the process is not inductive; instead, the theories predate the observations.

At the same time, the *Epidemics* also reveal features specific to the culture that produced them. In addition to allowing us to detect the range of ancient theories about the body that underpins the comments made about the progress of illness in a particular named individual, they also reflect the social context within which ancient medicine functioned. For example, although female patients are described, they are not directly named, instead featuring most commonly as "the wife of x." As David Schaps argued in 1977, women in ancient Greek culture were not mentioned in public, unless they were dead or of "ill repute."[3] This makes even more striking those passages in which a woman is named; for example, in *Epidemics* 6.8.32, Phaethousa and Nanno, who grow beards and then die, are doubly identified ("Nanno, wife of Gorgippos") or even triply identified ("Phaethousa of Abdera, wife of Pytheas").

Disorders affecting women feature in many Hippocratic treatises, along with issues about how a male physician can have access to the body of a female patient, and how far that physician should believe a woman's account of her health and illness. Most of the evidence for the male physician's engagement with such disorders, as well as for how women's bodies are interpreted,

---

2 V. Langholf, *Medical Theories in Hippocrates: Early Texts and the "Epidemics"* (Berlin: de Gruyter, 1990).

3 D. Schaps, "The Woman Least Mentioned: Etiquette and Women's Names," *Classical Quarterly* 27/2 (1977), 323–30.

comes not from the *Epidemics* but from the three volumes of *Gynaikeia*: "*Women's Matters*." Whereas the *Epidemics* accounts often close with the death of a specific patient, the various sections of the *Gynaikeia* sometimes end with general comments about the restoration of a healthy condition. For example, "If she becomes pregnant, she is healthy." Case histories, however, do not feature in the *Gynaikeia*; its women remain generalized, not specific.

There is one apparent exception to this rule of the absence of specific patients from the *Gynaikeia*, and it is another account of restoring health; and, here, also restoring fertility. In *Gynaikeia* 1.40 (8.96–98) we read about the patient Phrontis who examined her own vagina, recognized there was a problem (the verb is *egnô*) and reported it to the Hippocratic physician (*ephrase*, "she made it known"):

> Phrontis suffered those things that women suffer when the lochia are not flowing, and from this had pain in the private parts (*aidoion*), and having touched (it) she recognized that there was a blockage, and she made it known, and having been treated she purged and became healthy (*hygiês*) and fertile (*phoros*). If she had not been treated and if the lochia did not flow naturally, the ulceration would have been bigger and there would be a danger, if there was no treatment, that the ulcers would become cancerous. (Hippocrates, *Gynaikeia* 1.40, 8.96–98)

This section comes at the end of a passage describing how the "mouth of the *aidoion*" can become ulcerated after childbirth, as a result of the inflammation caused by the baby passing through the birth canal, so that the *aidoion* sticks together and causes closure; this provides the general context in which this isolated example of a patient features.

The great nineteenth-century French editor of the Hippocratic Corpus, Emile Littré, examining the variations in the manuscript tradition for the introductory part of this account, commented that "one easily recognizes that this concerns a particular observation which the author recounts" (*Gynaikeia* 1.40, 8.87, note). Manuscript C, downplayed by Littré but now considered the best surviving manuscript, adds the words "*hê gynê panta*," so that the start of the passage would read: "Phrontis. *The woman* suffered *all those things* that women suffer when the lochia are not flowing . . ."[4]

I am not alone in having previously presented Phrontis as a model for the ideal female patient, aware of her own body and when health becomes disease.[5] Aline Rousselle cited this example to support her claim that "in general, women examined themselves while they were in good health."[6] Three other passages, not mentioned by Rousselle, perhaps support this. In two of them, women examine the mouth of the womb and find it narrow and moist (*Gynaikeia* 1.59/L 8.118 and 2.155/L 8.330), and in another a woman examines herself and finds the mouth of the womb is closed or tilted (*Gynaikeia* 3.213/L 8.410).

Seeing Phrontis in this way, as a model patient, is consistent with the boundaries of Ann Hanson's category of "the experienced woman,"[7] as she has given birth; we know this because her problem is that the lochia (the discharge following birth) have not flowed as they should. Hanson argued that, while

---

4 H. Grensemann, *Hippokratische Gynäkologie* (Wiesbaden: Franz Steiner, 1982), 134.

5 H. King, "Medical Texts as a Source for Women's History," in A. Powell (ed.), *The Greek World* (London: Routledge, 1995), 199–218, at 203 and 214 n.30; King, *Hippocrates' Woman: Reading the Female Body in Ancient Greece* (London: Routledge, 1998), 48.

6 F. Pheasant, (trans.), *Aline Rousselle: Porneia: On Desire and the Body in Antiquity* (Oxford: Blackwell, 1988), 25.

7 A. E. Hanson, "The Medical Writers' Woman," in D. M. Halperin et al. (eds), *Before Sexuality: The Construction of Erotic Experience in the Ancient Greek World* (Princeton, NJ: Princeton University Press, 1990), 309–38, at 309–10; L. Dean-Jones, "Autopsia, Historia and What Women Know: The Authority of Women in Hippocratic Gynaecology," in D. Bates (ed.), *Knowledge and the Scholarly Medical Traditions: A Comparative Study* (Cambridge: Cambridge University Press, 1995), 41–58, at 55.

women's experience is denigrated as a reliable source in many types of ancient literature, the Hippocratic writers construct a "female counterpart to the idealized male patient, the intelligent layperson who works with the doctor to maintain health or to combat illness."[8] Women are graded on a scale based on their experience, and those who have such experience can be trusted as accurate witnesses to their own bodies and as partners in their own therapy, as in the description of inserting a tube so that milk can be poured into the womb, where this can be entrusted to the patient because "she herself will know where it is to be" (*Gynaikeia* 3.222/L 8.430). In this example too, the patient is clearly experienced, as she is married and has previously conceived. Her report that the womb is receiving the seed correctly can therefore be trusted.

But back to Phrontis, who appears to be "the only named woman in the Hippocratic gynaecological treatises."[9] Not only does her presence seem odd, as a named individual in a world of "wives of x," but her name itself does seem remarkably convenient for this story. Normally it is a male name, but another female Phrontis is known in classical Greek literature. In the middle of the fourth century BCE, Philiscus of Miletus wrote a funeral elegy for the Athenian orator Lysias. Here, Phrontis is the daughter of the Muse Calliope. Phrontis is called upon to give birth to a son, a hymn for Lysias that will spread his name across the world (preserved in Plutarch, *Lives of the Ten Orators*, 836c = fr. 1.5 West) [10]. David Leitao, who has examined the theme of "poetic paternity,"[11] the male who gives birth to songs or ideas, suggests that "this fictitious daughter of

8 Hanson, "The Medical Writers' Woman," in Halperin, 309–38, at 310.

9 L. M. V. Totelin, *Hippocratic Recipes: Oral and Written Transmission of Pharmacological Knowledge in Fifth- and Fourth-Century Greece* (Leiden: Brill, 2009), 116.

10 D. Leitao, *The Pregnant Male as Myth and Metaphor in Classical Greek Literature* (Cambridge: Cambridge University Press, 2012), 123.

11 Leitao, *The Pregnant Male as Myth*, 124.

Calliope is really just a personification of the poet Philiscus's own *phrontis.*"[12] Phrontis, literally "thought," can be used in the sense of "reflection;" but it can also mean "anxiety."

We should not automatically discount "Phrontis" as a real patient's name, of course; everyone has to be called something. But finding that the only named woman in the *Gynaikeia* is apparently called "Thought" should give us pause.[13] If she is not simply the ideal "thoughtful" patient, a further possibility is that this was not originally a name, but rather a symptom, or even the name of a disease. In his edition of the Hippocratic *Diseases* 2.72, Littré opened with "Phrontis, a dangerous disease," translating this as "worry" and adding his own diagnosis of "hypochondria" (L 7.108), but he also noted that the gloss "a dangerous disease" was not found in all manuscripts. Potter's 1988 edition and translation favored a different manuscript tradition here, one which omitted "a dangerous disease," and in addition he chose to emend the text to read "phrenitis" rather than "phrontis."

There is a common pattern in the Hippocratic Corpus by which diseases take their names from the part of the body in which they are situated,[14] and here both words come from the same root: the *phrên*, the part of the body responsible for reasoning, located at the midriff. Because it is a described as a thin and wide part with no cavity (*Sacred Disease* 17, L 6.392= 20, Loeb 2.180), the term is translated today as "the diaphragm."

12 Leitao, *The Pregnant Male as Myth*, 123.

13 There are many textual variants in the manuscript for the opening of this case. Even Littré, who was responsible for making this into a named woman, had his doubts about the text. Not all manuscripts even include the word "phrontis"; the earliest Latin translation, the 1525 complete Hippocratic Corpus by Marco Fabio Calvi, opened this section with "however for this woman the very same things will arise" (*Huic autem foeminae eadem venient*), using the sixteenth-century manuscript Vaticanus Graecus 278. See M. F. Calvi, *Hippocratis Coi medicorum omnium longe principis, Octoginta volumina*... (Rome: Franciscus Minitius, 1525), cxxi.

14 S. Byl and W. Szafran, "La phrénitis dans le Corpus Hippocratique: Etude philologique et médicale," *Vesalius* 2/2 (1996), 98–105, at 102.

Phrenitis was thought to be a dangerous acute condition characterized in the Hippocratic Corpus by fever and by what we would call delirium: disordered thought. In *Diseases* 2.72 the area of the *phrenes* swells and is painful, and the patient is afraid and has hallucinations. However, while Potter's emendation to "phrenitis" here makes sense, there is no manuscript justification for it.[15]

So, instead of *Gynaikeia* 1.40 offering us a unique reference to a named individual in the midst of a treatise about general conditions, or even a conveniently named ideal patient, we may here have a manuscript corruption in which the original was a description of "worry," as a further consequence of lochial retention, and "Phrontis" being an error for *phrenitis*. However, *phrenitis* is a condition of confused thought in which the patient goes "out of their mind": Phrontis thinks clearly and is very much in control. One of the symptoms of *phrenitis* is inability to speak[16]: yet Phrontis is able to communicate what is wrong with her.

While women patients do feature on many occasions in the Hippocratic Corpus, in some cases even having their words reported, we should be cautious how we handle the evidence in this particular case. Rather than a real, named individual who illustrates a general principle of health and disease, or an ideal "thoughtful patient" who provides a model for patient-physician interaction, these third-person descriptions could state what a generalized woman suffers, rather than anyone's specific experiences. Phrontis, whether real or ideal, does not provide the

15 G. C. McDonald, "Concepts and Treatment of Phrenitis in Ancient Medicine" (PhD diss., Newcastle University, 2009), 17 n. 9, notes that no other modern editor has made this amendment. Byl and Szafran, "La phrénitis dans le Corpus Hippocratique," note that retrospective diagnoses of *phrenitis* included, in the nineteenth century, encephalitis or meningitis and, in the early twentieth century, malaria.

16 Byl and Szafran, "La phrénitis dans le Corpus Hippocratique," 101.

complete picture. In other medical texts, women are doubted as reliable authorities on their own bodies. Deconstructing Phrontis suggests that we should treat our sources for the body with considerable caution, resisting our hope of connecting with real people, with real bodies just like ours.

CHAPTER FOUR

Health in Arabic Ethical Works

Peter Adamson

Conceptions of health in the medieval Islamic world often take their cue from Galen. This will come as no surprise as concerns medical literature. Less expected is Galen's crucial role as a source for the idea of "psychological," as opposed to bodily, health—that is, the health of the soul. Galen devoted several works to ethics, which were translated into Arabic. Particularly influential was a work called *On Character Traits*, lost in Greek but retained in the form of a summarizing Arabic paraphrase with the title *Fī l-Aḫlāq*.[1] Here Galen explicitly draws a parallel between the health of the soul and bodily health:

1 Arabic text in P. Kraus, "Kitāb al-aḫlāq li-Jālīnūs [*On Character Traits,* by Galen]," *Bulletin of the Faculty of Arts of the University of Egypt* 5 (1937), 1–51; J. N. Mattock (trans.), "A Translation of the Arabic Epitome of Galen's Book *Peri Ethon*," in S. M. Stern et al. (eds), *Islamic Philosophy and the Classical Tradition* (Oxford: Cassirer, 1972), 235–60. Revised English trans. in P. N. Singer (ed.), *Galen: Psychological Writings* (Cambridge: Cambridge University Press, 2013). Discussed in R. Walzer, "New Light on Galen's Moral Philosophy," *The Classical Quarterly* 43 (1949), 82–96. On Galen's influence on the Arabic ethical tradition see also G. Strohmaier, "Die Ethik Galens und ihre Rezeption in der Welt des Islams," in J. Barnes and J. Jouanna (eds), *Galien et la Philosophie* (Vandoeuvres: Fondation Hardt, 2003), 307–29. All translations are mine unless otherwise noted.

> Just as the body is beset by illness and ugliness (for example epilepsy or, for ugliness without illness, being a hunchback), so the soul is beset by illness and ugliness. Its illness is, for instance, anger; its ugliness, for instance, ignorance. (Galen, *On Character Traits*, 42–43)

The parallel suggests that we might conceptualize ethics as follows: just as the doctor can treat the body, helping the patient to maintain or restore its health, so the ethicist or ethical adviser can help people maintain or restore the "health of their body."[2] The example given here is that a soul that is prone to anger has a sort of illness; other examples we will meet later include sadness, weakness in the face of pleasure, and obsessive thoughts.

The task of treating such ailments of the soul was taken up in a group of works that form a kind of mini-genre in early Arabic philosophy. They draw on direct acquaintance with Arabic translations of Galen and base themselves partly or wholly around the notion that there could be a kind of medicine for the soul. In this chapter I will be looking at three such texts. In chronological order they are Abū Zayd al-Balḫī's *Maṣāliḥ al-Abdan wa-l-Anfus* (*Benefits for Souls and Bodies*, hereafter *Benefits*),[3] Abū Bakr al-Rāzī's *al-Ṭibb al-Rūḥānī* (*The Spiritual Medicine*),[4] and Miskawayh's *Tahḏīb al-Aḫlāq*

---

2 The parallel is not only Galenic but is also pervasive throughout antiquity. See J. Pigeaud, *La maladie de l'âme: Étude sur la relation de l'âme et du corps dans la tradition médico-philosophique antique* (Paris: Belles Lettres, 1981). Nonetheless, Galen seems to be the main source for the idea in Arabic texts.

3 Arabic text in two facsimile editions in the series overseen by F. Sezgin, *Abū Zayd al-Balḫī: Sustenance for Body and Soul* (*Maṣāliḥ al-Abdan wa-l-Anfus*) (Frankfurt a.M.: Institute for the History of Arabic Islamic Science,1984 and 1998). I will cite according to the page number of the 1984 facsimile. German translation of the section on the health of soul in Z. Özkan, *Die Psychomatik bei Abū Zaid al-Balḫī* (*gest. 934 AD*) (Frankfurt a.M.: Institute for the History of Arabic Islamic Science, 1990). On Abū Zayd see H. H. Biesterfeldt, "Abū Zayd," in U. Rudolph (ed.), *Philosophie der Geschichte der Philosophie: 8.–10. Jahrhundert* (Basel: Schwabe, 2012), 156–67.

4 Arabic text in P. Kraus (ed.), *al-Rāzī: Rasā'il falsafiyya* (*Opera philosophica*) (Cairo: Barbey, 1939), 15–96. English translation in A. J. Arberry (trans.), *The Spiritual Physick of Rhazes* (London: John Murray, 1950). French translation in R. Brague (trans.), *Al-Razi: La médecine spirituelle* (Paris: Flammarion, 2003).

(*The Refinement of Character Traits*, hereafter *Refinement*).[5] We can possibly trace all three works to the founding father of the Islamic philosophical tradition, al-Kindī. He was the teacher of al-Balḫī (850–934), and it may be that the latter was in turn a teacher of al-Rāzī (d. 925).[6] Certainly the ethical projects of al-Balḫī and al-Rāzī are strikingly similar. Where al-Balḫī's work is split into two sections, dealing respectively with physical and psychological medicine, al-Rāzī's *Spiritual Medicine* is explicitly presented as a companion piece to his encyclopedia on bodily medicine, the *Book for al-Manṣūr*. The later Miskawayh (d. 1030) accesses a wider range of sources and fuses the themes of Galenic ethics with passages drawn from Aristotle's *Nicomachean Ethics*. Yet he too looks back to al-Kindī, ending his *Refinement* with a lengthy quotation from his writings, and drawing on al-Kindī elsewhere when discussing the soul.[7]

It is therefore worth looking briefly at the work Miskawayh saw fit to quote as a conclusion to the *Refinement*. It is a nice example of "medicine for the soul," albeit that it does not seem to draw on Galen as our other authors do. This is an epistle by al-Kindī entitled *On Dispelling Sorrow*, which says about sadness or grief (*ḥuzn*) much the same thing that Galen remarked about anger. In this epistle al-Kindī says that sadness is an ailment of the soul:

5 Arabic text in C. Zurayk (ed.), *Miskawayh: Tahḏīb al-Aḫlaq* (Beirut: al-Nadin al-Lubnāniyya, 1966). English translation in C. Zurayk (trans.), *Miskawayh: The Refinement of Character* (Beirut: American University of Beirut, 1968).

6 The *Fihrist* (*List*) of the book merchant Ibn al-Nadīm reports that al-Rāzī studied with an al-Balḫī, but it is not clear whether this is the same man as our Abū Zayd. See further on this P. Adamson and H. H. Biesterfeldt, "The Consolations of Philosophy: Abū Zayd al-Balḫī and Abū Bakr al-Rāzī on Sorrow and Anger," in P. Adamson and P. E. Pormann (eds), *Philosophy and Medicine in the Formative Period of Islam* (London: Warburg Institute, 2018), 190–205.

7 See P. Adamson, "Miskawayh's Psychology," in P. Adamson (ed.), *Classical Arabic Philosophy: Sources and Reception* (London: Warburg Institute, 2007), 39–54, and P. Adamson and P. E. Pormann, "More than Heat and Light: Miskawayh's Epistle on Soul and Intellect," *Muslim World* 102 (2012), 478–524.

> Every pain for which one does not know the causes is incurable. We therefore ought to set out both what sadness is and what causes it in order to find a cure and to apply it with ease. (Sec. 1.1)[8]

Although al-Kindī is here concentrating on one particular psychological illness, he does at one point suggest a broader project of psychological medicine, such as is carried out by the later authors mentioned above:

> The welfare of the soul and curing it from its diseases is superior to the welfare of the body and curing it from its diseases, in the same way as the soul is superior to the body. (Sec. 4.1)

We have already seen how al-Balḫī and al-Rāzī explicitly planned their works around the parallel between medicine for body and for soul. This is less true of the *Refinement* in its entirety, but Miskawayh does use the theme to structure the sixth book of his work:

> We will discuss in this discourse the cure of the diseases which affect the soul of man and their remedies, as well as the factors and causes which produce them and from which they originate. For skilled physicians do not attempt to treat a bodily disease until they diagnose it and know its origin and cause. (Miskawayh, *Refinement*, 175, trans. Zurayk)

Notice that in none of these texts is the notion of "medicine for the soul" presented as a *metaphor*. Rather, the goal is to combat literal diseases of the soul.

8 Translation and section numbers from P. Adamson and P. E. Pormann (trans.), *The Philosophical Works of al-Kindī* (Karachi: Oxford University Press, 2012).

Medicine for Bodies and Souls

Why, then, would it make sense to see psychological deficiencies—especially ethical ones—as ailments and to envision a kind of medicine that treats the soul the way the more familiar kind of medicine treats the body? First and most obviously, we have the concept of health itself as a target of both bodily and psychological medicine. As noted in the introduction to this book, health is a normative notion. That might license the immediate inference that, insofar as the soul is in a good (or even ideal) state, we can think of the soul as "healthy." After all, we commonly extend the notion of health to the normative states of things other than human bodies, speaking for instance of "healthy trees," where this does not seem to be a mere metaphor. In the Arabic tradition, this consideration supported not only a parallel between "ethical treatment" and medicine but also between political rule and medicine. In particular, we find al-Fārābī repeatedly comparing the care that a good ruler has for his city to the doctor's care for his patients.[9] So far, though, we seem to have insufficient grounds for the idea that there is a "medicine" for souls. There are any number of arts that similarly operate with normative constraints, for instance carpentry (compare Plato's remarks on this art and the "form of Bed" at *Republic* 597a–d). But we do not find our authors saying that there is a carpentry of the soul or of the city.

Yet more specific parallels can be drawn. One is that medicine, both for bodies and for souls, has the twofold task of *preserving* and *restoring* health. This is of course a commonplace in medical literature. It is asserted for instance on the very first page of al-Rāzī's aforementioned medical encyclopedia for his patron al-Manṣūr ibn Ismāʿīl, governor of Rayy:

9 See for instance M. Mahdi (ed.), *Al-Fārābī: Kitāb al-Milla* (Beirut: Dār al-Mashriq, 2001), 56–7. A recent PhD has examined this topic in depth: B. El-Fekkak, "Cosmic Justice in al-Fārābī's Virtuous City: Healing the Medieval Body Politic" (PhD, diss., King's College, London, 2011).

> The art required for the care (*siyāsa*) of the body and its preservation is called "medicine." It has two parts: regimen (*tadbīr*) for the healthy body, so that its health continues, and restoration of the ailing body to a state of health.[10]

Al-Balḫī and Miskawayh explicitly extend the twofold purpose to the case of medicine for the soul:

> In the first treatise (*maqāla*) of this book, we have dealt with that which one needs to know and utilize in the regimen (*tadbīr*) of benefits for bodies: the preservation of their health, when it is present, and its restoration, when [health] has been ruined by the incidence of illnesses and maladies. . . . In this treatise, we intend to give an account of how to manage benefits for souls, how to keep their faculties in a good and balanced condition (*ʿalā sabīl al-ṣalāḥ wa-l-iʿtidāl*), and how to manage the removal of psychological affections (*aʿrāḍ nafsānī*) that befall them. (Al-Balḫī, *Benefits*, 269–70)

> As the medicine of bodies is divided primarily into two parts, the first to preserve health if it is present and the second to restore it if it is absent, so also we should divide the medicine of souls in this same way, trying to restore their health if it is missing and proceeding to preserve it if it is already there. (Miskawayh, *Refinement*, 176, trans. Zurayk)

The idea of taking steps now to prevent or minimize future psychological difficulties also dates well back to antiquity. One might think here, for instance, of the Cyrenaic technique of "pre-rehearsing"

10 Ḥ. al-Ṣiddīqī al-Bakrī (ed.), *Abū Bakr al-Rāzī: Al-Kitāb al-Manṣūrī fī l-ṭibb* (Kuwait: Maʿhad al-Maḫṭūṭāt al-ʿarabīya, 1987), 17–18.

possible future pains, a suggestion the Epicureans dismissed as counterproductive.[11]

Because both kinds of medicine are future directed and look to the preservation, not just the restoration, of health, both involve a regimen of recommended habitual behavior.[12] In the case of the body, this would include such things as a certain diet and exercise regime. It is less obvious what the regimen for the soul might be. Al-Balḫī distinguishes between "interior" and "external" remedies applied to the soul (*Benefits* 276, 283, 317–18). An external remedy might be listening to something calming (277, he presumably means something like music),[13] while an "interior" technique would be choosing "a moment when the soul is peaceful and at rest" to focus on useful thoughts. For instance one might think about the inevitability of suffering and hence the pointlessness of wishing it would never occur (277; also 283, 318). External influences can also involve this kind of consideration, as when we seek sound advice from others (283, cf. 285, quoted below, and 318). We will return to this idea that remedies for the soul would often involve some kind of thought process.

Another substantial parallel between the two types of medicine is that, in both cases, the healthy state being pursued is defined in terms of balance (*iʿtidāl*). As followers of Galen, our Arabic authors, of course, accepted the idea that bodily health involves a proper balance of the four humors. The soul too needs balance, in this case between three elements and not four: reason, spirit, and desire, the three aspects of soul established by Plato in the *Republic* and also accepted by Galen.

11 On this see T. O'Keefe, "The Cyrenaics on Pleasure, Happiness, and Future-Concern," *Phronesis* 47 (2002), 395–416.

12 The word *tadbīr* and other words of the same root are used to translate the Greek term δίαιτα, for instance in the Arabic translation of Galen's *The Dependence of the Soul on the Body*. See the glossaries in H. H. Biesterfeldt, *Galens Traktat "Dass die Kräfte der Seele den Mischungen des Körpers Folgen" in arabischer Übersetzung* (Weisbaden: Franz Steiner, 1972).

13 Compare the passage cited by F. Rosenthal, *The Classical Heritage in Islam* (London: Routledge, 1994), 265–66, where al-Rāzī is said to have described paintings in baths as having an effect on all three aspects of the soul. My thanks to Hinrich Biesterfeldt for this reference.

Galen himself already explained psychological health in terms of the balance of the three aspects of soul:

> The motions caused by the two bestial souls, insofar as they are imbalanced (*ʿalā ġayr iʿtidāl*), are unnatural to the human being. For they are imbalanced, and what is imbalanced departs from the proper condition of health (*ḫāriǧ ʿan istiqāmat al-ṣiḥḥa*). (Galen, *On Character Traits*, 27)

Al-Balḫī likewise compares the soul's health to humoral balance (*Benefits*, 276). Al-Rāzī and Miskawayh, meanwhile, base their ethical reflections explicitly on the tripartite soul.[14] Here is al-Rāzī ascribing to Plato a conception of what he calls "spiritual" medicine, which strives to confer balance on the soul:

> [Plato] holds that man should, by means of bodily medicine, which is the sort of medicine that is widely recognized (*maʿrūf*), and spiritual medicine, which is achieved by means of proofs and demonstrations, give equilibrium (*taʿdīl*, same root as *iʿtidāl*) to the actions of these souls, so that they may neither exceed nor fall short of what is intended. (*Spiritual Medicine*, 29)

Two caveats should be noted here, though. First, one might see the notion of "balance" or "harmony" between the three aspects as corresponding more narrowly to the virtue of justice rather than to psychological well-being in general. This would, of course, be close to

14 Al-Rāzī, *Spiritual Medicine*, 27, Miskawayh, *Refinement*, 15–16. Actually, as Miskawayh notes, it is not clear what terminology to use for the three aspects: he allows that both "souls" and "powers" would be possible terms (*Refinement*, 15). Al-Rāzī introduces the division by saying that Plato recognized three "souls," and himself frequently speaks of "rational soul," "desiring soul," and so on. I will continue to speak more neutrally of three "aspects." Of course the question of how to understand these three aspects in the Platonic text is a much-discussed one. See for instance J. Moss, "Appearances and Calculations: Plato's Division of the Soul," *Oxford Studies in Ancient Philosophy* 34 (2008), 35–68.

the discussion of justice in the *Republic*. Miskawayh duly equates justice with the peaceful cooperation (*musālama*) of the three aspects (*Refinement*, 18). A second caveat is that health in the soul is not strictly speaking "balance," at least, not in the sense that the three aspects are equal in strength, the way that the humors might be equally balanced in a healthy body.[15] Rather, the rational aspect of soul should dominate the other two aspects and especially the lowest, desiring or "bestial" soul. Thus Miskawayh goes on to add that there must be not only peaceful cooperation but that also the entire soul must capitulate (*istislām*) to the discerning faculty, which belongs to the rational soul (*Refinement*, 18). This would be as if health in the body were, for instance, that blood dominated the other three humors or that heat dominated cold, which is obviously not the case. One might nonetheless speak of "balance" in the soul (as Galen does) to mean that the lower parts have their *appropriate* degree of influence. This would mean that they still carry out their functions but are completely subordinated to the rational soul.[16]

A final reason, and perhaps the best reason, to envision medicine for souls as well as for bodies is that the cases of soul and body are not merely parallel. They are causally connected. For, as Galen famously argues in *That the Soul Depends on the Body*, the states of the soul depend on those of the body.[17] Thus a disease on the side of the body can lead to a disease on the side of the soul, and vice versa:

---

15 See further below, in the section called "Particularism," for a reason why this might be too simple a conception of bodily health.

16 The goal is certainly not that the lower parts of soul cease entirely to be active. As al-Rāzī points out, the desiring soul's function is to help keep the body alive through nutrition. This is reasonable enough in Platonic terms—if one had no desire for food, one might starve to death, and Plato himself speaks of some desires as "necessary" in this sense (*Republic* 558d–e). But we also see here an assimilation of Plato's tripartite soul (reason, spirit, desire) to that of Aristotle (reason, animal, vegetative), which is also evident in Miskawayh. See further Adamson, "Miskawayh's Psychology," 42.

17 One of Galen's best illustrations of this is drunkenness, where consumption of wine is able to affect even the workings of the rational soul (Kühn 6.777–78, 811–12). Compare al-Rāzī's polemic against the evils of excessive drink in *Spiritual Medicine*, which includes the remark that drinking can lead to the "loss of reason (*faqd al-ʿaql*)" (72).

> The human being is constituted from his soul and his body, and one cannot imagine him surviving without the union of the two . . . when the body ails and suffers, and is affected by harmful maladies, this hinders the faculties of the soul (including understanding (*fahm*), knowledge (*maʿrifa*) and others) from performing the activity in their [proper] way. (Al-Balḫī, *Benefits*, 273)

> As the soul is a divine, incorporeal faculty and as it is, at the same time, used for a particular constitution (*mizāǧ ḫāṣṣ*) and tied to it physically and divinely . . . you must realize that each one of them is dependent upon the other, changing when it changes, becoming healthy when it is healthy and ill when it is ill. (Miskawayh, *Refinement*, 175, trans. by Zurayk)

This gives us a strong reason to agree that deficiencies on the side of the soul are *literal* illnesses. After all such deficiencies result directly from, and can cause, bodily illnesses. In some cases it might even be reasonable to say that one and the same illness has manifestations at both the bodily and psychological level. Indeed al-Balḫī recognizes such illnesses. He remarks for instance that the obsessive thoughts called *wasāwis* are both a bodily *and* a psychological malady (*aʿrāḍ*, 323).

Objective and Subjective Deficiency

All this gives us strong grounds for the signature idea of Galenic ethics: medicine contains two parts, one for preserving and restoring the health of body, the other doing the same for the health of soul. If that is the case, though, then familiar tensions that arise when thinking about bodily health—such as discussed throughout this book—will also arise when considering psychological health. One such tension is between objective and subjective notions of health.

When we think about medicine in its restorative capacity, we typically imagine a patient who has presented with a complaint—the doctor's goal is to diagnose the source of the patient's suffering and treat it. But people who are ill or in bad bodily condition are frequently unaware of their parlous state. A person of a dissolute lifestyle might have badly balanced humors, yet think he is doing very well because his life is so pleasant (until, suddenly, it isn't). Drawing on Plato, al-Rāzī gives us a theoretical basis for understanding this phenomenon. In general, we are not aware of bodily health because it is simply the maintenance of a natural state, with no defect (*Spiritual Medicine*, 66). On the other hand, neither do we necessarily perceive our own defective states, especially at first—the slowness of changes from the neutral state makes it hard to tell that one is gradually becoming, for example, too dry or simply hungry, and thus occasions no pain even though what is happening could be harmful. The same is true in the other direction, as when a wound is healing slowly so that we do not feel pleasure. Only sudden changes away from or toward the natural state will give rise to pain and pleasure.[18]

With respect to the soul, it is perhaps even more obvious that people are imperfectly aware of their own deficiencies. The same dissolute person who is unaware that he is unhealthy in body may be quite satisfied with his way of life, unaware that it is vicious and harmful. At the same time, it is, of course, also common for people to find their psychological ailments agonizing. This tension accounts for a difference amongst our works of Galenic ethics. Al-Balḫī seems to direct most of his attention to people who have psychological *complaints*. For instance, people who are plagued by obsessive thoughts, or by grief or anxiety, are well aware that they have a problem. Perhaps for this reason, the closest thing he offers to a definition of psychological

18 For all this see P. Adamson, "Platonic Pleasures in Epicurus and al-Rāzī," in P. Adamson (ed.), *In the Age of al-Fārābī: Arabic Philosophy in the Fourth/Tenth Century* (London: Warburg Institute, 2008), 71–94.

health speaks only of the absence of maladies, with the soul being at rest (*Benefits*, 275), and not of virtue. One might hesitate, therefore, to think of al-Balḫī's *Benefits* as a work about *ethics*—it seems more like the work of an advice columnist, albeit one possessed of unusual literary sophistication. By contrast, al-Rāzī and Miskawayh clearly think they are doing ethics. Al-Rāzī is highly judgmental about people who display such shortcomings as gluttony or frequent drunkenness—he's even hard on people who fidget, as we'll see shortly. Miskawayh meanwhile feels no tension in weaving material from Aristotle's *Ethics* into a work with substantial borrowings from the Galenic ethical tradition.

We can rescue the unity of the genre by reminding ourselves that ancient ethics, and later the Arabic ethical tradition based on such Hellenic sources as Galen and Aristotle, are "eudaimonistic." What is at stake is the *flourishing* of the person, in this case the person's flourishing in respect of her psychological states. This means simply that the psychological states are as they should be. Just as on the bodily side, deficient states (that is, failures to flourish) may be subjectively perceived and cause distress, but they need not. If al-Balḫī focuses largely on cases where actual distress has arisen, this is more a matter of emphasis than a deep difference between his approach and that of al-Rāzī or Miskawayh. In fact some kinds of maladies covered in his *Benefits*, like anger, would not necessarily give rise to complaint on the part of the sufferer.[19] Conversely, al-Rāzī and Miskawayh deal with problems, such as sadness, that inevitably are bound up with a subjective experience of suffering. We can therefore say that perceived distress is a sufficient, but not necessary, condition for needing treatment by psychological medicine.

19 When he introduces the topic, he says merely that anger leads to *harm (aḏā*, 293) and not, for instance, pain.

Particularism

A second tension is the one that will occupy our attention for the rest of this chapter. Again, it is one that arises in both bodily and psychological medicine; I will introduce it by considering the bodily case. As mentioned, health in the body is often defined positively in terms of a physical "balance." In this sense, there is a universal notion of health that describes the ideal state for each and every human. On the other hand, a familiar claim in ancient medicine is that the ideal healthy state may vary from one person to another. It may be that your ideal state involves a higher degree of heat than mine, and even that a given person's healthy state differs at certain times of year or depends on the climate and other factors related to their location. This is why Galen is so insistent that the expert doctor will tailor his treatment to each patient he sees.[20] Let us call this aspect of Galenic (bodily) medicine "particularism."[21]

To what extent does particularism also apply to psychological medicine? Might the health of your soul differ from the health of my soul in significant ways? There are good reasons to expect our Arabic authors to think so. For one thing, the theme of particularism is very emphatic in Galenic bodily medicine. So it would

20 For an introductory discussion of this see P. van der Eijk, "Therapeutics," in R. J. Hankinson (ed.) *The Cambridge Companion to Galen* (Cambridge: Cambridge University Press, 2008), 283–303, especially 286–88. The possibility of individual variation in the ideal humoral balance is noted at 298, citing *De santitate tuenda* 1.1, 6.2, and 1.5, 6.13–15.

21 Here I should clarify that I am not using the word "particularism" with the sense it often has in contemporary ethics, namely that moral judgments do not rely on the application of universal principles. The notion of "bespoke" medicine I discuss here may seem close to that notion. Our texts do not, however, seem to rule out the idea that each individual's healthy state may be a function of the universal features that describe them—even if the combination of a given person's features is, as it happens, unique. (Thus, one can work out from universal principles that anyone with such-and-such symptoms, with such-and-such dispositions, in such-and-such a location, needs such-and-such treatment. This will be the right treatment for anyone who satisfies all the relevant criteria, even if in fact there is only one such person.) On contemporary moral particularism see the relevant entry in the online Stanford Encyclopedia: J. Dancy, "Moral Particularism," in E. N. Zalta (ed.), *The Stanford Encyclopedia of Philosophy*, 2013; B. W. Little and M. Little, *Moral Particularism* (Oxford: Oxford University Press, 2000).

rather undermine the parallel our authors want to draw with medicine for the soul if there were no grounds for particularism on the psychological side. Furthermore, at least in Miskawayh's case, there is the influence of Aristotle to consider. He famously cautions us not to expect too much precision in ethics, on the basis that in practical affairs we deal with particular situations:

> The whole account of matters of conduct must be given in outline and not precisely (τύπῳ καὶ οὐκ ἀκριβῶς), as we said at the very beginning that the accounts we demand must be in accordance with the subject-matter. Matters concerned with conduct and questions of what is good for us have no fixity (οὐδὲν ἑστηκός), any more than matters of health. The general account being of this nature, the account of particular cases (περὶ τῶν καθ' ἕκαστα) is yet more lacking in exactness. (*Nicomachean Ethics* 2.2, 1104a1–7)

The variety between particulars includes variation in ethical agents themselves, something Aristotle makes clear with his famous example of the outsized diet appropriate to Milo the wrestler (*Nicomachean Ethics*, 1106a26–b4).[22]

Miskawayh does not fail to recognize this aspect of Aristotle's teaching. Miskawayh makes it crystal clear we are dealing here with a difference in what ethics will mean to different people:

> The means of these extremes should be sought *separately for each person*. Our own task here is to note the all of these means [between extremes] and the rules governing them according to the requirements of our art, *and not for each individual* (*šaḫṣ*), for this

22 Translations taken from D. Ross (trans.), *Aristotle: the Nicomachean Ethics* (Oxford: Oxford University Press, 2009).

> would be impossible. The carpenter, the jeweler, and all the artisans acquire in their minds rules and principles only, for the carpenter knows the form of the door and that of the bed and the jeweler the form of the ring and that of the crown, in the absolute. Then each of them derives by these laws the individual things which he has in mind, but he cannot master all the individual things because they are infinite in number. . . . The art ensures the knowledge of principles only. (Miskawayh, *Refinement*, 25–26, trans. Zurayk, modified, *my emphasis*)

This seems to give us both a universal and particular element in ethics, just as we saw in bodily medicine. For both there will be general rules or principles, but the application of those principles to individuals will be, as Aristotle puts it, "inexact."

Furthermore, both Aristotle and Galen, of course, emphasize that people vary in terms of their inborn and acquired ethical tendencies, what Galen calls their "character traits"—which gives the title to the work already mentioned several times. In it he remarks:

> The praiseworthy states of the human soul are called "virtue," the blameworthy states "vice." These states are divided into two types. First, those that arise in the soul after the onset of thought, deliberation, and discernment (*al-fikr wa-l-rawiyya wa-l-tamyīz*)—these are referred to as knowledge, opinion, and judgment. Second, those that occur in the soul without thought, and these are referred to as character traits (*aḫlāq*). Some character traits appear in infants as soon as they are born, before the time of thought. (Galen, *On Character Traits*, 28)

> [Some character traits arise] prior to education (*ta'addub*). In general, none of the activities, affections (*'awāriḍ*), or character traits is present in any mature human unless it was already present in him

> at the time of his childhood. So it is false that all affections arise through opinion and thought, for opinion and thought are not affection, but rather either true or false belief. (Galen, *On Character Traits*, 30)

This again suggests that psychological medicine might need to tailor treatments to individual "patients," bearing in mind tendencies they may have had since birth.

But we need to be careful here. The last two points just considered (the one Aristotle illustrates with reference to Milo, and the one Galen illustrates with reference to babies) would support different kinds of particularism. Galen is not saying that each individual has a certain set of ethical predispositions that would be *ideal* for them, but simply that individuals vary in their character traits, ideal or not. The ethical significance of this would seem only to concern the steps one should take to bring someone closer to the ideal. For instance, there is no need to train a naturally abstemious person to avoid gluttony, and doing so could be harmful if it leads them too far in the opposite direction. Aristotle, by contrast, may at least be taken to say (and Miskawayh clearly did take him to say) that what is ethically ideal for one person may differ from the ideal for another person. In light of this we need to distinguish between two kinds of particularism that could be appropriate to psychological medicine: particularism concerning (1) the remedy offered, or (2) the goal sought. The analogous distinction in the physical case would be the difference between saying (1) different people need different kinds of treatment (e.g., the right drug or diet for one person may not be right for another person), and (2) different people will vary in terms of the state that counts for them as "health" (e.g., I might have an ideal degree of warmth that is different from yours).

With the exception of Miskawayh's remark quoted above, it turns out that there is remarkably little in our Arabic works of Galenic ethics to support particularism of type (2) regarding the soul. There is a good reason for this. As we have seen, two of our authors, al-Rāzī and Miskawayh, think that the health of the soul consists in the rational soul's domination over the body. This is not merely a general principle, but rather a robust conception of the good that does not vary from person to person. One might suppose that there could nevertheless be room for type (2) particularism, if one person's reason should rule his soul in light of one set of beliefs, while another person should have a different set of beliefs ruling her soul. But there is an obvious problem with that: surely we all should have the same beliefs, namely the ones that are true. The ideal case, then, will not be broadly the same but exactly the same for everyone: it will be for the rational soul to have true beliefs (or even better, knowledge) and to rule the rest of the soul in light of those beliefs.

Even here, one could find room for a kind of type (2) particularism, by suggesting that even if every ideal agent forms only true beliefs, each agent will have to form beliefs relevant to their own particular context. For instance I, but not you, am in a position to think, "I should not eat this almond croissant," because I am the one in a position to eat the croissant. This would, I think, be a good way of reconciling Aristotelian type (2) particularism with the Platonic conception of virtue as the rule of reason. But it is not one we find anywhere in our Arabic texts. As we will see in the next section, they think that the beliefs that psychological medicine inculcates are general ones, and they are intellectualist enough to think that bestowing health on someone is essentially a matter of inculcating the right beliefs. Nonetheless, they do give particularism an important place in their versions of Galenic ethics. It is always type (1) particularism, that is, particularism with regard to the kind of treatment each person needs to receive. In the final two sections of the chapter, I will explain why the treatment needs to be particular, even if the goal is always the same.

Cognitive Therapy

A striking feature of these works on Galenic ethics is the extent to which they describe the soul's health in purely intellectualist terms. This is faithful to Galen himself:

> When the soul reaches the aim which it seeks, namely the understanding (*maʿrifa*) of things, it is balanced and fine (*ḥasana*), but when it misses it, then it is perturbed and imbalanced. Thus the beauty of soul arises through knowledge, and its ugliness through ignorance. . . . The soul's understanding, through which it is fine, understands the elements from which the body is composed, and which generate, compose, and augment the affections of the soul. The discovery of their cures follows on the knowledge of this. Thus we never see a fine soul that is ill, as we see a very fine body affected by a severe illness. This cannot be otherwise, given that the knowing and beautiful soul first of all keeps itself healthy, and only then takes care of the body, because of its need to use it for its actions. But the body knows nothing of the essence of its own health, and cannot preserve it. (Galen, *On Character Traits*, 43)

Here we finally have an admission that psychological and bodily medicine do differ in at least one crucial respect. The body is not under the soul's control, the way that the soul is under its own control.[23] Rather, the soul is in charge of seeking the health of both itself and its body,

---

23 It may seem surprising that Galen, who was famously critical of the Stoic idea that the soul is rational through and through, should seem to be saying that knowledge is sufficient for good ethical condition. I do not, however, take him to be assuming that acquisition of true beliefs is by itself sufficient for the subduing of the lower soul. Rather, as we will see later in the paper, both Galen and his followers in the Arabic tradition think that the rational soul can be hindered if the lower soul is too powerful, something the Stoics would never have admitted since they didn't recognize a non-rational aspect of the human soul. What Galen is describing here, then, is a soul that is both knowledgeable *and* unimpeded by the lower aspects. Evidence for this can be found at *Character Traits* 39, where Galen says that someone who has achieved dominance of the rational soul can *then* go on to help others through advice and serving as a role model.

albeit that it prioritizes its own health—as we saw al-Kindī saying above, the soul's welfare is simply more important.

In keeping with this, all three of our authors concentrate much of their attention on persuading the reader (or hypothetical "patient") to modify their beliefs. Al-Balḫī often speaks as if the analogue to bodily medicine were specifically helping someone to have the right "thoughts":

> It was the habit of judicious kings to be attended by wise men, who could cure them of maladies of the soul (*aʿrāḍ nafsāniyya*) . . . through advice (*waṣāya*) and exhortations (*mawāʿiẓ*). . . . They depended on [the wise advisors] just as they depended on proficient doctors, who cured them of maladies of the body, since they knew enough to realize that they could not do without having both groups simultaneously, and that at some point there would be a need to get nourishment and medicine from both. (Al-Balḫī, *Benefits*, 285)

> We have said above that, just as bodily sicknesses are such as to be treated by means of bodily cures, so sicknesses of the soul are treated by means of psychological remedies, either by exhortations and reminders, or through thoughts (*fikar*), through which the person trains his soul and which he uses as a weapon and implement for keeping the pains of fears and sorrows away from himself. (Al-Balḫī, *Benefits*, 335–36)

Similar, albeit more technical, is al-Rāzī's conception of the process by which reason subdues the lower parts of the soul. He frequently speaks of the need to make "thought and deliberation (*fikr* and *rawiyya*)" lead the way, rather than letting desire push us into action. For instance:

> [The philosophers] offer a proof, based on the very constitution of man, that it is not constituted for preoccupation with pleasures and urges, but rather for the use of thought and deliberation (*fikr wa*

> *rawiyya*), given [man's] weakness concerning this [sc. the pursuit of pleasure] compared to irrational animals. (*Spiritual Medicine*, 24)

It's also worth noting that in a passage quoted above (*Spiritual Medicine*, 29), al-Rāzī says (on Plato's behalf) that medicine for the soul is "achieved through arguments and proofs (*bi-l-ḥujaj wa-l-barāhīn*)." One might suppose this means only that there are secure principles for psychological medicine, not that ethical agents are actually meant to rely on rationally proven beliefs when they deliberate. But as al-Rāzī says in the opening section of the *Spiritual Medicine*, it is reason (*ʿaql*) and reason alone that leads us to a good life. The intellectualist consequences are spelled out in later passages of the work, for instance when he remarks that whereas desire pushes us to choose on the basis of what is agreeable, *ʿaql* chooses on the basis of argument (*ḥujja*, 89). Or again:

> Know that the judgment (*ḥukm*) of reason (*al-ʿaql*) that the state of death is better than that of life is in accordance with [reason's] conviction (*iʿtiqād*) in the soul, but it can happen that one continues to follow desire in respect of this. For the difference between desirous opinion and intellectual opinion is that one chooses, is influenced by, follows, and adheres to desirous opinion without clear argument (*ḥujja*) or evident rationale, but only out of some sort of inclination (*mayl*) towards this opinion, and agreement and affection for it in the soul. Whereas one chooses intellectual opinion through clear argument and evident rationale, even if the soul finds it hateful and deviates from it. (*Spiritual Medicine*, 94)

One might therefore take the *Spiritual Medicine* to be a long sequence of "arguments" and "proofs" that will help the rational soul of the reader to have the right beliefs, which can then give rise to appropriate actions. Actually, as we'll see below, this would be to underestimate the complexity of al-Rāzī's strategy, but his emphasis on reasoning does provide us with good grounds to see him as upholding an intellectualist ethics.

That Miskawayh would be no less intellectualist is hardly surprising, given his broad allegiance to the tradition of late ancient Platonists like Plotinus. Like al-Rāzī, he defines man's perfection in terms of "discrimination and deliberation" (*tamyīz* and *rawiyya*, at *Refinement*, 12). Furthermore, he seems to put a high degree of trust in the power of belief and knowledge to guarantee good action:

> When a person *realizes* (*ʿalama*) that [passions and lusts] are not virtues but vices, he will avoid them and be loathe to be known for them. But if he *believes* (*ẓanna*) that they are virtues, he will pursue them and become accustomed to them (*ṣārat la-hu ʿāda*). (Miskawayh, *Refinement*, 10, trans. Zurayk, *my emphasis*)

> Wisdom is the virtue of the rational and discerning soul. It consists in the knowledge of all existents *qua* existing, or if we wish to say so, the knowledge of things divine and human. This knowledge bears the fruit of understanding which of the possible actions should be performed and which should not be. . . . In order to grasp the essences of these divisions [sc. of the virtues], we must have recourse to their definitions, for it is by the knowledge of definitions that we understand the essences of things sought, which are always constant. This is demonstrative knowledge which never changes and is not impaired by doubt in any way, for just as the virtues which are in essence virtues do not become, under any condition, other than virtues, so also is the knowledge of them [always the same and never changing]. (Miskawayh, *Refinement*, 18–19, trans. Zurayk)

How can he say these things, though, given that we saw him following Aristotle in acknowledging a gap between universal knowledge and the indefinite multiplicity of particular cases? As it turns out, he can be remarkably insouciant about this difficulty:

> Man attains his perfection and is able to perform his own distinctive activity when he knows all the existents. By this I mean that he knows their universals and their definitions which are their essences, not their accidents or their properties which render them infinite [in number]. For if you know the universals of existents, you come also to know their particulars in a certain way, since particulars do not depart from their universals. (Miskawayh, *Refinement*, 41, trans. Zurayk)

This seems to take most of the wind out of the sails of type (2) particularism: Miskawayh is the only one of our three authors to recognize the problem of applying reason to particulars in the ethical sphere, and now it looks as if he doesn't think it is much of a problem after all.[24]

Curing Sorrow

To illustrate how the intellectualism of these works is carried out in practice, I would like to consider a specific psychological "ailment," namely sorrow or grief (*ḥuzn, ġamm*). This case will incidentally allow us to see the close textual relationship between all four texts. The founding text for the treatment of sorrow in the Arabic tradition is, of course, the aforementioned *On Dispelling Sorrow* by al-Kindī. It puts forward many considerations about why one should never succumb to sorrow, providing everything from stringent philosophical argument to memorable anecdotes about Alexander the Great and Socrates. But one particularly central idea is the following one:

> Sensible things which we love and seek out, can be interrupted by anyone and seized by any power. It is impossible to protect

24 For an interesting and more detailed discussion of the relation between intellect and particulars, see Miskawayh's *On Soul and Intellect*, trans. in Pormann and Adamson, "More than Heat and Light," sec. 1.

> them, and one cannot be sure that they do not perish, fade away, or change. . . . If we want something corruptible to be not incorruptible . . . then we want from nature something which is not natural. He who wants something unnatural wants something which does not exist. Someone who wants what does not exist will seek in vain. And he who seeks in vain, will be distressed. (Al-Kindī, *On Dispelling Sorrows* 1:4–6, trans. Adamson and Pormann)

Al-Kindī wants us to see that if we make our happiness dependent on the continuation of perishable things, we are guaranteeing our future unhappiness. This is a familiar point from ancient ethics, common to both Stoics like Epictetus, who inferred that we should value only our own will, and Platonists, who inferred (as al-Kindī does) that we should instead value eternal, intelligible things. But there is a more distinctive idea at work here too: wanting the permanence of impermanent things is *incoherent*, since one is both valuing the nature of what one wants and rejecting that nature insofar as it guarantees transience. This would be like wanting to have a bath without wanting to get wet.

Whatever we might make of this argument, it was apparently a big success with later readers. We find it in all three works I discuss in this chapter:

> Another [remedy] is to think about the constitution and construction of this world (*al-dunya*): that it does not belong to anyone to spend their whole life having what they want and love, so that they would never lose what they love or miss out on what they seek. This being so, everything he loves and any felicity that does come to him is an advantage (*fāʾid*) and luxury (*ġinya*). (Al-Balḫī, *Benefits*, 319)

> When a reasonable person examines and considers that in this world which is subject to generation and corruption, and sees that their elements change and flow with no stability or permanence as individuals—rather they must all vanish . . . he should reckon

> the period of their continuation to be a bonus (*faḍl*) for him, and his savoring of them a profit (*ribḥ*), since they inevitably disappear and cease to be. When someone longs for the permanent continuation of [these things subject to corruption], he wants what cannot possibly exist. But someone who longs for what cannot exist thereby brings grief upon himself, and turns away from his reason towards his desire. Furthermore, grief and sorrow do not last concerning things that are unnecessary for continued life, because they are quickly replaced and come around again. (Al-Rāzī, *Spiritual Medicine*, 67–68)

> If [someone] knows that in this world of generation and corruption nothing is stable or continuous—there is stability and continuity only in the world of the intellect—then he will not crave for what is absurd, nor pursue it. If he does not pursue it, he will not feel sorrow in missing out on what he desires. (Miskawayh, *Refinement*, 217)

We know that Miskawayh was acquainted with al-Kindī's epistle since, as mentioned, he quotes from it at the end of the *Refinement*. It stands to reason that al-Balḫī was also familiar with his master al-Kindī's writings. Al-Rāzī presents more of a challenge when it comes to determining historical influence; here I will content myself with noting the striking similarities between the passages just quoted from him and al-Balḫī (especially the idea that the term of survival for any worldly good is a kind of "profit").

But the main point is not one about historical influence. It is that the repeated use of this argument points to a shared intellectualism. All our authors seem to presuppose that one can dispel or avoid sorrow about worldly things simply by realizing the implications of the perishability of those things. To put this another way, *insofar as one is rational*, one will not experience such sorrow. For, as al-Kindī already pointed out, the sorrow implicitly involves something "absurd": wanting the perishable to be imperishable. This is stated explicitly in Miskawayh, and it seems to be the same reason that al-Rāzī considers sorrow an abandonment of reason (the sorrowful person "wants what cannot possibly exist"). The obvious rejoinder to this whole line of thought is that one could well understand the impossibility of a state of affairs and still wish it would come about. One might wish that a dead loved one would come back to life, for instance, without being under any illusions about the impossibility of such an event. More generally, it would seem to be simply false that laying out arguments—no matter how compelling—is sufficient to change humans' desires and behaviors.

Galen acknowledges this in *On Passions of the Soul*, complaining that he has often persuaded people with good advice, only to see them lapse into their old ways thereafter (ed. Kühn, 5.52).[25] Of course, someone who is in the business of writing a book of ethical advice is not apt to stress the inefficacy of ethical advice. But al-Rāzī, at least, displays a rather subtle approach to this problem, by mentioning "second-best" considerations that might have more impact on a vicious person than more perspicuous arguments. For instance, despite believing that reason leads us to accept the soul's immortality, he gives a battery of arguments against the fear of death that would work for someone who thinks they will cease to exist upon the death of the body (*Spiritual Medicine*, 93).[26] Similarly, he at one point recalls trying to persuade a glutton that overeating will ultimately cause them pain greater than the pleasure they take in eating. This is not the real reason to avoid overeating (the real reason is that it involves domination of desire over the rational soul). But it has a better chance of being effective than the real reason:

---

25 At *On Character Traits*, 31, he similarly remarks that judgments that arise from habit and nature are difficult to eliminate merely through rational argument.

26 By contrast Miskawayh (at *Refinement*, 209) simply says that the fear of death is occasioned by false belief.

> This and other such remarks are of more benefit to someone who has not engaged in philosophical training (*riyāḍāt al-falsafa*) than proofs (*ḥujaj*) built on philosophical foundations (*uṣūl falsafiyya*). (Al-Rāzī, *Spiritual Medicine*, 71)

Particularism and the Lower Soul

Obviously, then, the intellectualist account cannot be the whole story. Rational arguments, naturally enough, can affect only the rational soul—the lower soul and especially the desiring soul remain unmoved by them. Galen says as much:

> The vegetative soul is not susceptible to being educated (*taʾdīb*) nor does it steer a straight course through training and discernment (*tamyīz*); it is improved and subdued only through taming (*qamʿ*). (Galen, *On Character Traits* 42)

In this he is followed by Miskawayh who likewise says that the "bestial" soul is not "susceptible to education (*adab*)" (*Refinement*, 54). In a sense this leaves the intellectualism of our authors standing: psychological health will still be understood as the correct functioning of reason. It is just that this correct functioning can be *undermined* by the contrary impulses of the lower soul. Again, Galen makes this particularly clear:

> Since errors arise from false opinion, while affections arise from irrational impulse (τὰ μὲν ἁμαρτήματα διὰ [τὴν] ψευδῆ δόξαν γίγνονται, τὰ δὲ πάθη διά τιν' ἄλογον ὁρμήν), I judged that one should *first* (πρότερον) free oneself from the affections. For we are probably led by these too somehow into false opinion. (Galen, *Passions of the Soul*, ed. Kühn, 5.7, trans. Singer, modified, *my emphasis*)

Notice that Galen already makes the crucial inference: even if health lies in knowledge, as we have seen him claim, health presupposes more than straightening out one's beliefs. It also demands an *antecedent* process, in which the lower soul is brought to heel so that it cannot interfere with reason's functioning.

There are two potential sources of trouble in the lower soul. First, a person may simply be born with bad tendencies (e.g., toward anger or a certain kind of desire). Second, regardless of their innate dispositions, they may have been habituated to allow their lower soul too much free rein. Only once these sources of vice are eliminated can we be sure that true beliefs will lead to happiness. Meanwhile, good nature and habituation will be of no avail if one's beliefs are false. Galen lays this out in a crucial methodological remark:

> In this book I base all that I investigate on what is evident in small children, in order more easily to discriminate the purely bestial motions from what has some admixture of the opinions and doctrines that belong to the rational soul . . . Among people, some live their life with what is naturally appropriate for them, avoiding what is not, without deliberation (*rawiyya*). Others apply deliberation and thought to the natures of things, so as to form the view that the appropriate thing is to follow what is naturally appropriate, and avoid what is not, or the other way around. So develops the way of life (*sīra*) of each of them, from that moment on, and for the rest of their lives they are led to actions through natural inclination and by acquired belief. . . . You must consider those who are past the age of childhood, regarding their actions and the causes of those actions. You will find that the cause of some is character (*ḫulq*) and of others belief (*ra'y*). That which derives from nature or habit is caused by character, but that which derives from thought and consideration is caused by belief . . . The character trait comes to be through constant habit. (Galen, *On Character Traits*, 30)

This passage confirms a threefold analysis of the potential sources of vice: in the non-rational soul, innate dispositions and habit form our "character." Good character is a necessary condition for happiness, simply because *bad* character will undermine the rational soul. But it is not a sufficient condition, since even a person of good character will need to have true beliefs (or ideally knowledge), such as the beliefs propounded by Galen and the authors of our Arabic works in Galenic ethics.

The slipperiness of this distinction may account for the fact that our authors seem torn between two accounts of virtue. We have seen both Galen and Miskawayh defining virtue (what Galen calls the "beauty" or "fineness" of the soul in a passage cited above, *On Character Traits*, 43) in purely intellectualist terms as true belief, knowledge, or wisdom. But virtue is frequently explained in terms of the balance of the soul or taming the irrational soul. The soul's health is likewise identified with such a balance, or even, as in al-Balḫī, in terms of the mere absence of psychological maladies. Strictly speaking though, the absence of such maladies and/or domination by the irrational soul are not constitutive of virtue, but the mere removal of an impediment. Still, if we want a formula that captures both the positive and negative sides of the conception of psychological health that runs through this Galenic tradition, this would be not be hard to give. We could say that it is the good functioning of reason, both in terms of its domination of (i.e., it is not being undermined by) the lower soul and in terms of its having knowledge. This is, perhaps, why al-Rāzī begins his *Spiritual Medicine* with a passage in praise of the usefulness and importance of reason (*ʿaql*) in human life.

Meanwhile, the capacity of the lower soul to undermine reason gives us a place for particularism in Galenic ethics. For even if, as I have argued, the positive functioning of the rational soul is the same for everyone (that is, we should all have the same beliefs—e.g., that worldly things are not to be valued, or death not to be feared), the threat posed to reason by the lower soul will vary from person to person. Since mere rational argument will not work on the desiring soul (as we have seen it is immune to "education," *adab*), we need another strategy: habituation. This provides Miskawayh with the opportunity to integrate Aristotelian with Galenic ethics. He defines character (*ḫulq*) as "a state of the soul which causes it to perform its actions without thought or deliberation" (*Refinement*, 31, trans. Zurayk) that can come from two sources. On the one hand it may arise "naturally, on the basis of [bodily] mixture (*mizāǧ*)," on the other hand through habit (*ʿāda*). This gives us two reasons why ethical treatment needs to be tailored to each individual person: the variation in their bodies, and the variation in their preexisting habitual dispositions.

I will return to the issue of bodily mixture in the next section; now let us briefly consider the point about habituation. The role of habit in ethical life provides a confirmation of the idea that medicine for souls, like medicine for bodies, has to do with a good "regime"—for instance, one that will habituate the soul to restrain its desires. A good example is the way we fear what is unfamiliar, a point made by Galen (*On Character Traits*, 32–33) and echoed by Balḫī (fear results both from being "ignorant of the nature (*anniyya*) of things" and lacking "experience (*taǧriba*) of the senses"—one might see here an invocation of both rational judgments and non-rational habituation). Miskawayh does something especially interesting with the theme of habituation, by repeatedly associating it with the law (*šarīʿa*) of Islam:

> It is the law that sets the youth on the straight path, habituates them to admirable actions, and prepares their souls to receive wisdom, to seek the virtues, and to reach human happiness through sound thought and upright reasoning (*qiyās*). (Miskawayh, *Refinement*, 35)

Notice again the way that habituation, which reforms the lower soul, is a first stage that removes obstacles that could hinder the acquisition of wisdom.[27]

It should be noted that this process of habituation concerns especially the desiring soul. The role of the spirit is more complicated. On the one hand, this aspect of the soul should certainly not be allowed to dominate. Anger is one of the psychological maladies discussed in our texts and can obstruct the development of reason.[28] On the other hand, as Plato said and is especially emphasized by al-Rāzī, one can use the spirit's sense of indignation against one's own desiring soul. He tells a story about a king who is stung by criticism of his fidgeting:

> This man's rational soul influenced his irascible soul, by means of rage and haughtiness, so that resolve became vigorous and secure in his rational soul, such that it exercised a strong influence on him and became a reminder, making him attentive whenever he started to become neglectful. Upon my life, the irascible soul was made solely in order to help reason against the desiring soul. (Al-Rāzī, *Spiritual Medicine*, 78)

This is a reminder that the goal of "balance" in the soul is not to make the lower souls completely inert or ineffective. Rather, we need only train the lower souls not to obstruct reason. In the case of the spirit, the well-trained lower soul can be positive help to reason.

---

27 Cf. *Refinement*, 49, 129. In the latter passage he even says that the best case scenario is being exposed to the law in childhood and then studying philosophy, and finding it to agree with that to which one has been "previously habituated."

28 Miskawayh, *Refinement*, 196, 205; cf. Balḫī, *Benefits*, 295 which speaks of waiting for a restful moment to combat anger with thought, *fikr*.

Particularism and Bodily Influence

Finally let us consider the second cause of particular ethical traits, which is the human's underlying physical state. Al-Balḫī frequently invokes "nature" to explain differences between people's souls, saying for instance that people may be "by nature (*bi-ṭibāʿihī*)" prone to anger or easily frightened "because of a delicate nature (*riqqat al-ṭabʿ*) and quickly changing soul" (*Benefits*, 293–94, 306). A particularly illuminating passage on this reads:

> Each person is not affected by [psychic maladies, *aʿrāḍ*] to the same extent, for they differ in terms of which maladies befall them. This is because each of them takes on [maladies] in accordance with his bodily mixture (*mizāǧ*), and how strong or weak is his basic composition. Some are quick to anger, while others come to anger slowly. Similarly, in some people there is much fear and anxiety about what is frightful, while others are steadfast and composed. Similarly what applies to (*aḥkām*) women and children and those who have weak natures (*ṭabāʾiʿ*), is different from what applies to strong men, with regard to the extent to which each of the [maladies] affects them. (Al-Balḫī, *Benefits*, 271)

The relevance of "bodily mixture" is linked to the fact that we share ethical dispositions with certain animals (e.g., *Benefits,* 338, regarding the trait of sociability). This is probably based on Galen's consideration of children and animals at such passages as *On Passions of the Soul* (ed. Kühn, 5.38–40). But more generally the influence of mixture is established in Galen's treatise *That the Soul Depends on the Body*, which of course makes a powerful case that ethical traits and other psychological states are at least partially caused by bodily states.

Again, Miskawayh finds an opportunity here to harmonize Galen with Aristotle, since Aristotle famously says that anger is associated with the boiling of blood around the heart (*On the Soul*, 403a31).

Miskawayh repeats this assertion and then remarks, "people differ in this respect according to bodily mixture," (*Refinement*, 193–94). Similarly al-Balḫī says that obsessive thoughts are caused by an excess of bile (*mirra*).[29] The importance of mixture also explains why climate is an important influence on character (Miskawayh, *Refinement*, 47, 175). This may be linked to Hippocratic ideas about climate, but the main source for this in Arabic was probably Ptolemy,[30] as used early on by al-Kindī.[31] Finally, it should be noted that just as the body can affect the soul, so the soul can affect the body and its mixture. Thus al-Rāzī remarks that envy causes "prolonged sadness (*ḥuzn*), worry (*hamm*), and obsessive thoughts (*fikar*). Upon the incidence of these symptoms in the soul, [the body] undergoes prolonged sleeplessness and bad diet, which are followed by poor coloring, bad appearance, and the disruption of the [humoral] mixture" (*Spiritual Medicine*, 51). As al-Balḫī remarks, "the soul of every living thing is powerfully disposed towards the body in which it resides, and has a harmony with it" (*Benefits*, 353).

Conclusion

The upshot of all this is that psychological medicine, just as much as bodily medicine, must take account of the particular characteristics of each "patient." Each of us is born with certain natural tendencies caused by our bodies, and these tendencies are then modified through habit (and other factors that affect the mixture itself, like diet). The

29 At *Benefits*, 344 he mentions an alternative possible cause, namely a demon (*šayṭān*)—here one might think of Christian ascetic literature such as Evagrius, but the idea is also attested in the *ḥadīth*. At 345–46 he again invokes bodily mixture while discussing obsessive thoughts.

30 *Tetrabiblos* 2.2.

31 See *On the Proximate Agent Cause of Generation and Corruption*, sec. 7, in Adamson and Pormann (trans.), *The Philosophical Works of al-Kindī*. Al-Kindī here alludes to the Galenic claim that the states of soul depend on those of the body, showing that he realizes the Galenic theory can explain the phenomena mentioned by Ptolemy (that climate has an effect on things like hair, as well as psychological traits).

advice, anecdotes, and arguments gathered in Arabic works like the ones we have surveyed will not, then, work equally well on everyone. Someone who really wants to ensure their psychological health should not only read such books but also seek individual advice tailored to their needs. This is perhaps why Galen recommends that we call on the services of a friend to criticize our particular shortcomings.[32] Though this outside perspective can be useful, ultimately all of us must, to some extent, be our own doctors.[33] For this, if for no other reason, if you want to be happy, you must know yourself.[34]

32 *Passions of the Soul*, ed. Kühn, 5.9–10; cf. al-Rāzī who explicitly refers to Galen on this topic at *Spiritual Medicine*, 35, and Miskawayh who also knows Galen's discussion and points out that an enemy would be better for this job than a friend (*Refinement*, 189).

33 Cf. al-Balḫī's contrast of external and interior medicine, discussed above.

34 This paper was written with the support of the Leverhulme Trust, which is here gratefully acknowledged.

Reflection

THE RATIONALITY OF MEDIEVAL LEECHBOOKS

Richard Scott Nokes

The very word "leech" for a medieval physician sounds romantic, in the way that practices from the past can seem exotic and alien. I must admit, that even after years of studying medieval Anglo-Saxon medical books, the first time I entered the British Library manuscript room to examine *Bald's Leechbook*, I had the sensation of being involved in some occult activity, discovering arcane mysteries from ancient texts. The room was lit with electric lights, not candles, appointed with computers, not mystical symbols, and scholarly dress tended toward jackets and ties, not robes and pointy hats. Yet I couldn't help feeling as though I were in some horror movie, in which uttering an incantation out loud could unleash an unspeakable curse. Of course, my fantasies were an example of our modern prejudices, not a representation of the reality of the Middle Ages. Our romantic ideas about the medievals obscure the truth that medicine has always largely been a pragmatic affair. While the *Leechbook* claims that afflictions in the natural world occasionally have supernatural causes, the treatment nearly always privileges the natural.

The *Leechbook* is actually three different medical texts that have been bound together into a single manuscript (Royal 12.D.xvii), containing hundreds of remedies, transmitted over several centuries. The first two sections (*Leechbook I* and *II*) are commonly called *Bald's Leechbook*, because a colophon in the manuscript lets us know that "Bald is the owner of this book, which he ordered Cild to write." The third book is generally just called *Leechbook III* and contains much material duplicating what we find in the previous two books, suggesting it had a history of independent transmission before settling into its home in *Bald's Leechbook*.

The perennial debate regarding the contents of the *Leechbook* and similar medieval texts is to how rational or irrational they are, and whether they are properly considered magic, medicine, religion, or some amalgam of all those categories. One of the most influential works of scholarship on these texts, J. H. G. Grattan and Charles Singer's book *Anglo-Saxon Magic and Medicine*, marks this ambiguity in its title.[1] Grattan and Singer are not, however, ambiguous regarding the rationality of these texts, famously writing:

> Surveying the mass of folly and credulity that makes up the Anglo-Saxon leechdoms, it may be asked, "Is there any rational element here? Is the material based on anything that we may reasonably describe as experience?" The answer to both questions must be, "very little."[2]

At the opposite end of the spectrum is the article, "A Re-Assessment of the Efficacy of Anglo-Saxon Medicine."[3] In an act of scholarly jujitsu, the authors demolish the assertions by Godfrid Storms regarding the magical elements of a remedy by

1 J. G. H. Grattan and C. Singer, *Anglo-Saxon Magic and Medicine: Illustrated Specially from the Semi-Pagan Text Lacnunga* (London: Oxford University Press, 1952).

2 Grattan and Singer, *Anglo-Saxon Magic and Medicine*, 92.

3 B. Brennessel, M. Drout, and R. Gravel, "A Reassessment of the Efficacy of Anglo-Saxon Medicine," *Anglo-Saxon England* 34 (2005), 183–95.

experimentally testing the salve prescribed and demonstrating that it actually worked.[4] Of course, most scholars fall between these two poles; some considering remedies with incantations as magical charms, and some considering remedies with exclusively natural elements as medicine. Still others diminish the distinctions between magic, medicine, and religion, as Karen Jolly does in her book *Popular Religion in Late Saxon England*.[5] But all the debate regarding the rational/irrational and medicine/magic has distracted from the fundamentally pragmatic nature of the *Leechbook*. Medicine is, by its very nature, a practical affair dealing with real human needs, needs that have not really changed in millennia. Although the treatments for ailments and broken limbs may have developed over time, the human experience of breaking a limb today is basically the same as a broken limb 5,000 years ago . . . or in this case, 1,000 years ago.

Like many other ancient medical manuals, the *Leechbooks* are organized in a roughly head-to-toe fashion: *Leechbook I* focusing on exterior ailments, and *Leechbook II* focusing on internal ailments. Some other medical manuals, such as the *Old English Herbarium* and *Medicina de Quadrupedibus*, are organized according to what we might call the "active ingredient" of the remedy—herbal ingredients in the former, and animal ingredients in the latter. We do not find, say, sections on ailments that are caused by being elfshot (pierced by the tiny, invisible arrows of supernatural beings) or by witchcraft. Instead, they are organized entirely according to their physical elements. Nor is this surprising. One would presumably seek the help of a leech for a specific affliction, some particular symptom. When the leech Bald was looking at a patient, although he might have wondered whether

4 G. Storms, *Anglo-Saxon Magic* (Halle: Jijhoff, 1948).

5 K. L. Jolly, *Popular Religion in Late Saxon England: Elf Charms in Context* (Chapel Hill: University of North Carolina Press, 1996).

the illness was caused by an unbalance of humors, or witches, or elves, doubtless his primary thought was, "This man is feverish and vomiting. How can I heal him?" When using any of these books, the medieval leech was searching for a way to heal the afflicted.

Modern scholars use the *Leechbooks* differently. We are looking for a window on the past, and generally more interested in the ways medieval medical practice differed from modern practice. As a result we tend to focus on the strange and exotic; scholarship becomes strongly biased toward novelty. A good example of this bias is the dozen-or-so extant Anglo-Saxon charm texts. These texts are widely known and much studied by medieval scholars, who have been drawn by their magical-sounding incantations. In my experience, scholars are often unaware that any other remedies even existed, let alone that volumes and volumes of remedies are extant. In no way are these charm texts representative, in terms of either content or number. Yet they are the first texts that spring to mind when we hear the word "leechbook."

Still, we should not ignore the supernatural elements of medieval medicine. Although the priorities of the remedies found in the *Leechbooks* are pragmatic, behind the practical treatments there is clearly an understanding that supernatural forces must play some part in some illnesses. There is a simple reason why the *Leechbooks* do not focus on these forces: what, in the end, could be done about them beyond entreating God for help? Instead, the *Leechbooks* focus on what can actually be done by the leech to aid the afflicted. For that reason, the dichotomy between "rational" medieval herbal remedies and "irrational" medieval charms and prayers is a false one. As modern scholars, we remove ourselves from the very function of these remedies with a set of wildly foolish anachronistic assumptions: if the leech believed that an illness was spread by invisible creatures called "elves," that leech was bound by irrational superstition. If the leech had only believed that an illness was spread by invisible creatures called "microbes," that leech would have been rational and wise.

For the medieval leech, focusing on elves would be irrational, not because belief in elves is superstitious, but because the patient has come to the leech for a cure. The rational approach is to affect a cure for the ailment, not to offer the suffering patient a thesis on the potential causes. So, for example, with the charm "For Waterelf Disease," found in *Leechbook III*, the remedy opens, "If a man has waterelf disease, then his fingernails will be dark and his eyes teary and he will look down . . . " This is then followed by an herbal remedy (mixed in ale and holy water), and an incantation for the leech to sing three times. A modern scholar's focus on elves, holy water, and incantations cause us to see this as irrational, but it is entirely rational and pragmatic. After naming the disease, the remedy immediately lists the symptoms, followed by a cure.

In *The Greeks and the Irrational*, E. R. Dodds describes an encounter with a young man who remarks that he is unmoved by Greek art because "it's all so terribly *rational*, if you know what I mean."[6] By contrast, modern scholars frequently dismiss medieval science as terribly irrational. Yet the remedies labeled "irrational" by modern scholars were seen as a bridge between the physical and supernatural worlds. While the medieval leech certainly acknowledged the power of supernatural forces, he and his patients still had to dwell in the natural world. Given this reality, it speaks poorly of modern scholars that we have so little studied the efficacy of the remedies in the medieval leechbooks. Instead, we tend to dismiss them as pseudo-scientific, because on rare occasions they blame illnesses on supernatural forces. We miss our opportunity to cross the bridge offered by these charms.

6 E. R. Dodds, *The Greeks and the Irrational* (Berkeley: University of California Press, 1951), 1.

CHAPTER FIVE

Health in the Renaissance

Guido Giglioni

It will not be out of place, at the beginning of this chapter, to offer a few introductory remarks on the Latin terminology of health. Although the centuries between the Middle Ages and the early modern period saw the increased use of national vernaculars in science, the language of health remained heavily indebted to the original Latin matrix, all the more so since during the late Middle Ages and throughout the Renaissance, Greek and Arabic sources had become available in Latin translations and led to a wealth of linguistic solutions concerning the terminology of health. The Latin word *salus* included the meanings of "health," "safety," and "salvation." As such, it was a word laden with medical, political, and religious connotations. For instance, dwelling on the relationship between health and salvation, Petrarch (1304–1374) brought into sharper focus the root of salvation (*salutifera radix*) rediscovered through a Platonic meditation on death.[1] Pietro

1 E. Carrara (ed.), "Petrarch: Secretum," in G. Martellotti et al. (eds), *Petrarch: Prose* (Milan: Ricciardi, 1955), 30.

Pomponazzi (1462–1525), to give another example, underlined the political sense of *sanitas*, when in his *De immortalitate animae* he referred to the Platonic and Aristotelian view of the statesman as the physician of the souls, concerned with the right behavior of his people, and not with their knowledge or culture.[2] As for the strictly medical meanings of *salus, sanitas*, and *valetudo*, the latitude of health (discussed later) was ample, for doctors and laymen were fully aware that one could have good or bad health, undergo various degrees of chronic dysfunction, and oscillate between states of valetudinarian and hypochondriac impairment.[3]

In what follows, I examine the principal changes that the idea of health underwent during the Renaissance (spanning, roughly, from the second half of the fourteenth century to the beginning of the seventeenth). The guiding thread in my analysis of the interrelated concepts of soundness, safety, and salvation will be provided by the prophylactic framework of the so-called six non-naturals (environmental conditions, nutrition, physical exercise, evacuations, sleeping habits, and emotions), that is, those factors that, placed at the juncture of both nature and culture, had been regarded, since antiquity, as capable of affecting human health in crucial ways. Furthermore, because of their epistemological and methodological flexibility, the six non-naturals have the advantage of bringing to the fore the many intersections between the physical, social, and political spheres of human experience. For this reason, Renaissance hygiene can be seen as the ideal vantage point to investigate the relationships between individual bodies and the body politic; and the interplay of bodily, mental, and spiritual conditions in the definition of well-being; not to mention the many bio-ethical questions related to such issues as good life, longevity, and

2 G. Morra (ed.), *P. Pomponazzi: Tractatus de immortalitate animae* (Bologna: Nanni and Fiammenghi, 1954), 186–88.

3 On medical Latin in the early modern period, see G. Giglioni, "Medicine," in P. Ford, J. Bloemendal, and C. Fantazzi (eds), *Brill's Encyclopaedia of the Neo-Latin World* (*Macropaedia*) (Leiden: Brill, 2014), 679–90.

vitality. I refer to a variety of authors and sources, but three names in particular stand out: Marsilio Ficino (1433–1499), Alvise Cornaro (1484–1566), and Girolamo Cardano (1501–1576). Ficino, a physician and a priest known for his recovery and reinterpretation of the Platonic tradition, also wrote one of the most popular texts of psychosomatic prophylaxis in the early modern period. Cornaro, a layman, was the author of a bestselling guide to healthy life in the Renaissance, the *Discorsi intorno alla vita sobria* (*Discourses about Sober Life*), in which he argued that everyone could lead a healthy life without the need to resort to physicians and medicines.[4] Cardano, finally, was a professor of medicine at the universities of Pavia and Bologna and a renowned practitioner. He wrote an important manual of hygiene, *De sanitate tuenda* (*How to Preserve One's Health*, 1560), where, as we will see, he scrutinized the matter of a healthy life (*salubritatis materia*) and its natural and cultural implications.

The Historical Background

How Renaissance people perceived the vicissitudes of their health, both at a learned and a popular level, was marked by a number of significant variables. Here I mention the new appreciation of Galen's works, a remarkable progress in anatomical techniques and discoveries, the spread of new epidemic diseases, and a distinctively antimedicalizing attitude among humanist writers.

At the end of the Middle Ages, while commenting on Galen's influential definition of medicine from his *Ars medica* (*Medical Art*) as "knowledge of things healthy, unhealthy and which are neither healthy, nor unhealthy (*scientia salubrium, et insalubrium, et neutrorum*),"[5] scholastic doctors sparked an intense debate over the

---

4 In A. Di Benedetto (ed.), *Prose di Giovanni della Casa e altri trattatisti cinquecenteschi del comportamento* (Turin: UTET, 1970), 355–420.

5 K. G. Kühn (ed.), *Galen: Claudii Galeni Opera omnia*. 20 vols (Leipzig: Knobloch, 1821–33), vol. 1, 307–8.

categories of health (*salubre*), illness (*morbosum*), and the intermediate condition known as "neutrality" (*neutrum*).[6] In the course of the sixteenth century, especially after the Aldine presses had issued the 1525 *Editio princeps* of Galen's works, a more radical and philologically more sophisticated approach to his oeuvre—especially to such influential treatises as *Quod animi mores corporis temperamenta sequantur* (*The Habits of the Minds Follow the Temperaments of the Body*) and *De placitis Hippocratis et Platonis* (*The Opinions of Hippocrates and Plato*)—resulted in successive waves of naturalism of a distinctively medical flavor. The two best known cases are that of the Spanish physician Juan Huarte de San Juan (c. 1530–1592), whose *Examen de ingenios para la ciencias* (*An Appraisal of Mental Skills for the Different Branches of Knowledge*, published originally in 1575 and later reissued in 1594 in a revised form) provided a thoroughly physicalist understanding of human life and health; and the Italian philosopher Bernardino Telesio (1508–1588), who devoted a large part of his work to formulating a new concept of natural health and social wholesomeness.

Another aspect of Galen's medicine that underwent renewed scrutiny between the end of the Middle Ages and the Renaissance was his view of health as a condition defined by the changeable and individual nature of human beings. During the sixteenth century, university professors of medicine who were open to both exegetical and anatomical innovation, such as Giambattista da Monte (1498–1552), Matteo Corti (1475–1544), and Girolamo Cardano, debated whether health was an ideal and timeless norm of perfection against which doctors were supposed to measure the condition of individual bodies or rather a pragmatic and flexible criterion, known as "latitude of health" (*sanitatis latitudo*), which depended on factors such as age, environment, diet, and sex. Stressing the clinical and empirical foundations of medicine, Da Monte, besides editing Galenic texts, promoted

---

6 T. Joutsivuo, *Scholastic Tradition and Humanist Innovation: The Concept of Neutrum in Renaissance Medicine* (Helsinki: Finnish Academy of Science and Letters, 1999).

anatomical investigations both in Ferrara and Padua. He also presented the art of restoring health as knowledge of degrees rather than a science of immutable essences.[7]

Despite his predominance in the medical panorama of the Renaissance, Galen was not the only force to cause a reorientation in ideas and values about human health. Indeed, a series of important anatomical discoveries accelerated a long-term shift from a temperamental and humoral consideration of health—grounded in ideas of well-balanced mixtures of humors and cosmological harmonious correspondences—to one based on structures, obstructions, and tensions. The typically digestive and humoral view of life processes, marked by varying degrees of slowness and porosity (a porosity that was deemed to facilitate exchanges between solid, fluid, and airy states of matter, between physical and mental faculties, and between the natural and moral orders of life) gradually gave way to a new medical framework in which life was defined by speed and mechanical motion: organs were regarded as neatly defined structures that were constantly being filled and emptied through channels and pores, while mental operations were described in terms of accumulation and release of tension.

To this intellectual milieu of general Galenic reappraisal, we should add two more factors: one material, the other intellectual; that is to say, the spread of new pestilences, and the influence of the humanist movement. Unknown epidemic diseases, such as syphilis, smallpox, and sweating sickness, acted as powerful catalysts in the field of public health and urged policymakers to engage in farsighted sanitary policies, which in the end led to the birth of the hospital in the modern sense.[8]

---

7 On Da Monte's innovations in clinical medicine, see J. J. Bylebyl, "The School of Padua: Humanistic Medicine in the Sixteenth Century," in C. Webster (ed.), *Health, Medicine and Mortality in the Sixteenth Century* (Cambridge: Cambridge University Press, 1979), 335–70.

8 J. Henderson, *The Renaissance Hospital: Healing the Body and Healing the Soul* (New Haven, CT: Yale University Press, 2006). On the impact that new diseases had on sanitary strategies and institutions, see C. Cipolla, *Public Health and the Medical Profession in the Renaissance* (Cambridge: Cambridge University Press, 1976); J. Arrizabalaga, J. Henderson, and R. French (eds),

Renaissance humanism, on the other hand, exercised a significant influence on the evolution of medical ideas and practices by bringing to the fore several ethical issues underpinning contemporary health plans and by criticizing forms of extreme medicalization. Petrarch, it's worth remembering here, had already condemned scholastic doctors for their reductionist attitude about human health and their attempts to replace ethical and rhetorical competences with various forms of medical treatment.[9] Moreover, humanist doctors (such as Antonio Musa Brasavola) and humanists interested in medical topics (such as Desiderius Erasmus) reinforced the already prevalent assumption that health was a characteristic trait of individual bodies, immersed in the life of larger social and political bodies, exposed to the vicissitudes of historical change.[10] Here humanist preoccupations with historicism and cultural relativism met with contemporary anxieties about the definition of natural power. This had obvious consequences for how doctors and laymen undertook the task of protecting and promoting human health. The account of health as a state "according to nature" was being increasingly questioned in favor of a broader, more pragmatic notion of health understood as the condition whereby humans were not impeded in their activities and aims, in keeping with conjectural, pragmatic, and utilitarian views of knowledge. According to Galen's standard definition, health consisted in a balanced proportion between warm, cold, moist, and dry qualities. The best combination

---

The Great Pox: The French Disease in Renaissance Europe (New Haven, CT: Yale University Press, 1997); R. French and J. Arrizabalaga, "Coping with the French Disease: University Practitioners' Strategies and Tactics in the Transition from the Fifteenth to the Sixteenth Century," in R. French et al. (eds), *Medicine from the Black Death to the French Disease* (Aldershot: Ashgate, 1998), 248–87.

9 P. G. Ricci (ed.), *Petrarch: Invective contra medicum*, in Martellotti et al., *Petrarch: Prose*, 682–84. For the case of a physician dealing with the rhetorical aspects of his profession (Girolamo Cardano), see G. Giglioni "The Many Rhetorical Personae of an Early Modern Physician: Girolamo Cardano on Truth and Persuasion," in S. Pender and N. S. Struever (ed.), *Rhetoric and Medicine in Early Modern Europe* (Farnham: Ashgate, 2012), 173–93.

10 On medicine and humanism, see Bylebyl, "The School of Padua"; V. Nutton, "The Rise of Medical Humanism: Ferrara, 1464–1555," *Renaissance Studies* 11 (1997), 2–19; N. G. Siraisi, *History, Medicine, and the Traditions of Renaissance Learning* (Ann Arbor: University of Michigan Press, 2007).

of these qualities, especially warmth and moisture, resulted in the best temperament, that is, the most harmonious bodily condition (Galen, *Opera omnia*, 6:1–2; Galen, *Hygiene*, 5).[11] This situation, however, remained largely hypothetical, for actual temperaments were never fully balanced in all their qualitative degrees. Moreover, since the inevitable formation of superfluous substances and impurities prevented the body from being fully repaired through food, drink, respiration, and pulse, a complex system of evacuations and perspirations was needed to expel all that had not been completely assimilated by the organism (Galen, *Opera omnia*, 6:59–68; Galen, *Hygiene*, 38–41). In this sense, the precarious equilibrium of temperamental and individual health was crucially related to the various stages of digestion ("concoction," to use the technical term). The delicate interplay of health, digestion, and evacuations was traditionally associated with the doctrine of the so-called six non-naturals.

At the end of the fifteenth century, the growth of the print trade combined with a more diffused level of literacy favored a wider spread of a particular kind of literature concerned with medical advice. According to Sandra Cavallo and Tessa Storey, this development and a number of other crucial factors helped create a more proactive attitude toward the elimination of diseases and the preservation of health. These factors were the commercialization of new drugs, which more often than not were manufactured with substances coming from the East and the Americas; the spread of the Paracelsian view of illness, which became quite influential during the sixteenth and seventeenth centuries; and a certain appeal exercised by charlatans and their empiric remedies. A "culture of prevention" spread in Renaissance Italy, where physicians (not only in the courts, but also in urban settings) were often consulted as advisers in matters of health maintenance and the prevention of illness. Greater attention to hygiene, both

11 See C. Spon (ed.), *G. Cardano: Opera omnia*, 10 vols. (Lyon: Jean-Antoine Huguetan and Marc-Antoine Ravaud 1663. Repr. Stuttgart-Bad Cannstatt: Frommann, 1966), 6:23b.

personal and public, physical and emotional, was being incorporated into new models of conduct, influenced by more genteel lifestyles, changes in household management, a diffused interest in how to employ leisure time, not to mention that in Catholic countries the post-Tridentine Church emphasized norms of bodily decorum and emotional restraint.[12] It is a clear sign of the new orientation in health care that in his book on life *De vita libri tres* (*Three Books on Life*), the fifteenth-century philosopher and physician Marsilio Ficino (1433–1499) felt the need to include special advice for people living in cities.[13] In *De la institutione di tutta la vita de l'homo nato nobile e in città libera* (*The Lifelong Education of a Gentleman Born in a Free City*) published in 1542, Alessandro Piccolomini (1508–1578) complained that in the new urban reality human well-being had been reduced to a medical and natural philosophical issue, without paying due attention to the soul and its spiritual needs.[14] In his *Essayes* (published in 1597 and again, in a much enlarged edition, in 1612), Francis Bacon (1561–1626) warned that health, both physical and mental, was the first thing that men of action risked losing: "Certainly, Men in Great Fortunes, are strangers to themselves, and while they are in the pulse of businesse, they have no time to tend their Health, either of Body, or Minde" (Bacon, *The Essayes or Counsels, Civill and Morall*, 34).

The Six Non-Naturals

It is evident that, since antiquity, the discourse on health was not confined only to ways of curing illnesses and restoring the balance

12 S. Cavallo and T. Storey, *Healthy Living in Late Renaissance Italy* (Oxford: Oxford University Press, 2013), 4, 7–9.

13 C. V. Kaske, and J. R. Clark (ed. and trans.), *M. Ficino: Three Books on Life* (Binghamton, NY: Center for Medieval and Early Renaissance Studies, 1989), 220.

14 Eugenio Refini, "De la institutione di tutta la vita de l'homo nato nobile e in città libera," in *Vernacular Aristotelianism in Renaissance Italy Database* (VARIDB), https://vari.warwick.ac.uk/items/show/4920.

of lost soundness but also involved ways of preserving a condition of well-being, both physical and mental. Starting with Hippocrates, the art of preserving health was associated with a series of measures concerning prevention, prophylaxis, and hygiene. Preventive measures included a broader outlook about the relationship between life and death; notions of old age and prolongation of life; general considerations of how human beings should promote vital expansion and ethical fulfillment; and, finally, on a broader scale, ways of maintaining the social and political order. In his treatise on how to preserve one's health, Galen had distinguished between two different strategies for dealing with health: preservation and restoration. In keeping with the same division, in his *De sanitate tuenda* (1560), Cardano divided medicine into two principal parts: "one protects health when this is present; the other restores it when it is absent." Health needed to be protected, Cardano explained, for no single body in nature, however strong and wholesome, could last long without medical assistance. In Cardano's opinion, the part of medicine devoted to prophylaxis and prevention was more important than the one aiming at recovering lost health. He thought it was better "not to fall ill for a long time rather than to heal people that have fallen ill." The reason was that "those who fall ill do not know when their disease will end, and, in case they recover their health, they will not go back to their previous level of wholesomeness (*integritas*)" (Cardano, *Opera omnia*, 6:10–11).

Since Hippocratic times, and continuing with such seminal texts as Aristotle's *Parva naturalia* and Galen's *De sanitate tuenda*, the overlapping fields of therapy and prevention became populated with three different kinds of ontological entities: immutable natural principles (the "naturals," κατά φύσιν); pathogenic factors of all sorts (the "unnaturals," παρά φύσιν); and various intermediate conditions that were subject to a limited amount of change and therefore were considered not completely natural (the "non-naturals," οὐ κατά φύσιν μέν, οὐ μήν ἤδη παρά φύσιν). The medical category of non-naturals originated therefore from secular attempts to adjust cultural habits to biological processes (Galen, *Opera omnia*, 6:27–28, 57; Galen, *Hygiene*, 19, 35).[15] Besides the seven "naturals" (elements, temperaments, humors, organs, faculties, actions, and spirits), tradition had gradually systematized the number of the non-naturals into six: air, food and drink, physical exercise, sleep and waking, evacuations, and emotions. By non-naturals, doctors referred to those factors—both natural and cultural—that were thought to affect all aspects of human life. Unlike therapy, which addressed abrupt and exceptional events (the unnaturals, i.e., illnesses), health regimens centered on ordinary events of everyday life. It was by using the non-naturals as a template that doctors elaborated a set of precautions and devices, largely on an empirical and practical basis, to maintain relatively stable standards of a healthy life.

The increasing emphasis placed on variables such as change, contingency, individuality, environment, and habit—a typical aspect of Renaissance culture, reinforced by the poignant sense of the past diffused by the humanists—confirmed the view that health depended on the correct administration of the six non-naturals: where to live, what to eat and drink, how to regulate the cycle of sleep and waking, how many hours to devote to exercise and leisure time, how to purge the body and the mind of physical and emotional pollutants, and finally how to take advantage of the emotional energy provided by the passions without suffering mental and physical exhaustion. Because of their dual status—natural and cultural—the six non-naturals represented the prophylactic and ecological conditions of an embodied mind (or ensouled body), situated in a complex environment and exposed to natural and social constraints at once. Human

---

15 On the origin and evolution of the "six non-naturals," see L. J. Rather, "The 'Six Things Non-Natural': A Note on the Origins and Fate of a Doctrine and Phrase," *Clio Medica*, 3 (1968), 337–47; L. García-Ballester, "On the Origins of the 'Six Non-Natural Things,' in J. Kollesch and D. Nickel (eds), *Galen und das Hellenistiche Erbe* (Stuttgart: Steiner, 1993), 105–15; K. Albala, *Eating Right in the Renaissance* (Berkeley: University of California Press, 2002); Cavallo and Storey, *Healthy Living in Late Renaissance Italy*.

beings were regarded as truly cultural-biological engines, capable of absorbing energy from the environment (through eating, drinking, and breathing), of processing that energy through physical and mental exercise (including sleeping, dreaming, imagining, and feeling), and of releasing it back into the environment in the form of excretions, evacuations, and cultural artifacts. For Renaissance human beings living in a culture of non-naturals at every social level, food was both a material and spiritual experience. Moreover, because of the key role played by the environment, the six non-naturals underpinned a view of health that was natural and social at the same time, private and public. To recapitulate, we can say that by relying on a constant triangulation between the spheres of nature (life), un-nature (illness), and non-nature (health), the art of preserving health took account of three main variables: the individual, shaped by a specific proportion of qualities and humors (*temperamentum* or *complexio*), in particular by a definite ratio between innate heat (*calor innatus*) and original moisture (*humidum radicale*); the impact that external influences from the environment, above all air and food, exercised on the individual; and, finally, the influence of social and cultural factors such as habits, customs, laws, ethical precepts, religious practices, fashions, and ways of managing the household economy.

From a more specifically philosophical point of view, the balance of nature, un-nature, and non-nature was predicated on the interplay of life and death. Though there is a natural tendency toward decay and dispersal, living beings had also been described since ancient times as driven by an original appetite for self-preservation that manifested itself through the biological functions of eating, drinking, and breathing in and out (the heartbeat was a function of the digestive system and not of the then-unknown circulation of the blood; Galen, *Opera omnia*, 6:6–7; Galen, *Hygiene*, 7). Premodern and early modern regimens of prophylactic measures can be seen as all-encompassing models of socio-bio-ethical conduct centered on the functions and the effects of the digestive system. According to Cardano, to give an example, the part of medicine that dealt with the preservation of human health had a physical counterpart in the very function of nutrition and metabolism, for our bodies are always "in a state of continual change" (Cardano, *Opera omnia*, 6:30b). In order for a body to be able to reconstitute its daily loss of vital matter, Cardano went on, "sagacious nature" had provided the body with "a faculty of preservation" whose principal instrument was the very "sensitive" mouth of the stomach (Cardano, *Opera omnia*, 6:20a).

In writing his own art of preserving health, Cardano was following a well-trodden path in physiology. The foundations remained the classical principles of Galenic anatomy before William Harvey (1578–1657) discovered that blood circulated incessantly inside the body at a very high speed. The substance of the living body was warm and moist. As such it was subject to produce waste that needed to be constantly replaced through food and drink (*alimentum*). Sleeping was the state that was more conducive to digestion (*concoctio*), and it was a means of restoring life based on that particular relation of inverse proportionality that, in Cardano's opinion, connected the operations of conscious activity (*facultas animalis*) to the unconscious processes of vegetative life (*facultas vitalis et naturalis*). As a result, when the mind was at rest, the bodily functions grew more active, and vice versa, when the mind was at its most alert, the vegetative operations slowed down (Cardano, *Opera omnia*, 6:21a). Moreover, a large part of the vital economy of the body rested on the way in which waste materials were handled by the secretive and excretive faculties (*excrementorum* [*ut ita dicamus*] *oeconomia*), while illnesses were caused by accumulation of superfluous matter (Cardano, *Opera omnia*, 6:11, 30a). Warmth, finally, had a central role in maintaining the health of the whole mechanism, especially through the many exchanges between internal (*proprius*) and external (*ambiens*) heat, exchanges that to a large extent relied on the air as a vehicle of breathed-in and breathed-out substance (Cardano, *Opera omnia*, 6:31b).

Considered as a general biological process affecting several parts of the body at various levels and in different periods, digestion (*coctio*) was therefore a central feature in ancient, medieval, and early modern medicine, in the domains of both physiology and pathology, and within pathology, in the complementary spheres of physical and mental illnesses. The same held true of hygiene and prevention, so much so that the six non-naturals can be primarily described as ways of identifying and managing the innumerable operations and variables affecting the economy of the digestive functions: above all nourishment, drinking, and evacuations, but also motion, thinking, and socializing. Sleeping, for instance, was closely related to digestion, for it was during sleep that the most critical phases of concoction designed to restore the life of the organism were supposed to occur, and so was moderate exercise, both physical and mental (walking in a garden, for instance, while admiring the colors and harmonies of the landscape), for they were operations that facilitated digestion. Likewise, passions could have both positive and negative consequences on the principal digestive phases. Diagnoses of melancholy and hypochondria often suggested disorders that affected the faculties of digestion more than the nervous system. The environment as a whole, finally, and not just the breathed air, was perceived in terms of one external force capable of influencing metabolism in all its stages. Echoing Galen and Avicenna, Ficino repeatedly stressed the importance of digestion for a healthy life (*digestio vitae radix*), not only because undigested or poorly digested nutritive matter (*cruditas*) was behind a large number of illnesses, but also because the seemingly simple act of digesting food did not stop at the operations performed in the stomach and the intestines. Since antiquity, digestion had been regarded as a large and complex vital enterprise, consisting of many stages, called "first digestion" in the stomach, "second digestion" in the liver, third in the veins, fourth in the various parts of the body, fifth in the left ventricle of the heart, and sixth in the brain (Ficino, *Three Books on Life*, 172–74; Helmont, *Sextuplex digestio alimenti humani*). The brain completed the digestive process by elaborating animal spirits, the main tools for imaginations and thoughts.

Prophylaxis: A Subtle Art of Paradoxes

Early modern physicians were well aware that the best regimen was the one adjusted to the specific needs of each individual. Prophylactic norms could not be indifferently applicable to every different situation. This was yet another reason, unlike the naturals (which, in keeping with persisting assumptions of Hippocratic and Aristotelian ontology, were considered to be unmodifiable conditions of biological development), the six non-naturals were subject to a certain margin of manipulation and could be used to ameliorate the physical and mental condition of individual human beings. Everyone, Cardano argued, should choose the regimen (*victus ratio*) that they found appropriate for themselves, especially those who were not lucky to be healthy and strong by nature, and therefore needed instructions on how to conduct their lives (Cardano, *Opera omnia*, 6:22b). This was an area where the six non-naturals allowed further scope for intervention, that is to say, the area between nature and technology, on the one hand, and between nature and habit formation, on the other hand. Alvise Cornaro, the author of one of the most celebrated guides to healthy living in the Renaissance, the *Discorsi intorno alla vita sobria*, never tired of repeating that "method" (*arte*) and "practice" (*uso*) were forces capable of curing natural faults (*vizi e mancamenti naturali*) and mitigating the pressure of natural determinants (Cornaro, *Discorsi intorno alla vita sobria*, 361, 363, 383). In his observations and suggestions, Cornaro mentioned Galen and was well aware of the tradition of the six non-naturals. He recommended a controlled use of food and drink (*non mangiare se non quanto digerisce il mio stomaco con facilità*), of sleep (*non impedir i miei sonni ordinarii*) and sex (*uso del matrimonio*), a careful attention to environmental conditions (*non stanziar in mal aere, non patir dal vento, né dal sole*) and to the effects of the passions (*la malinconia, e l'odio, e le altre perturbazioni dell'animo; gli accidenti dell'animo*; Cornaro, *Discorsi intorno alla vita sobria*, 363–64, 376). Predictably, though, food and drink—which he called "the two orders of the mouth" (*i due ordini della bocca*)—were the most important non-naturals, the one to which all the others were supposed to be reduced (Cornaro, *Discorsi intorno alla vita sobria*, 363, 384). Cornaro died when he was about eighty-years-old. The story goes that, suffering from gout at an early age and fearing he would die of sudden death, he decided to change his lifestyle drastically and follow a most austere regimen of food control. A friend of Pietro Bembo, Sebastiano Serlio, Pietro Aretino, and Sperone Speroni, he recounted his personal experience in his *Discorsi*, four tracts published between 1558 and 1565. They praised moderation (*sobrietà*), which was seen as the "true medicine for both the soul and the body" (Cornaro, *Discorsi intorno alla vita sobria*, 380–81). The work enjoyed immense success and was translated into French, German, and English. Cardano mentioned on many occasions the "little book" that everyone was reading at the time (*qui in omnium manibus est*; Cardano, *Opera omnia*, 6:15b).

Being both individual and holistic, regimens of health required sensible application. During the Renaissance, the art of preserving health was still seen as a most subtle exercise of judgment, through which precepts of general knowledge (medicine, but also meteorology, geography, astrology, physiognomy, ethics, and economy) were applied to concrete and particular situations. By undergoing a whole range of adjustments and negotiations, subtlety was mostly required in addressing various sets of biological polarities (loss and repair, short and long life, moisture and heat, putrefaction and exsiccation). These polarities lay at the very foundations of prophylactic and preventive strategies. In *De sanitate tuenda*, Galen had described the condition of health as the result of a precariously balanced relationship between variables of opposite nature. As mentioned, according to the principles of his medicine, perfect health could only work as a regulative principle, for in real terms perfection remained unachievable. Galen had assumed that any living body was in the end bound to lose its principal vital constituents, due to the organism's inability to restore the vital frame in its entirety and complexity, while an increasing quantity of excretions (*excrementa*) and unrecyclable wastes (*superfluitates*) kept building up inside the body (Cardano, *Opera omnia*, 6:30a). Already in the Middle Ages, a number of physicians interested in the use of chemical remedies had begun to look for ways of overcoming this biological impasse through the progress of technology (especially in the domain of chemistry), but it would be only with such militant and confident advocates of medical progress as Francis Bacon and René Descartes (1596–1650) that the indefinite extension of one's life was presented as a plausible desideratum, both philosophically and technologically.[16]

Another polarity affecting the health of human beings was caused by the air, which was supposed to prey on moisture (*consumere humidum*) while extending the life of the individual (*producere vitam*; Cardano, *Opera omnia*, 6:32a). It was not easy to find the right balance between air, heat, and moisture, given the difficulty of establishing a correct proportion between the extremes of putrefaction and exsiccation, firmness and fineness, contraction and relaxation. The always shifting equilibrium could be maintained by tinkering with the sources of innate heat (*calor innatus*) and original moisture (*humidum radicale*). From this point of view, the easiest way of keeping the balance of heat and moisture in check was through monitoring the condition of one's blood. A constant attention to the quality of blood, for instance, was essential in Ficino's health regimen: "Let not the blood be fiery," he cautioned the reader of his *De vita*, "not watery, but airy—not airy like a too-dense air, lest it be too much like water, nor like a very subtle air, for fear it may easily kindle into fire" (Ficino, *Three Books on Life*, 177). The reason why Ficino's materia medica was particularly rich in spices

16 G. Giglioni, "The Hidden Life of Matter: Techniques for Prolonging Life in the Writings of Francis Bacon," in J. R. Solomon and C. Gimelli Martin (eds), *Francis Bacon and the Refiguring of Early Modern Thought: Essays to Commemorate the Advancement of Learning (1605–2005)* (Aldershot: Ashgate, 2005), 129–44.

(*aromatica*) was because in his eyes these substances contained an ideal proportion of heat, moisture, and viscosity. As such, they were especially beneficial for human life (Ficino, *Three Books on Life*, 192). In being the biological cornerstones of human health, both physical and mental, innate heat and original moisture produced opposite effects that the good physician was able to turn into complementary actions. While humidity was the material that fueled the activity of heat, heat was the force that moderated the tendency to putrefaction embedded in humidity. "We start to dry up as soon as we are born," said Cardano at the beginning of his *De sanitate tuenda* (Cardano, *Opera omnia*, 6:10). He summed up centuries of medical investigations by stating that the main aim of a healthy regimen was to prevent excessive internal moisture from initiating processes of putrefaction and external sources of heat (*calor ambiens*) from preying on innate heat (*calor proprius*; Cardano, *Opera omnia*, 6:29b, 31b).

Of the foundational polarities, the most striking of all was perhaps the one between the length and the healthiness of life (between its energy and meaning, as it were). It's certainly no accident that a good number of physicians and natural philosophers shared the belief that a long life was not necessarily a healthy life. Cardano explained that the goal of enjoying good health (*ratio valetudinis*) often did not coincide with that of living a long life (*ratio vitae*; Cardano, *Opera omnia*, 6:12). Bacon, as we will see, despite advocating the need to extend the limits of human existence, made clear that longevity did not come at the detriment of pleasure and virtue. Cornaro agreed with many early modern physicians and philosophers that an unbalanced temperament (*mala* or *trista complessione*) could be transformed into an opportunity for human beings to become more aware of their limitations and to live a more meaningful life. On the contrary, he considered a sound temperament (*perfetta complessione*) at birth to be the main reason behind a poor health (*mala condizione*) in old age, so much so that they had to be rightly blamed for that outcome (*essi stessi ne sono cagione*; Cornaro, *Discorsi intorno alla vita sobria*, 360, 369, 382, 397). Huarte de San Juan went so far to argue that, in some circumstances, temperamental dysfunctionality could be the starting point for a better life.[17]

To recapitulate, regimens of health were practical, individual, and embodied (affecting both individual and social bodies). Being a discipline that was constantly engaged in finding the correct balance between sets of interrelated polarities, the art of preserving health required flexible epistemological canons. Cardano referred to the expertise concerned with healthy living indifferently as *scientia, ars*, and *disciplina* (Cardano, *Opera omnia*, 6:15ab). He described the "science" of health as a type of knowledge about human bodies (*subiecta*) in which substances and situations needed to be constantly evaluated in order to decide whether they were worth using or avoiding (*res adhibendae et fugiendae*), in what ways (*modi*) and in relation to what causes (*causae*; Cardano, *Opera omnia*, 6:11). It was therefore a kind of practical knowledge demanding attention to a dizzying amount of variables (*tanta rerum multitudo ac varietas*): nutrition, digestion, quantity and quality of food, the order to be followed in assuming food, age, sex, passions, geographic regions, and weather (Cardano, *Opera omnia*, 6:11–2, 29b–30a). Within such a complex tangle, Cardano listed seven principal objectives: to preserve one individual's specific temperament; to eat the right food; to favor evacuations in all their forms, from sperm and urine to sweat and tears; to maintain the organs in their proper condition; to improve the quality of the surrounding air; to protect the body from all possible sources of "external violence"; and, finally, to take care of all those functions that, as explained by Aristotle in *Parva naturalia*, were "common" to both the body and the soul (Cardano, *Opera omnia*, 6:30ab). It was precisely at the intersection of corporeal and mental well-being that the "science" of health was confronted with its biggest challenges.

---

17 G. Serés, (ed.), "J. Huarte de San Juan," in *Examen de ingenios para las ciencias* (Madrid: Cátedra, 1989), 174–82.

The Health of the Mind

As one of the six non-naturals, the changeable states of the mind (its "accidents") testified to the fact that health (*salus* understood as soundness, safety, and salvation) was inextricably natural, spiritual, and political. Mariano Santo (1488–1560), a physician from Barletta (in Apulia), who studied anatomy in Rome and practiced surgery in Venice and wrote commentaries on Avicenna, wrote a speech devoted to the arts of healing in which he defined medicine as a pillar of human society in that it preserved the bond between the soul and the body. Santo championed a loosely medicalizing program, with strongly anti-Aristotelian tones. It was heavily indebted to that particular accord between medicine and polity that Plato outlined in his *Timaeus* (86b–88d). Happiness could not simply be regarded as a property of the soul, as Aristotle argued in the *Nicomachean Ethics* (1177a–1178a); rather, in the true spirit of Galenic medicine, it could only be a prerogative of embodied souls. Christ himself, added Santo, had announced his coming into the world as a physician (*se medicum confiteatur*) and as God made human (*deus humanatus*; Santo, *Oratio de laudibus medicinae*, 316^{r}). In his commentary on Avicenna's *Canon*, he countered Heinrich Cornelius Agrippa (1486–1535), who in *De incertitudine et vanitate scientiarum* (*On the Uncertainty and Vanity of the Sciences*, 1527) had rejected medicine as unreliable learning. In contrast, Santo embraced the Avicennian consideration of the medical art as a form of all-inclusive knowledge culminating in the "most pure" and "most holy" medicine governing "the intellect and the will" (Santo, *Commentaria in Avicennae textum*, 4).

In the "Proem" to *De vita*, Ficino defended the idea that there were different kinds of medicine, distinguished according to their different ways of engaging with the physical and the spiritual aspects of human life. First of all, he differentiated an "Apollonian" therapy, based on the wise use of herbs and songs, from a "Dionysian" cure, which relied on wine and a most joyful condition of serenity (*securitas laetissima*;

Ficino, *Three Books on Life*, 102). Ficino was convinced that the pursuit of wisdom required both the health of the body (*sanitas corporis*), epitomized by Hippocrates, and the health of the mind (*sanitas mentis*), symbolized by Socrates. And yet, seen from the point of view of "Apollo," Socrates was, in fact, more effective than Hippocrates, for the mind remained ontologically superior to the body (Ficino, *Three Books on Life*, 160). It was however Christ who, in Ficino's account, fulfilled both types of health, physical and spiritual (*vera utriusque sanitas*; Ficino, *Three Books on Life*, 106, 160). Christian religion combined mental serenity with pious devotion: "as soon as the mind is purged of all fleshly perturbations through moral discipline and is directed towards divine truth (i.e., God himself), truth from the divine mind flows in and productively unfolds the true reasons of things" (Ficino, *Three Books on Life*, 163). This was also the reason why faith and prayers were able to enhance the effects of administered drugs in the most powerful way (Ficino, *Three Books on Life*, 202).

Both Santo and Ficino were able to appeal to the medical tradition, for Galen had already defended the physicians' ability to contribute to the mental well-being of humans. In many of his works, he had described mental health as the province of both medicine and philosophy. Since the "habit of the mind," he argued in *De sanitate tuenda*, was impaired "by faulty customs in food and drink and exercise and sights and sounds and music," the "hygienist" (ὁ τὴν ὑγιεινὴν μετιών) was supposed to be skilled in all these matters and "to mould the habit of the mind" (Galen, *Opera omnia*, 6:40; Galen, *Hygiene*, 26). Moreover, as one of the six non-naturals, the passions of the soul (*accidentia animi*) were universally regarded as a crucial factor among the many variables affecting human health. Ficino was particularly concerned with the ways in which emotional imbalances could affect the stability of the psycho-somatic compound (*imaginationis motus; laboriosus animi corporisque motus; anxietas; ira; solitudo et maeror*; Ficino, *Three Books on Life*, 168, 172, 188, 211). The old adage about keeping "a sound mind in a sound body" (*mens sana in corpore sano*) was at the core of his medical philosophy (Ficino, *Three Books on Life*, 184).[18] The proverb was all the more appropriate when the lives of philosophers and men of letters were at stake, for the minds of such people are exposed to all sorts of professional hazards. Ficino complained that scholars had yet to find their doctor (*solus litterarum studiosis hactenus deest medicus aliquis*), for they were prone to neglect their health; above all, they were particularly heedless of the principal instrument of their intellectual activities (*instrumentum ingenii*), that is, the *spiritus*, understood in medical terms as the most refined part of the blood (Ficino, *Three Books on Life*, 108, 110, 146, 152). To remedy this situation, Ficino's directions about the healthy life of the body and the mind hinged on the natural proximity between spirits, odors, and air. As the vehicle of sublunary and celestial qualities, air mediated between bodily spirits and the faculties of the soul; "pure and luminous air, odours and music" were especially beneficial to the life of the intellectuals (*ingeniosi*). While investigating the relationship between food and the soul, Ficino insisted on the kinship of *spiritus* and *odor*. An odor, he said, is sublimated food: "we call an odour that vapour into which digested food is subsequently transformed" (Ficino, *Three Books on Life*, 223).

We can sum up Ficino's position on health preservation by saying that mental soundness consisted for him in a very delicate balance of humoral components, degrees of temperature, and levels of density and rarefaction (Ficino, *Three Books on Life*, 118–22). Phlegm and yellow bile were important, but it was the melancholic humor (black bile) that had the greatest impact on the health of the mind. In *De vita*, he compared black bile and the *spiritus* produced by melancholic

18 On Ficino's medicine of the mind, see P. O. Kristeller, *The Philosophy of Marsilio Ficino* (New York: Columbia University Press, 1943), 351–401; N. L. Brann, *The Debate over the Origin of Genius during the Italian Renaissance* (Leiden: Brill, 2001); M. J. B. Allen, "Life as a Dead Platonist," in M. J. B. Allen and V. Rees (eds), *Marsilio Ficino: His Theology, His Philosophy, His Legacy* (Leiden: Brill, 2002), 159–78; G. Giglioni, "Coping with Inner and Outer Demons: Marsilio Ficino's Theory of the Imagination," in Y. Haskell (ed.), *Diseases of the Imagination and Imaginary Disease in the Early Modern Period* (Turnhout: Brepols, 2011), 19–51.

reactions occurring in the blood to aqua vitae. As such, they produced the most refined fuel, the one, that is, belonging to the thinking activity, capable of collecting and concentrating the mind (*in suum centrum animum colligit*; Ficino, *Three Books on Life*, 120). Excesses in mental activity could affect one's health in two ways, following the well-known patterns of exsiccation and refrigeration: concentration was likely to dry up the brain, while the overuse of spirits made the blood thick and cold (Ficino, *Three Books on Life*, 114, 134).[19]

With respect to the six non-naturals in general, Ficino warned about three specific "monsters" and how they could impair mental health: sex (*Venereus coitus*), excess of food and drink (*vini cibique satietas*), and sleep deprivation (*ad multam noctem frequentius vigilare*; Ficino, *Three Books on Life*, 122–23, 182, 188, 217). He recommended scholars to shun Saturn, a mythological emblem symbolizing the "secret and too constant pleasure of the contemplative mind," for in that space of mental brooding Saturn was deemed to devour "his own children," that is, one's own thoughts (Ficino, *Three Books on Life*, 213). In this domain, Ficino's dietetic instructions were directed at strengthening the faculties of the soul, by making the senses more alert, corroborating memory, refreshing the imagination, and sharpening the mind. Although, in line with the principles of Galenic psychiatry, Ficino's characterization of mental prophylaxis hinged on the notion of material temperament (especially black bile, as just noted), he nevertheless did not downplay the contribution of moral philosophy (*disciplinae moralis instituta*) and the extensive benefits that could derive from relaxing activities and spiritual ways of comforting both the senses and the imagination:

> I advocate the frequent viewing of shining water and of green or red colour, the haunting of gardens and groves and pleasant walks along

19 On Ficino and melancholy, see the classic R. Klibansky, E. Panofsky and F. Saxl, *Saturn and Melancholy: Studies in the History of Natural Philosophy, Religion and Art* (London: Nelson, 1964).

> rivers and through lovely meadows; and I also strongly approve of horseback riding, driving, and smooth sailing, but above all, of variety, easy occupations, diversified unburdensome business, and the constant company of agreeable people. (Ficino, *Three Books on Life*, 135–37)

To this list he added music, songs, games, moderate laughter, and devices aimed at recovering memories from one's childhood (Ficino, *Three Books on Life*, 188, 212–24). Because of the connections they established between sensible qualities and mental patterns, synaesthetic associations were especially appreciated by Ficino. His analysis of the color green, for instance, remains masterful:

> The frequent use of green, since it recreates the spirit of sight, which is in a way the principal part of the animal spirit, refreshes also the animal spirit itself. And we will also remember that if the colour green, which among the colours is the middle grade and the most tempered, is so good for the animal spirit, much more will those things which through their qualities are the most temperate will help the natural and vital spirits and conduce greatly towards our life. (Ficino, *Three Books on Life*, 206)

Given the general coordinates of Ficino's philosophy, the mind's *sanitas* had necessarily a positive effect on the well-being of the human beings on two levels, individual and cosmological. From an individual point of view, the mind was able to secure a condition of sound health by expanding the scope of self-knowledge (*unusquisque se cognoscat*). This meant that, by acquiring a richer experience of their inner life, all people were in the position of becoming their own best physicians (*suique ipsius moderator ac medicus esto*; Ficino, *Three Books on Life*, 216). With respect to cosmological life, Ficino's model of temperamental medicine rested on ideal correspondences between states of corporeal harmony and their symbolical counterparts more than on relationships between bodily humors and sensible qualities. Heavens, human bodies, and animal spirits were related to each other because they were the most "temperate"—that is, balanced—things in the universe. By means of analogical correspondences, the spirit thus became suited to receive energy and knowledge from celestial things (*spiritus per temperata coelestibus conformatur*; Ficino, *Three Books on Life*, 206). "I advise you to observe what Jupiter the even-handed taught Pythagoras and Plato: to keep human life in a certain equal proportion of soul to body and to nourish and augment each of the two with its own proper foods and exercises" (Ficino, *Three Books on Life*, 212).

For all these reasons, Ficino's guidelines for preserving health provided an original synthesis of cosmological, theurgic, and religious teachings, unified by the need to achieve a stable condition of mental soundness. Likewise, the immensely popular *Discorsi* by Cornaro owed part of their success to the way in which the Venetian nobleman combined medical advice, ethical directions, and religious devotion. He described his "pious" medicine (*santa medicina*) as a lifestyle in which the soul was allowed to dwell in a good body (*buona stanza nel corpo*), in peaceful harmony with the humors, the senses, and the appetites (Cornaro, *Discorsi intorno alla vita sobria*, 380). He stated that he was able to enjoy "two lives at the same time," the earthly one through the senses (*con l'affetto*), and the heavenly one through the intellect (*col pensiero*; Cornaro, *Discorsi intorno alla vita sobria*, 400). For Cornaro, a frugal life invigorated both a healthy brain and a serene mind (*un cervello purgato* and *alte e belle considerazioni delle cose divine*; Cornaro, *Discorsi intorno alla vita sobria*, 372–73). He reiterated the commonsense assumption that everyone was able to cure himself in the best manner (*l'uomo non può essere medico perfetto di altri, fuor che di sé solo*) by highlighting the irreducibly individual character of human health, seen as a unique combination of qualities, temperaments, and faculties (Cornaro, *Discorsi intorno alla vita sobria*, 368).

Other Renaissance authors followed a different route to defend the legitimacy of mental health. Rather than blaming the senses and the appetites, authors such as Telesio and Bacon warned against assigning too conspicuous a role to the sphere of the intellect in one's cognitive and ethical life. Their philosophical programs entailed a momentous rehabilitation of the senses—from sense perception (discernment) to good sense (judgment). Telesio, it should be said, acknowledged that not everything in the mind could be reduced to the senses, for that part of the soul whose functions developed from the seed (*educta e semine*) was in fact the material substratum onto which God had directly and immediately grafted an individual and immaterial soul. It was on this basis that Telesio managed to differentiate between the souls of nonhuman and human animals. The former were mere corporeal *spiritus*, the latter, specific "forms" imprinted by God on the corporeal *spiritus*. In the work he wrote to rebut Galen's division of the faculties of the soul, entitled *Quod animal universum ab unica animae substantia gubernatur* (*The Living Being as a Whole Is Governed by the One Substance of the Soul*), Telesio argued that the *spiritus* was able to perform all the biological and cognitive functions but not to account for some higher tendencies of a moral and religious order (such as self-sacrifice and disinterested acts of altruism), which transcended the level of purely biological self-preservation (Telesio, *Quod animal universum*, 188; Telesio, *De somno libellus*, 380).

Telesio was aware that the great majority of doctors, both the ones belonging to the traditional camp and the follower of new trends, shared the view that a change in bodily conditions, induced through a careful use of diet, drugs, and a variety of material stimulations, could affect the state of the mind. More complex, however, was the question concerning the type of mental operations that were able to alter the body. Telesio's solution to this problem—a bold solution indeed—derived from his original views about the soul. In his opinion, every living being, including man, was governed by one sentient entity. Regardless of whether people called it *anima, universitas,* or *substantia*, the underlying meaning remained the same, that is, the most rarefied and active part of the vital fluid, the *spiritus*. In *De rerum natura* (published in three different versions in 1565, 1570, and 1586), he described the *spiritus* as a continuum of pneumatic energy, diffused in every part of nature and in each single organism, capable of perceiving and reacting to all the stimuli received from the outside. Telesio labeled this spirituous core as a "fully perceptive spirit" (*spiritus omnino omniscius*), that is to say, a material substance that at any time was aware of its surrounding reality down to the last detail. He thus viewed the soul as a collective pneumatic organization capable of connecting the knowledge that pertained to *spiritus* as a whole (*universitas spiritus*) and the knowledge coming from local and peripheral aggregations of *spiritus* to the functioning of the organs and each bodily part. Being in charge of the vital organization of the body, the *spiritus* was also able to enact the best conduct of life (*ratio vivendi*; Telesio, *De rerum natura,* 3:348, 352; Telesio, *Quod animal universum*, 215–16).

This model had important consequences in relation to notions of physical health, physical pleasure, and the physiological processes underlying the activity of knowledge. Given the all-pervasive activity of the fully sentient—*omniscius,* in fact—*spiritus,* Telesio intertwined very closely cognitive operations, physiological processes, and feelings of pain and pleasure. In doing so, he tied perception, pleasure, and happiness together, for it was through sentient reactions that the primordial forces of nature—heat and cold—could immediately probe and sense any advantage or harm coming from external things. These unmediated perceptions were always accompanied by feelings of pleasure or pain. Pleasure and pain, in turn, were acts of knowledge. Laughter, for instance, was described by Telesio as a motion that signaled the extent to which *spiritus* was taken "by the greatest desire to preserve itself and at the same time to draw pleasure from it" (Telesio, *Quod animal universum,* 214). Given the pneumatic dimensions of Telesio's prophylaxis, it is easy to understand why air was for him the most important non-natural, for the *spiritus* inside the body of living beings constantly interacted with the *spiritus* outside them, that is, air (Telesio, *Quod animal universum*, 196). It was certainly no accident that Telesio had written a short treatise dealing with the effects of the atmosphere on human health, *De iis quae in aere fiunt et de terraemotibus* (*On Atmospheric Phenomena and Earthquakes*).[20]

To a certain extent, Bacon shared Telesio's central argument in *De rerum natura* that the health of the mind depended crucially on the health of the senses. Bacon's medicine of the mind, as we will see in the next section, implied a rediscovery of the unadulterated life of the senses, and through the senses, of the innermost appetitive life pervading the whole body of nature. As would also happen with René Descartes (see the chapter by Gideon Manning in this volume), Bacon thought that the notion of health had to be dramatically reinterpreted by exploring the nature of matter and life along radically and experimentally new lines. In his *Historia vitae et mortis* (1623), he rejected the foundations of traditional and institutional medicine (*turba medicorum*), that is, the already mentioned notions of original moisture (*humor radicalis*) and natural heat (*naturalis calor*), in favor of a view of vital phenomena seen as processes that could be perpetually repaired and restored (Bacon, *The Instauratio Magna*, pt. 3, 144). Distancing himself from Ficino's and Cornaro's advice, Bacon thought that the mind was not the cure, but a substance in desperate need of a cure, for left to its own devices—mainly, imaginations and

20 On Telesio's views on nature, self-preservation and health, see N. Badaloni, "Sulla costruzione e la conservazione della vita in Bernardino Telesio (1509–1588)," *Studi Storici*, 30/1 (1989), 25–42; M. Mulsow, *Frühneuzeitliche Selbsterhaltung: Telesio und die Naturphilosophie der Renaissance* (Tübingen: Niemeyer, 1998); G. Giglioni "Spirito e coscienza nella medicina di Bernardino Telesio," in G. Ernst and R. M. Calcaterra (eds), *"Virtù ascosta e negletta": La Calabria nella modernità* (Milan: Angeli, 2011), 154–68; "Introduzione," in G. Giglioni (ed.), *Bernardino Telesio: De rerum natura iuxta propria principia libri IX* (Rome: Carocci, 2013), xi–xxxii; Giglioni, "Medicine."

passions—the mind would be trapped in a world of phantasms (*idola*) and lose touch with reality (nature).[21]

Conclusion: Long Life, Healthy Life, Meaningful Life

Another development concerning the art of preserving health that gathered particular momentum during the early modern period was the almost obsessive attention with which physicians, philosophers, and divines focused on the complex interplay of vitality, longevity, and lifestyle. Galen had already demonstrated that, in order for human beings to enjoy good health, they needed to watch carefully the way in which they aged, in addition to control the evacuation of excrements and the replacement of wastes (Galen, *Opera omnia*, 6:8–9; Galen, *Hygiene*, 9–10). Ficino referred to a Chaldean precept (*regula Chaldaeorum*) intimating that humans slowly cleansed extraneous humors (*peregrinos humores imbibitos corpori expurgare gradatim*) to recover their lost youth (Ficino, *Three Books on Life*, 218). By and large, growing old was a key variable in any health regimen, for while death could not be avoided, it could be deferred. As stated by Cardano, all living bodies had a natural tendency to last and postpone decay (Cardano, *Opera omnia*, 6:30b). Accordingly, and following a long tradition, he divided the art of prophylaxis by referring to two different aims: preservation of good health (*ad servandam bonam valetudinem*) and prolongation of life (*ad producendam vitam*; Cardano, *Opera omnia*, 6:12).[22]

As explained at the beginning of this chapter, all the operators in the field of "health protection" (*sanitas tuenda*) described the functions of

21 On Bacon's idea of "medicining the mind" and its legacy in the seventeenth century, see G. Giglioni, "Medicine of the Mind in Early Modern Philosophy," in J. Sellars (ed.), *The Routledge Handbook of the Stoic Tradition* (London: Routledge, 2016), 189–203.

22 On aging in the Renaissance, see C. Skenazi, *Aging Gracefully in the Renaissance: Stories of Later Life from Petrarch to Montaigne* (Leiden: Brill, 2013).

life in terms of natural heat, and, to be effective, they all agreed that heat required a constant supply of energy, a "primordial" or "radical" source of moisture (the *humidum radicale*). The medical tradition had described this fuel in terms of an oily (*pinguis*) and airy (*aerius*) moisture. Summing up the way in which Aristotle had explained the processes of life and death in the *Parva naturalia* (464b–467a; 478b), Ficino reminded his readers that death could derive from either "resolution" (*resolutio*) or "suffocation" (*suffocatio*), respectively depending on whether the natural fuel was lacking or the heat was quenched by an excess of fluids or putrefaction (Ficino, *Three Books on Life*, 168, 216). In discussing the natural sources of death, Galen too had emphasized the role of exsiccation. While growth relied on a balanced relationship of moisture and dryness, old age was determined by a slow but unstoppable (and in the end deadly) increase of dryness (Galen, *Opera omnia*, 6: 4–6; Galen, *Hygiene*, 7).

As argued by Cornaro in his *Discorsi*, to look at aging as a process that could be gradually delayed was tantamount to perceiving the final term of one's life as the painless outcome of a natural and slow consumption of vital moisture (*bella e desirabil morte è quella che ci dà la natura per via di risoluzione*; Cornaro, *Discorsi intorno alla vita sobria*, 379, 383). A "good" death was therefore evidence of a "good" life, extended through a long series of ethically meaningful acts. As pointed out, Cornaro's approach to a healthy life was based on few simple precautions centered on common sense and moral commitments, such as pursuing order and virtue (*la virtù dell'ordine*), applying oneself to routine tasks (*la forza dell'uso*), establishing good habits (*l'uso negli uomini col tempo si converte in natura*), and practicing a frugal lifestyle (*la natura si contenta di poco*; Cornaro, *Discorsi intorno alla vita sobria*, 357, 361, 366, 371). The best antidote to avoid an early death, he stated, consisted in "a sober and orderly life," premised on a relentless exercise of virtue (Cornaro, *Discorsi intorno alla vita sobria*, 360). He recounted how nothing had better kept him away from the grip of death than the "great order" he had followed for many years. Cornaro believed in the power of virtue to impart structure and purpose to one's mortal life, for "it is impossible in nature that he who leads an orderly and continent life may fall ill or die of unnatural death before his time comes" (Cornaro, *Discorsi intorno alla vita sobria*, 367, 398).

Moral and religious issues, therefore, played a key role in Cornaro's account of healthy longevity. Significantly, he identified three specific "vices"—or better three "cruel monsters"—which more than anything else could hamper the course of human life: social conformism (*l'adulazione e le cerimonie*), the spreading of heterodox views in religion (*il viver secondo l'opinione Luterana*), and succumbing to the empire of bodily appetites (*crapula*). These "monsters" had adulterated, respectively, "the honesty of civic life, the religion of the soul and the health of the body." Cornaro's principal aim behind his treatises was to defeat the third "monster." In his diagnosis, *crapula*, that is, debauchery and intemperance, derived from gluttony (*vizio della gola*), which in turn was the symptom of a distorted use of the sense faculties (*senso* and *appetito*, or *vivere secondo il senso*; Cornaro, *Discorsi intorno alla vita sobria*, 358, 362). He was particularly concerned with three desires: sex (*desiderio della concupiscenza*), glory (*desiderio degli onori*), and wealth (*desiderio della roba*). More than any other appetite, they could expose human life to serious mortal dangers (Cornaro, *Discorsi intorno alla vita sobria*, 373). The solution lay in a life modeled on the ideals of natural simplicity (*la semplicità della natura*), moderation (*la santa continenza*), and reason (*la divina ragione*), achieved by following an extremely frugal diet in which people were advised to eat only the amount of food that was strictly necessary and thus easily digestible. Most of all, they needed to mistrust taste as a reliable indicator of nourishing and healthy food (*quello che sa buono, nutrisce e giova*) and rather embrace that piece of practical advice which suggested always to stop eating before one felt full (*non saziarsi di cibi è uno studio di sanità*), for any surplus of food would inevitably be converted into peccant humors (*tristi umori*; Cornaro, *Discorsi intorno alla vita sobria*, 359, 362–63, 383–86).

Cornaro described in glowing terms long life as a good life: not a *vita morta*, but a *vita viva* (Cornaro, *Discorsi intorno alla vita sobria*, 374, 378). He looked at nutrition as the foundation of long-lasting order, at both an individual level and a social level. Other authors preferred to stress the power that virtue and thinking had in prolonging life. Ficino, for instance, agreed with many philosophers and physicians since antiquity that a long life depended on a prudent use of judgment (*prudentis iudicii perspicacia*; Ficino, *Three Books on Life*, 166). Cardano, too, listed health among the crowning achievements in a long and prosperous life, full of wisdom and intellectual accomplishments, in which death was expected to arrive in the most natural of ways, "without pain" (Cardano, *Opera omnia*, 6:15b). In this respect, the art of preserving health had clear ethical overtones:

> We are led astray by the abundance and variety of things, for in our greedy attraction for flavours we are all driven to titillate our palate, without paying attention to any difference, not the ones concerning the surrounding air and physical exercise, nor the ones regarding the other six non-naturals, where we look for opportunities to be healthy (*salubritatis materia*). The result is that the course of our life is short and precarious. And this happens not because it is something that is inherently determined by nature (as some like to think), but because it is procured as a result of our mistakes. (Cardano, *Opera omnia,* 6:11)

As noted, the time-honored doctrine of the six non-naturals combined a strong faith in the prophylactic effects of prudent action with close attention to the power of natural constraints. There was an evident recognition that human "art" could play a significant role in changing the conditions of one's life. Within this framework, ethical prescriptions were seen as capable of preserving and improving human health. Unassisted by the art of prevention, human beings were bound to lead a disorderly, indeed deranged, life (*inordinate, imo insane*), breathing polluted air (*aër pravus*), in the grip of "sorrows, anxieties, insomnia, sex and purposeless exertions," prone to excessive eating and drinking, at the mercy of time and chance, and finally dying a death that was "unnatural," "unexpected," "violent," or "due to illness" (Cardano, *Opera omnia*, 6:11).

And yet long life was not necessarily synonymous with happy and meaningful life. Precisely because control over the biological conditionings of one's existence implied the social and ethical use of the non-naturals, a desire to postpone the end of one's life had always to be accompanied by a parallel desire to make the right decisions. If in some circumstances life was shortened by bad habits and dietary preferences, which would dissipate the reserve of natural heat, dissolve the vital moisture and dry up the spirits, in other cases a short life lived with intensity could add meaning and vitality precisely by accelerating the course of the principal vital processes. This means that the pursuit of happiness could sometimes be at odds with the attempt to live a long life. Indeed, more often than not, a sickly life lasted longer than a healthy one. Valetudinarianism was certainly not the symptom of a happy existence, and yet it led its devotees to a long life. Ficino noticed how weak and frail people, who obsessively took care of their lives, were capable of living longer than healthy but imprudent people (Ficino, *Three Books on Life*, 166).

Cardano was therefore right in pointing out that means of assistance for a sound life (*auxilia salubritatis*) were not the same as those that lead to a long life (*auxilia longitudinis vitae*; Cardano, *Opera omnia*, 6:32a). In the end, whatever doubts one might have had about the best relationship between long life and good life, the fact remained that a long life risked re-enacting Tithonus's predicament, the situation symbolized by a human awarded with the divine gift of immortality but without eternal youth. If it was true that, since time immemorial, the long life of the biblical patriarchs and pious anchorites had been taken as a proof that longevity, moral perfection, and emotional contentment were all parts of one comprehensive understanding of healthy life, a renewed sense of natural order and spiritual immanence induced more than one thinker to question the contribution to happiness provided by physical and spiritual maceration (Cornaro, *Discorsi intorno alla vita sobria*, 402). Bacon insisted on keeping "healthy life" and "long life" as separate categories: people could increase the level of activity and nimbleness in their spirits and in so doing shorten the course of their life, or conversely, they could prolong life, and thus damage their health. Given the complexity of the question and the superiority of the active over the contemplative life, in his *Historia vitae et mortis* Bacon went so far to deny that physical health could be taken as normative: "The duties of life are more important than life itself" (Bacon, *The Instauratio Magna*, pt. 3, 240).[23] His main preoccupation concerned the question of how to live a healthy and long life that was also worth living. In his program for the reformation of human learning, this was one of the reasons behind his decision to link the idea of physical health to the complementary ones of methodical soundness (medicining of the mind) and religious salvation (*salus*).[24]

23 See also Spon, *Cardano: Opera omnia*, 6:15a.

24 Research leading to this chapter was supported by the ERC Grant 241125 MOM.

Reflection

EARLY MODERN ANATOMY AND THE HUMAN SKELETON

Anita Guerrini

The Roman physician Galen wrote in his instruction manual *On Anatomical Procedures*: "As poles to tents and walls to houses, so are bones to living creatures." In his short treatise *On Bones for Beginners* he added that bones were the "hardest and driest parts of the living body and, as one might say, the earthiest. . . . All else depends on or is attached to them." Therefore knowledge of the skeleton must precede any other exploration of the body.[1] They were the beginning of anatomical knowledge and the conclusion of the process of dissection.

The human skeleton has had multiple meanings in history: medical, scientific, and symbolic. These perceptions have shifted over time and place, and as anatomical study rose to prominence in early modern Europe, they continued to coexist. Between the sixteenth and the mid-eighteenth centuries, a critical juncture in the history of anatomy, the skeleton became an object of scientific regard while retaining long-held symbolic

1 C. Singer (trans.), *Galen: On Anatomical Procedures* (Oxford: Oxford University Press, 1956), 2, 5; C. Singer, "Galen's Elementary Course on Bones," *Proceedings of the Royal Society of Medicine* 45 (1952): 767–76, at 768. This is a translation of Galen's *De ossibus ad tirones.*

and emotional connotations as symbols of death and as relics.[2] The skull held particular significance: to anatomists, it held the brain and therefore the senses, and to Christians it was the seat of the soul.

The skeleton first became a focus of rational study with the ancient Greeks; Aristotle described the bones of "blooded" (i.e., vertebrate) animals in *History of Animals.*[3] By the time Galen wrote 500 years later, knowledge of the skeleton was a customary aspect of medical education. But when dissection ceased after the fall of Rome, skeletons ceased to be scientific objects for several centuries, while retaining their symbolic resonance. Skeletons had appeared in frescoes and mosaics at Pompeii as reminders of mortality, and Christian iconography adopted this symbolism. Images of the "Dance of Death" in which skeletons danced with the living emerged with the plagues of the fourteenth century, and the figure of the *transi*, or decomposing corpse, adorned tombs from the late fourteenth century into the seventeenth.[4]

The first printed image of the human skeleton was a "Dance of Death" in the *Nuremberg Chronicle* of 1493 (plate 3).[5]

The first printed anatomical representation of the skeleton also dated from 1493, indicating its reemergence as a scientific object.[6] Skeletons, complete and in parts, were prominent in the illustrated textbooks of anatomy that began to appear in the sixteenth century, which often employed familiar tropes of mortality.

---

2 On the definition of "scientific object," see L. Daston, "Introduction: The Coming into Being of Scientific Objects," in L. Daston (ed.), *Biographies of Scientific Objects* (Chicago: University of Chicago Press, 2000), 1–14.

3 A. L. Peck (trans.), *Aristotle: History of Animals*, vol. 1 (Cambridge: Harvard University Press, 1965), 516a8–516b31.

4 H. Weaver (trans.), *Philippe Ariès: The Hour of our Death* (New York: Alfred A. Knopf, 1981), 113–16; K. Cohen, *Metamorphosis of a Death Symbol: The Transi Tomb in the Late Middle Ages and the Renaissance* (Berkeley: University of California Press, 1973).

5 Weaver, *Ariès, Hour of our Death,* 116–18; R. Saban, "Les premières représentations anatomiques des squelettes humain imprimées en Alsace au XVe siècle," *113e Congrès nationale des sociétés savantes 1988, Questions de l'histoire de la médecine* (1991), 27–46, at 29; J. de Vauzelles, *Les simulachres et historiées faces de la mort, autant élégamment pourtraictes, que artificiellement imaginées* (Lyon: Soubz l'eseu de Cologine, 1538).

6 Saban, "Premières représentations anatomiques," 30–32.

Berengario da Carpi in 1523 showed a skeleton standing over an open sarcophagus, while a skeleton held an hourglass in Felix Platter's textbook sixty years later. Into the eighteenth century, anatomists employed these symbols and others such as winding sheets and scythes, much like the artistic convention known as the "vanitas" genre (plate 4). The title page of a 1615 book on bones by the Leiden anatomist Pieter Pauw (1564–1617) featured skeletons engaged in a dance of death.[7]

The order of dissection in these textbooks most often began with the skeleton, as Galen had advised, even though it logically would appear last and not first, a product of dissection rather than its origin. André du Laurens (1558–1609) explained this practice in Galenic terms in his often reprinted 1593 text: the bones were the most similar of the parts of the body, being made all of one substance; they were the most dry and earthy (following the cosmic order); they gave form to the rest of the body. Bones also provided exceptionally good evidence of the divine plan of the body.[8] In practice, this meant that the anatomy theater required an assembled skeleton before any dissection took place, and by the end of the seventeenth century, several manuals detailed the construction of a skeleton.

The first works wholly devoted to osteology, the science of the bones, appeared in the sixteenth century. In 1556 a corrected Latin translation of Galen's *De ossibus ad tirones* appeared, edited by the renowned Paris anatomist Jacques Dubois, known as Sylvius (1478–1555). Dubois noted in his preface that although Galen had used monkey skeletons, now human bones could be examined.[9] The preliminary matter of the much-reprinted *Alphabet anatomic* of the Montpellier surgeon Barthélémy Cabrol (1529–1603)

---

7 *Primitiae anatomicae de humani corporis ossibus* (Leiden: Iusti a Colster, 1615).

8 P. Pauw, *A. du Laurens: Historia anatomia humani corporis* (Paris: Excudebat Iametus Mettayer and Marcus Ourry, 1600), 50–51, 65–87.

9 J. Dubois [Sylvius, pseud.], *Iacobi Sylvii . . . Commentarius in Claudij Galeni de Ossibus ad Tyrones libellum, erroribus quamplurimis tam Graecis quàm Latinis ab eodem purgatum* (Paris: Petrum Drouart, 1556), 3–4.

included a sonnet to Cabrol and his skill in uncovering the skeleton, comparing the body to a house with the skeleton as the foundation. Cabrol listed "ostéologie" on his title page among the subjects he treated, possibly the first use of that word in French. Expanding on the architectural metaphor, he stated that the skeleton sustains the body "as pillars do a house." His work was not illustrated, but the 1633 Dutch translation by Vopiscus Fortunatus Plemp included an engraving of Pauw's Leiden anatomy theater with a skeleton presiding.[10]

Bones also held increasing interest to those concerned with human and animal generation and development. The 1573 essay on the development of the bones of the fetus by Volcher Coiter (1534–1576) was only the first of several treatises on this topic, and his anatomical tables published the previous year had included the first scientific illustration of a fetal skeleton.[11] Fetal skeletons continued to be a focus of interest: Coiter's work was reprinted in 1659, and Amsterdam anatomist Frederik Ruysch (1638–1731) employed them in anatomical dioramas on "vanitas" themes, enacting small but heartfelt dramas on the theme of death.[12] William Hunter (1718–1783) noted that a complete fetal or infant skeleton was highly prized.[13]

Yet because they were hidden from view, bones served in this period as imperfect markers of health. While abnormalities of bone structure and development were widely documented—particularly within the genre of the "monstrous"—what are now

10 B. Cabrol, *Alphabet anatomic* (Tournon: C. Michel and G. Linocier, 1594), "Au dit Sieur Cabrol sur son livre des os. Sonnet," 5. In ARTFL's *Dictionnaires d'autrefois* the term "osteology" only appears in 1762. B. Cabrol, *Ontleeding des menschelycken lichaems. Eertijts in't Latijn beschreven door Bartholmaeus Cabrolius* (Amsterdam: Cornelis van Breugel voor Hendrick Laurentsz, 1633).

11 V. Coiter, *Externarum et internarum principalium humani corporis partium tabulae* (Nuremberg: Gerlatzenus, 1572).

12 F. Ruysch, *Thesaurus anatomicus,* in *Opera Omnia,* 4 vols. (Amsterdam: Jansson-Waesberg, 1720–33).

13 W. Hunter, Lecture notes 1775–76 fol. 257, Library, Royal College of Surgeons of London.

recognized as bone diseases attracted little medical attention. The hunched back of certain forms of tuberculosis was considered to be congenital before Percivall Pott (1714–1788) recognized its connection to tubercular infection. Rickets and other effects on the bones and joints, because of vitamin deficiency, were recognized, but their causes were unknown; and treatments, when they existed, were ineffective. Skilled surgeons could set simple fractures and dislocations, but compound fractures, because of the dangers of infection, often resulted in amputation.

According to the renowned Paris surgeon Pierre Dionis (1643–1718), the most common surgical operation for illnesses of the head was trepanation: cutting out a piece of the skull. In ancient times, it had been used to treat a variety of ailments, including headaches, but by the late seventeenth century, it was used mainly in cases of head trauma and particularly skull fracture, and also, although rarely, in cases of hydrocephalus.[14] Trepanation was performed with a circular, hand-cranked drill on a patient who (unlike in most surgical operations of the time) was often unconscious; indeed, Dionis believed that loss of consciousness as a result of a blow or a fall always required trepanation. The anatomical and spiritual significance of the skull made this a particularly fraught operation, and Dionis spent several pages in his surgical textbook detailing the procedure and how to determine if it was necessary.[15]

By the end of the seventeenth century, the term "osteology" was well established in a number of European languages, as was the study of bones, including the comparative study of human and animal bones. When a new edition of the massive compendium *Bibliotheca anatomica* (first published in 1685) appeared in 1711, almost the entire first volume consisted of works on the bones, with citations from a dozen authors, most of them dating from

---

14 P. Dionis, *Cours de chirurgie* (Paris: Chez la Veuve d'Houry, 1708), 349.

15 Dionis, *Cours de chirurgie*, 335–64.

after 1650. The Paris anatomist Joseph-Guichard Duverney (1648–1730) gave a separate course on the topic beginning in the 1680s. Works on osteology multiplied into the eighteenth century; at least twenty appeared between 1650 and 1750. Nonetheless, although its identity as a scientific object was now firmly established, the skeleton retained its symbolic and emotional resonance. When the many charnel-houses in Paris began to be cleared in the late eighteenth century and their bones transferred to the empty quarries below the city, the bones were carefully arranged to form an aesthetically pleasing and emotionally resonant façade of arm and leg bones and skulls (see plate 5). The lintel above the entrance to these catacombs read "Arrête! C'est ici l'Empire de la Mort" (Stop! Here is the empire of death).[16]

---

16 P. Koudounaris, *The Empire of Death* (New York: Thames and Hudson, 2011), 132–33.

CHAPTER SIX

Health in the Early Modern Philosophical Tradition

Gideon Manning

It shows great prudence and virtue... not cowardliness and fearfulness, to set store by one's health.

BARTOLOMEO PASCHETTI, *Del conserver la sanità* (1603)[1]

Introduction

The early modern period, which in this chapter refers to the seventeenth century, is of particular importance in charting the terrain covered in this book as a whole.[2] Though this is the period covering the so-called "scientific revolution," it does not represent a complete break with the past. Far from it, for no less than in earlier periods,

1 Cited in S. Cavallo and T. Storey, *Healthy Living in Late Renaissance Italy* (Oxford: Oxford University Press, 2013), 277.

2 T. M. Lennon, "Bayle and Late Seventeenth Century Thought," in J. P. Wright and P. Potter (eds), *Psyche and Soma: Physicians and Metaphysicians on the Mind-Body Problem From Antiquity to Enlightenment* (Oxford: Oxford University Press, 2000), 197–216, uses the analogies of navigation and topography to good effect in discussing mind-body dualism, and I follow his lead to discuss the concept of health.

seventeenth-century philosophers aspired to the good life and, as in earlier periods, the idea of the good life was linked to tranquility, happiness, and virtue, all of which were conceived through the medical idiom of health.[3] Additionally, a healthy body was generally thought to facilitate a healthy mind, and even if it was possible to exercise enough discipline of mind to assure a good life independent of one's bodily states, the advantages offered by a healthy mind were often thought to include a healthy body. In other words, ancient wisdom was still very much alive in the early modern period.

Taking a closer look at the medical topography where the concept of health is most at home, there are landmarks in the early modern period that look surprisingly familiar, that is to say, contemporary. There was, for example, a thriving genre of vernacular medical writing that details what measures can be taken at home, such as Philibert Guibert's *L'Apothicaire du médecin charitable* (1625), as well as numerous general medical self-help guides, including Bartolomeo Paschetti's *Del conserver la sanità* (1603) and Thomas Tryon's *The Way to Health, Long Life and Happiness* (1697), the latter of which even advocates a vegetarian diet. Moreover, the consolidation and expansion of anatomical investigation, as well as the lasting innovations and discoveries of a cadre of "great doctors," such as Santorio Santorio, William Harvey, Gasparo Aselli, Thomas Willis, Thomas Sydenham, and Marcello Malpighi, all took place in the seventeenth century with long-term consequences for the concept of health. There are also several medical firsts that stand as the entry point to what today are densely populated lands, including the first extended study of population demography in John Graunt's *Natural and Political Observations . . . upon the Bills*

---

3 For an account of what made early modern life worth living, see K. Thomas, *Ends of Life: Roads to Fulfillment in Early Modern England* (Oxford: Oxford University Press, 2009). A philosophical introduction to the early modern good life can be found in J. Cottingham, *Philosophy and the Good Life* (Cambridge: Cambridge University Press, 1998), whereas M. L. Jones, *The Good Life in the Scientific Revolution: Descartes, Pascal, Leibniz and the Cultivation of Virtue* (Chicago: University of Chicago Press, 2006) offers a presentation of the good life tied specifically to early modern science.

of Mortality (1662), which analyzed recorded causes of death in different English locales, and the first human blood transfusions, which occurred almost simultaneously in the late 1660s at Britain's Royal Society and France's Académie des Sciences.[4]

There are equally familiar landmarks in the philosophical and scientific landscape of the time, though historians have rarely discussed how the concept of health relates to this terrain. Easily recognized are philosophers like René Descartes, Pierre Gassendi, Thomas Hobbes, Baruch Spinoza, John Locke, Nicholas Malebranche, and Gottfried Wilhelm Leibniz, all of whom sought to replace or supplement the philosophical teaching of "the Schools" with an alternative more in keeping with the "new science" developed by figures such as Francis Bacon, Galileo Galilee, Descartes, Jan Baptiste van Helmont, Robert Boyle, Robert Hooke, Christian Huygens, Leibniz, and Isaac Newton. Together, early modern philosophy and science precipitated a crisis about the nature of substance, space, causation, motion, force, and the laws of nature, all of which was bundled in a rhetoric of innovation that continues to inform our scientific ideals.

All these facts are well known, but the reason early modern philosophy is of particular importance to the history of the concept of health is because the landmarks just referred to are often blamed for medicine having lost its way. This is especially true of the contributions made by Descartes. For example, the neurologist Antonio Damasio believes Descartes neglects the "psychological consequences of diseases of the body proper ... [and the] body-proper effects of psychological conflict."[5] Mark Sullivan notes the consensus that "the source of medicine's ills is . . . René Descartes," and Kay Toombs explains

---

4 Any good survey of early modern medicine will discuss the figures and topics mentioned above. See, e.g., A. Wear, "Medicine in Early Modern Europe," in L. I. Conrad et al. (eds), *The Western Medical Tradition: 800 BC to AD 1800* (Cambridge: Cambridge University Press, 1995), 215–362.

5 A. Damasio, *Descartes' Error: Emotion, Reason and the Human Brain* (New York: Avon Books, 1995), 251.

that Descartes's "paradigm is incomplete."[6] Put simply, though with a bit more detail, Descartes's famous dualism between mind and body and his causally monistic view of life, where living things are presented in the common yet purely material terms of machines and their dispositions, set the agenda for much of the seventeenth century and have had disastrous consequences in the form of an impersonal scientific medicine ever since.[7] Thus, to the extent that we have run aground in modern medicine and in contemporary thinking about health, the early modern period stands out as the period during which we took a wrong turn, thanks largely to Descartes.

Such accusations obscure the many ways in which medicine and the multiple meanings of "health" and "healthy living" influenced the early modern philosophical tradition, and vice versa. Yet Descartes does provide an excellent vantage point for surveying the concept of health within the early modern philosophical tradition. For one, he is an example of a philosopher with an account of the "medicine of the mind" directed not just at the avoidance of error but also the general Stoic ideal of tranquility, happiness, and controlling the passions; and he repeatedly claims an interest in the "medicine of the body," going so far to offer advice and even record his own remedies. Additionally, and this much his contemporary critics get right, Descartes's influence was profound and nearly immediate, so much so that by focusing on Descartes and the Cartesian aftermath, we can glimpse how seventeenth-century physicians and philosophers struggled with the implications of the "new science" and the appropriate way to understand the relationship

---

6 M. Sullivan, "In What Sense Is Contemporary Medicine Dualistic?" *Culture, Medicine and Psychiatry* 10/4 (1986), 331; and S. K. Toombs, "Illness and the Paradigm of Lived Body," *Theoretical Medicine* 9/2 (1988), 201; respectively.

7 T. M. Brown, "Descartes, Dualism, and Psychosomatic medicine," in W. F. Bynum et al. (eds), *The Anatomy of Madness: Essays in the History of Psychiatry*, vol. 1 (New York: Tavistock, 1985), 40–62, rejects the conclusion that Descartes is the source of these and related errors. Cf. I. Switankowsky, "Dualism and Its Importance for Medicine," *Theoretical Medicine* 21 (2000), 567–80. Perhaps the most interesting question to ask here is why this fiction persists or how it arose; the answers are not at all obvious.

between mind and body when our health is at stake. For this reason Descartes, and a handful of medical Cartesians, will serve as the primary focus of this chapter.

The next section of this chapter contains a series of reminders about the concept of health relevant to the early modern period. Here the goal is to survey the broader landscape of the seventeenth century and the context in which Descartes's work belongs, where it is just one part of a larger whole. In the third section of the chapter, the emphasis will be on resurrecting a more accurate account of Descartes's relationship with medicine and the concept of health. This section identifies Descartes's three forms of medical advice: biomechanic, psychosomatic, and naturaopathic.[8] In addition, it will highlight a conceptual difficulty Descartes recognized; namely, a dispositional or functional account of health—we are healthy when we function normally—cannot be justified outside of references to our lived and embodied experience as composites of mind and body. The concluding section traces several interpretative choices made by medical Cartesians representative of the last half of the seventeenth century—Henricus Regius, Jacques Rohault, Johannes De Raey, Johannes Clauberg, Friedrich Gottfried Barbeck, and Tobias Andreae—who, in their effort to reconcile medical practice with dualism and mechanism, did not always agree with one another about what therapeutic strategies to adopt.[9]

8 I have borrowed the first two labels from D. Des Chene, "Life and Health in Descartes and After," in S. Gaukroger et al. (eds), *Descartes' Natural Philosophy* (New York: Routledge, 2002), 723–35, and the third from S. Voss, "Descartes: Heart and Soul," in Wright and Potter, *Psyche and Soma*, 173–96.

9 These figures were first called to my attention in the groundbreaking scholarship of T. Verbeek, *Descartes and the Dutch: Early Receptions to Cartesian Philosophy, 1637–1650* (Carbondale: Southern Illinois University Press, 1992); Verbeek, "Tradition and Novelty: Descartes and some Cartesians," in T. Sorell (ed.), *The Rise of Modern Philosophy: The Tension Between the New and Traditional Philosophies from Machiavelli to Leibniz* (Oxford: Clarendon, 1993), 167–75; and F. Trevisani, *Descartes in Germania: La ricezione del Cartesianesimo nella Facolta filosofica e medica di Duisburg (1652–1703)* (Milano: F. Angeli, 1992); expanded in Trevisani, *Descartes in Deutschland: Die Rezeption des Cartesianismus in den Hochschulen Nordwestdeutschlands* (Wien: LIT, 2011). For English-language readers, portions of *Descartes in Germania* are summarized in Des Chene, "Life and Health," and J. E. H. Smith, "Heat, Action, Perception: Models of Living Beings

Finding and Defining Health

Defining health in the early modern period is not especially easy. Sometimes it is understood as the absence of pain or disease, as a kind of symptom-free existence. At other times, it is conceived as a balance of traditional humors or an ordering of the chemical elements of the body. And at still other times it is described as a robustness of function in the face of external pressures.[10] Health can be mechanically defined or chemically defined, or it may not be defined at all. By analogy to the concept of a healthy body, the healthy mind is in turn conceived as free from error or vice, well ordered, spiritually pure, or resolute in the pursuit of truth and virtue, or, again, may not be defined at all. In fact, during the seventeenth century, health's companion concept of disease appears far more often precisely because there were so many ways one could be sick, and it was sickness and its discomforts more than health that required attention and treatment.[11]

---

in German Medical Cartesianism," in M. Dobre and T. Nyden (eds), *Cartesian Empiricisms* (Dordrecht: Springer, 2013), 105–24.

10 This second positive conception of health—not as a balance but as the ability to resist environmental insults—finds expression in the philosophical project of Baruch Spinoza, who links *conatus* ("appetite" or "striving") with the essence of finite things. A. Gabbey speculates that Spinoza's *conatus* may derive, at least in part, from the medical tradition: see "Spinoza's Natural Science and Methodology," in D. Garrett (ed.), *The Cambridge Companion to Spinoza* (Cambridge: Cambridge University Press, 1996), 168. Spinoza's views of finitude, human vulnerability, and health are discussed in A. Schmitter, "Responses to Vulnerability: Medicine, Politics and the Body in Descartes and Spinoza," in S. Pender and N. S. Struever (eds), *Rhetoric and Medicine in Early Modern Europe* (Surrey: Ashgate, 2012), 141–71. For Spinoza's connection to medicine and the medical tradition more generally, see W. Aron, "Baruch Spinoza and Medicine," *The Hebrew Medical Journal* 2 (1963), 255–82. I thank Raphael Krut-Landau for helpful discussions of Spinoza's views.

11 Helen King has observed that it "is much easier to talk about disease than health" because disease comes in many forms and is invariably noticed, whereas health "lives in the shadow of disease" often only coming to our attention when it is gone: H. King, *Health in Antiquity* (London: Routledge, 2005), 3. King's observation applies as much in the early modern period as to the classical period. Citing but one example, Montaigne asks rhetorically, "How much more beautiful health seems to me after the illness" when "the beautiful light of health" returns: D. M. Frame (trans.), *Montaigne: The Complete Works: Essays, Travel Journal, Letters* (New York: Knopf, 2003), 1021; cited in M. Schoenfeldt, "Aesthetics and Anesthetics: The Art of Pain Management in Early Modern England," in J. F. van Dijkhuizen and K. A. E. Enenkel (eds), *The Sense of Suffering Constructions of Physical Pain in Early Modern Culture* (Leiden: Brill, 2009), 26.

At a time when malnutrition in all its forms was common, and poor sanitation and hygiene were the norm, diseases and infections were spread rapidly in the early modern period, especially among the poor and urban masses.[12] Chronic diseases were also a fact of early modern life, ranging from persistent skin diseases of all kinds, to dysentery, gout and kidney stones. Epidemics of plague also continued to affect Europe throughout the seventeenth century, with unfathomably horrible outbreaks in Naples in 1656, when more than 300,000 of a city of 500,000 died, and in London in 1665, the plague to which we owe Daniel Defoe's gripping *History of a Plague Year* (1720). Taking these facts together, it is hardly a surprise that early modern patients did not have great confidence that medicine could heal them or that anxiety about loss of health was a constant theme that occupied medical writers, poets, dramatists, and visual artists throughout the seventeenth century.

Equally unsurprising, references to health and disease have a prominent role in surviving diaries and early modern correspondence, where discussion in a letter could easily turn to personal health or the health of family and friends.[13] This included dramatic first-person accounts of pain and even anguish and loss, and equally obvious efforts at objective description, sometimes even by patients, of crippling discomfort. It mattered, of course, whether one's correspondent was a physician or healer, and whether the author or recipient of the letter was a woman

---

12 Public health efforts became especially efficacious in the eighteenth century (see Tom Broman's contribution in the present volume). For the prehistory of modern public health, see the early chapters of D. Porter, *Health, Civilization, and the State: A History of Public Health from Ancient to Modern Times* (London: Routledge, 1999). For the existence of health boards that exercised broad powers in the early modern world, see C. Cipolla, *Public Health and the Medical Profession in the Renaissance* (Cambridge: Cambridge University Press, 1976), which also demonstrates that urban planning from the late medieval period did, in fact, lead to an increased focus on sanitation and clean air to maintain healthy living conditions.

13 For discussion of physicians as part of the Republic of Letters, and not simply answering medical queries from patients, see H. Steinke, and M. Stuber, "Medical Correspondence in Early Modern Europe. An Introduction," *Gesnerus* 61 (2004), 139–60 and the references they provide; for a slightly earlier period, see N. Siraisi, *Communities of Learned Experience* (Baltimore MD: Johns Hopkins Press, 2013).

or a man, but, in all these cases, references to health and disease were common.[14]

The correspondence of the philosopher Lady Anne Conway, and those in her circle, is an especially clear example of how health and disease enter early modern letter writing. From her teens until her death at the age of forty-eight, Conway suffered debilitating attacks of pain. In her own letters, she referred to her "afflictions," but, as Sarah Hutton has noted, Conway tended to present herself as resolved to endure, exposing little of her true suffering. The correspondence of her family presents a different picture. In 1658, Conway experienced acute and prolonged pain for more than seven weeks. At the time, her husband Lord Conway wrote to his brother-in-law George Rawdon, "her sighs, and grones come so deep from her, that I am terrifyed to come neere her."[15]

Conway's correspondence with her mentor and confidant, the Cambridge philosopher Henry More, shows his personal concern for Conway's health, but also his efforts to find a cure for her physical and psychological suffering. Securing Conway's health, in other words, was a significant topic of More's correspondence with her, and the health he wished for her was the absence of pain. It was More, for example, who arranged for the famous physician Francis Mercury van Helmont (son of Jan Baptiste van Helmont) to attend to her. It was also More

---

14 The wealth of scholarship concerning woman and their indispensable role in early modern medicine from the household to the anatomical theater continues to grow. For several exemplary recent studies, see K. Park, *Secrets of Woman: Gender, Generation, and the Origins of Human Dissection* (Brooklyn: Zone Books, 2006); E. Leong, "Making Medicine in the Early Modern Household," *Bulletin of the History of Medicine* 82 (2008), 145–68; A. Rankin, *Panaceia's Daughters: Noblewomen as Healers in Early Modern Germany* (Chicago: University of Chicago Press, 2013); and M. DiMeo, "'Such a Sister Became Such a Brother': Lady Ranelagh's Influence on Robert Boyle," *Intellectual History Review* 25 (2015), 21–36. O. Weisser, *Ill Composed: Sickness, Gender, and Belief in Early Modern England* (New Haven, CT: Yale University Press, 2016) is an eye-opening account of the role gender plays in the experience of illness.

15 Lord Conway to Major Rawdon, August 17, 1658; cited in S. Hutton, "Making Sense of Pain: Valentine Greatrakes, Henry Stubbe and Anne Conway," in L. Jardine and G. Manning (eds), *Testimonies: States of Mind and States of the Body in the Early Modern Period* (Dordrecht: Springer, forthcoming).

who put her in touch with Frederik Clodius, who nearly poisoned Conway using the then popular chemical medicine of mercury.[16]

More's own medicine for Conway was psychosomatic. It amounted to the (sadly common) advice offered to anyone who suffered chronic pain. More counseled "patience and fortitude, the reading of spiritual guides . . . and strengthening [Conway's] mind with philosophy and religion."[17] There was obviously more than a hint of piety and morality in this advice. Elsewhere More wrote, "the Diseases of the Body are, for the most part, from the Vices of the Mind," believing that "a purifi'd Mind goes a great way to the purging and purifying of the Body" so much so that "there is no Remedy so powerful . . . as a severe application of Virtue and Piety."[18] Consistent with this, More's advice to Conway was medicine for the mind and not an intervention in the body, at least not directly. At most it was a medicine meant to affect Conway's experience of her body, and only secondarily would this change of mind affect her body.

While one of Conway's physicians, van Helmont, thought that many diseases resulted from the mind's disordered passions, and so could be treated directly in the way More advised, this was not clearly More's view nor the view of most other physicians or philosophers at the time, although they certainly did believe that the mind affected the body, and vice versa.[19] Conway would eventually reject the

---

16 Mercury was a common chemical medicine, especially popular as a cure for "the pox" or "French disease," as it was known outside of France, which appears to have been a virulent form of syphilis.

17 Hutton, "Making Sense of Pain." For more on Conway's life and the Platonic character of her philosophy, see S. Hutton, *Anne Conway: A Woman Philosopher* (Cambridge: Cambridge University Press, 2004); and C. Mercer, "Platonism in Early Modern Natural Philosophy: The Case of Leibniz and Conway," in C. Horn and J. Wilberding (eds), *Neoplatonic Natural Philosophy* (Oxford: Oxford University Press, 2012), 103–26.

18 H. More, *An Account of Virtue, or, Dr. Henry More's Abridgment of Morals Put into English* (London: Printed for Benj. Tooke, 1690), 147–48; cited in Hutton "Making Sense of Pain."

19 For more on Van Helmont, see G. B. Sherrer, "Philalgia in Warwickshire: F. M. van Helmont's Anatomy of Pain Applied to Lady Anne Conway," *Studies in the Renaissance* 5 (1958), 196–206; S. Hutton, "Of Physic and Philosophy: Anne Conway, Francis Mercury van Helmont and Seventeenth-Century Medicine," in A. Cunningham and O. Grell (eds), *Religio Medici: Medicine and Religion in Seventeenth-Century England* (Surrey: Aldershot, 1996), 218–46; Hutton, *Anne Conway*, 140–55. One of Descartes's followers, Tobias Andreae seems to share a similar view to

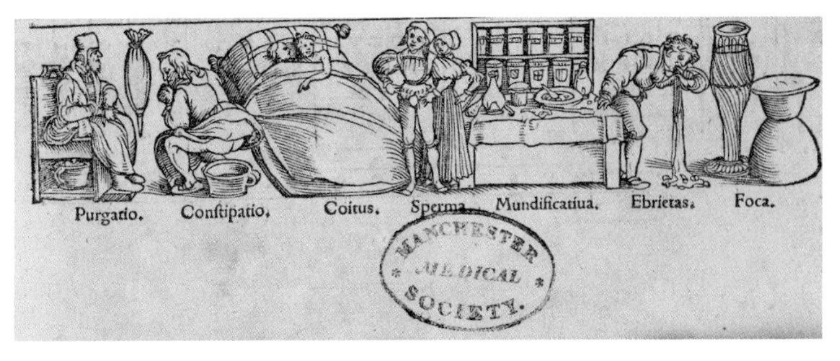

PLATE 1 Detail of a woodcut, illustrating the Latin translation of the *Almanac of Health* (Strasburg: Apud Ioannem Schottum librarium, 1531), p. 103. © University of Manchester

PLATE 2 Avicenna, *Canon*, MS Tritton 12, fol. 1b. Royal College of Physicians, London.

PLATE 3 *Liber chronicarum*, "Dance of Death," 1493. Wellcome Library, London.

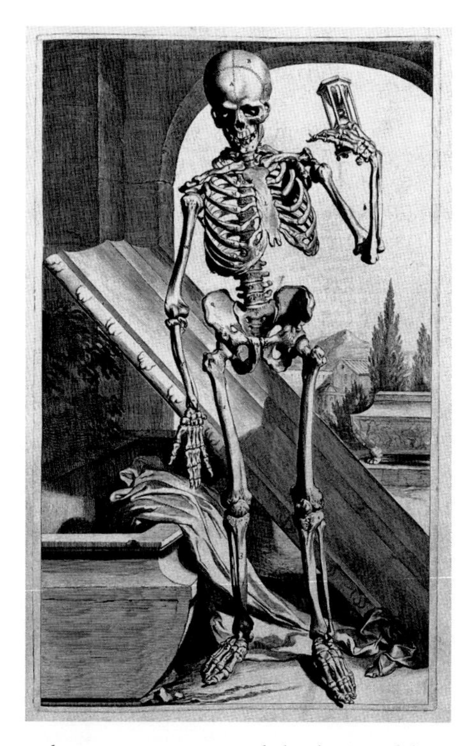

PLATE 4 *Anatomia humani corporis*, engraved plate by Gerard de Lairesse, 1685. Wellcome Library, London.

PLATE 5 Paris Catacombs (bones deposited 1787–1859). Wikimedia Commons.

PLATE 6 William Hogarth, *Marriage à la Mode: The Lady's Death*, ca.1743. National Gallery, London.

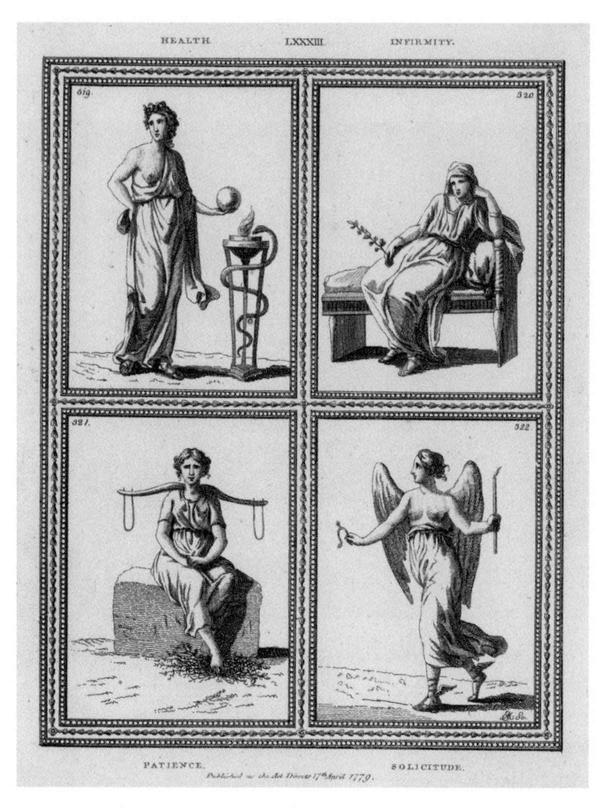

PLATE 7 George Richardson, *Iconologia*: "Health," "Infirmity," "Patience," and "Solicitude," 1779.

PLATE 8 *Gutsave Courbet, Le Déséspère* (*The Desperate Man*), 1845.

consolations of philosophy, in the end experiencing the terrible indifference of family and friends that those with chronic disease still find today after long years of suffering.

It would compound the injustice of Conway's illness to propose it as the key to understanding her life and work. Yet her life, and her philosophy's strong rejection of dualism between mind and body, reminds us that medicine and medical care in the early modern period was the most obvious real-world test of a philosopher's or scientist's beliefs about the natural world. Did they know enough to secure their own health and longevity or the health and longevity of others? This question was not out of place in the early modern period. Indeed, philosophers and scientists alike were not just consumers in the medical marketplace, as Conway was in her efforts to cure her pain. They doctored themselves and their acquaintances, as More tried to do for Conway, and they frequently saw themselves as contributing to the advancement of medical care of both the mind and body.

And this reminds us of something else. Early modern philosophy was as much a way of life, with practical implications actively pursued and explicitly advertised, as it had been in earlier times.[20] In the seventeenth-century English context, for example, speculative and experimental science served as a means to both correct the mind's defects and secure its perfection. This image of the medicine of the mind derives from a number of sources, but two traditions enjoying newfound vitality in the sixteenth century are especially important, namely Stoicism and the Christian emphasis on our fallen state.[21]

---

van Helmont's (on which see the conclusion of this chapter), but even though others can be found expressing similar views, we should not necessarily infer that they viewed the mind as the source of all disease or that a change of mind could cure all the body's diseases.

20 For accounts of early modern philosophy as a way of life, see the references in note 3; S. Corneanu, *Regimens of the Mind: Boyle, Locke, and the Early Modern Cultura Animi Tradition* (Chicago: University of Chicago Press, 2011); and the essays in C. Condren et al. (eds), *The Philosopher in Early Modern Europe: The Nature of a Contested Identity* (Cambridge: Cambridge University Press, 2006).

21 Stoic influence on early modern philosophical thought is discussed in the essays collected in J. Miller and B. Inwood (eds), *Hellenistic and Early Modern Philosophy* (Cambridge: Cambridge

While the immediate religious and moral value of engaging in philosophy or science was being declared, their indirect advantages were also being cited. It was believed that the new science, as a science of the natural world, would eventually reveal many secrets, including secrets about the human body. An improved medicine of the body would follow this newfound knowledge and enable further cultivation of the mind once freed from the body's daily and sometimes painful distractions. Ultimately, this cycle would lead to the realization of goods like happiness and tranquility during our earthly lives.

Along with such optimism came the belief that a new method for avoiding errors and discovering truths was required if we were to achieve these goals. As early modern philosophy came to be associated with these new methods, it was the latter (the method), as a form of education and cultivation, either in the guise of a traditional logic, a logic of invention, or a method of interpretation, that took on the role of a "medicine of the mind." Francis Bacon is an important case in point. In the first instance, he describes his moral philosophy as a "medicining of the minde" in the *Advancement of Learning* (1605), something he reiterated in his later *De Augmentis scientiarum* (1623).[22] "Medicining of the body," wrote Bacon, it is appropriate "first to know the diuers Complexions and constitutions, secondlye the diseases, and lastlye the Cures." And so too "in medicining of the Minde, after knowledge of the diuers Characters of mens natures, it foloweth in

University Press, 2003). For the Stoic background to the early modern medicine of the mind in particular, see Guido Giglioni's chapter in the present volume, as well as his "Medicine for the Mind in Early Modern Philosophy," in J. Sellars (ed.), *The Routledge Handbook of the Stoic Tradition* (London: Routledge, 2016), 189–203; D. Jalobeanu, "Francis Bacon and Justus Lipsius: Natural Philosophy, Natural Theology and the Stoic Discipline of the Mind," in J. Papy and H. Hirai (eds), *Justus Lipsius and Natural Philosophy* (Bruxelles: Wetteren Universal Press, 2007), 107–21; and Corneanu, *Regimens of the Mind*. The significance of religion to early modern science would be hard to overstate. For entry into this unsurprisingly large body of literature that includes discussion of original sin, see P. Harrison, *The Fall of Man and the Foundations of Science* (Cambridge: Cambridge University Press, 2007).

22 I thank Soranna Corneau for correspondence about Bacon's medicine of the mind. Any misunderstandings are my own. For more on Bacon see Guido Giglioni's chapter in the present volume.

order to know the diseases and infirmity[i]es of the mind, which ar no other the*n* [*sic*] the perturbations and distempers of the affec-tions [*sic*]."[23]

The precise details of Bacon's early medicine of the mind are not easily summarized. Its goal is to teach virtue by curing us of the distempers and passions that distract our intellect. Whether the medicine of the mind alone can deliver us to a virtuous life is ambiguous; it looks more like a preparation, a ground clearing for the diligence and continuous labor needed to achieve virtue. Still, the Stoic rhetoric Bacon adopts represents the best state of the soul (virtue) by analogy to the ideal state of the body, namely health.[24] The analogy even extends to discussing vice, which Bacon presents as comparable to the diseases of the body. Each of these diseases is a disturbance, the former of the healthy mind, the latter of the healthy body, and each can be avoided or corrected with the appropriate regimen, either a medicine for the mind or a medicine of the body.[25]

Bacon goes on to extend much of the same language from his moral philosophy to describe his cure for our false beliefs, that is, his method in natural philosophy. Bacon's response to error required long-term care and an effort to fight what he identified first in the *Advancement of Learning* and later in the *New Organon* (1620) as the obstacles to human reasoning: the "Idols" or "Illusions" of the Tribe, Cave, Marketplace, and Theatre.[26] The first of these impediments, the Idols

23 M. Kiernan (ed.), *The Oxford Francis Bacon*, vol. 4 (Oxford: Oxford University Press, 2000), 149.

24 Similarly, in *De Augmentis* the good of the mind is presented as analogous to the good of the body consisting in "health" (absence of perturbations), "beauty" and "strength": J. Speeding, R. Ellis, and D. Heath (eds), *The Works of Francis Bacon*, 15 vols. (London: Longman, 1860), 5:30.

25 Other self-styled reformers in the seventeenth century, including the German university professor and Calvinist minister Johann Heinrich Alsted, would describe philosophy, and not method per se, as "a universal medicine... by means of which the diseases which the mind suffers in knowing things, words, and modes can be removed"; cited and translated in H. Hotson, *Johann Heinrich Alsted 1588–1638: Between Renaissance, Reformation, and Universal Reform* (Oxford: Clarendon Press, 2000), 73.

26 L. Jardine, and M. Silberthorne (eds), *Francis Bacon: The New Organon* (Cambridge: Cambridge University Press, 2000), 40 (I.34).

of the Tribe, include limitations owing to human nature in general, such as the limitations of sense experience. The second, Idols of the Cave, are more personalized limitations of mind and body, including those errors that result from individual prejudices, habits, or styles of reasoning. Idols of the Marketplace derive from our common language, which does not reliably distinguish phenomena. Finally, Idols of the Theatre are like fictional worlds built using systems of philosophy and traditional (though inadequate) rules of demonstration.

In the *New Organon* these four sources of error are described as being in need of a "remedy," a word Bacon uses repeatedly throughout the text.[27] It is his view that conducting experiments will ameliorate the errors of the senses, and in his more polemical moments he declares that the Idols of the Cave and Marketplace "can be eliminated" while those of the Theatre can "in no way" be eliminated unless a new method is introduced into philosophy.[28] We must "indict [the Idols]," Bacon says, "and . . . expose and condemn the mind's insidious force." Exposing and restraining the mind is necessary "in case after the destruction of the old, new shoots of error should grow and multiply from the poor structure of the mind itself." This is precisely why constant vigilance is needed, to avoid the return of error and to keep the mind well ordered. Bacon's constructive proposal is to "fix and establish for ever the truth that the intellect can make no judgement except by induction in its legitimate form."[29] And this is precisely the remedy or medicine the *New Organon* will provide us. It delivers a method that limits the mind to the proper form of induction.

---

27 Additional examples of medical language in the *New Organon* include referring to the method Bacon advocates as offering "true helps" (*vera auxilia*) (I.9); "treatment and cure" (*remedium & medicinam*) (I.94); "ministration" (*ministramus*) and "regulation" (*regimus*) (I.126).

28 Jardine and Silberthorne, *Bacon: The New Organon*, 19.

29 Jardine and Silberthorne, *Bacon: The New Organon*, 19. See also I.68 where Bacon declares the Idols "must be rejected and renounced and the mind totally liberated and cleansed of them, so that there will be only one entrance into the kingdom of man, which is based upon the sciences" (56).

But if Bacon was optimistic about the cures available in the early years of the seventeenth century, others were less so later in the century. Leibniz, who thought he might successfully reconcile the Catholic Church with Protestantism, so himself far more an optimist than most, would lament in 1670: "we are ignorant of the medicine of bodies and of minds."[30] And Henry More, whom we met earlier, identified a "disease incurable," namely, atheism, or "perfect *Scepticisme*." A "thing rather to be pitied or laught at," thought More, "then seriously opposed."[31] Pursuing the analogy between the medicine of the mind and the medicine of the body just a bit further, More's "disease incurable" reminds us that there is a limit to what the medicine of the mind could cure or even ameliorate, just as there were diseases of the body that had no remedy.

The aim of this section has been to show that there are many lands where the concept of health can be found. It is present in what we today would characterize as philosophical and medical texts, but it is also a part of social and religious ritual and constituted, in part, by the spaces in which we live. It is in literature and art, and religion, but also relevant to economic and political events effecting tens of thousands. In other words, were this an entire volume about the early modern period, all these contexts would deserve attention, and this diversity should not be forgotten in spite of the narrow focus required here.

Descartes and His Medical Philosophy

Nearly everything mentioned about health and the medicine of the mind and body in the previous section applies to Descartes. His

30 L. E. Loemker (ed. and trans.), *Philosophical Papers and Letters: A Selection* (Dordrecht: D. Reidel, 1969), 132. For the view of Leibniz as an optimist and "conciliatory eclectic," see C. Mercer, *Leibniz's Metaphysics: Its Origins and Development* (Cambridge: Cambridge University Press, 2001).

31 A. Gabbey, "'A Disease Incurable': Scepticism and the Cambridge Platonists," in R. H. Popkin and A. J. Vanderjagt (eds), *Scepticism and Irreligion in the Seventeenth and Eighteenth Centuries* (Leiden: Brill, 1993), 89.

correspondence is replete with medical discussion, and he freely offers advice to his correspondents about their health or his desire to cure them of their ailments.[32] Descartes also developed his own remedies, one of which, for constipation, Leibniz thought enough of to have transcribed and preserved in its only surviving copy.[33] Uniquely among the standard bearers of early modern philosophy, Descartes was also offered a position on the medical faculty at a major European university.[34] Indeed, so central were the notions of health and disease to his projects that in 1637, in the concluding section of the *Discourse on Method*, he wrote of a commitment "to devote the rest of [his] life to nothing other than trying to acquire some knowledge of nature from which we may derive rules in medicine which are more reliable than those we have had up till now" (*Writings*, 1:43).[35]

In this section, we will proceed to examine, first, Descartes's commitment to medicine and his therapeutic view of philosophy. Next we will consider evidence of his interest in the medicine of the body. Finally, after considering the relationship between the medicine of the body and the medicine of the mind, we will identify the three forms of therapeutic advice present in his correspondence. In spite of his many innovations, Descartes's medical advice remained quite traditional; this mix of innovation and tradition is a persistent mark of the early modern concept of health.

---

32 The first volume of Descartes's correspondence published in 1657 came with a subheading indicating that "les plus belles Questions de la Morale, Physique, Medecine, & des Mathematiques" were discussed in the volume.

33 The recipe is cited and translated in J. E. H. Smith, "Early Modern Medical Eudaimonism," in P. Distelzweig et al. (eds), *Early Modern Medicine and Natural Philosophy* (Dordrecht: Springer, 2015), 325–41, at 325–26.

34 For discussion of the University of Bologna's efforts to hire Descartes, see G. Manning, "Descartes and the Bologna Affair," *British Journal for the History of Science* 47 (2014), 1–13.

35 Descartes's work is cited in the text using the now standard Cambridge translations of J. Cottingham et al., *Descartes' Philosophical Writings*, 3 vols (Cambridge: Cambridge University Press, 1985–1991), hereafter abbreviated *Writings*.

Medicine for the Mind

The position Descartes takes in his first publication, the *Discourse*, involves more than a commitment to preserving the health of the body or even learning how to reclaim it once lost, both common goals defining a physician's activities. Just prior to the passage quoted earlier, Descartes elaborates his motivation. "Even the mind depends so much on the temperament and disposition of the bodily organs that if it is possible to find some means of making men in general wiser and more skillful than they have been up to now, I believe we must look for it in medicine" (*Writings*, 1:43). For Descartes, medicine of the body is the key to our moral and intellectual improvement. Elsewhere in the same work, he wrote, "we might free ourselves from innumerable diseases, both of the body and of the mind, and perhaps even from the infirmity of old age, if we had sufficient knowledge of their causes and of all the remedies that nature has provided" (*Writings*, 1:143).

It would be a mistake to dismiss Descartes's sentiment from the *Discourse*. In 1645 Descartes wrote to the Marquis of Newcastle, "the preservation of health has always been the principal end of my studies." Two years later, in the French letter preface to the *Principles of Philosophy*, the principal benefit of philosophy—which includes not just metaphysics but also natural philosophy, the two disciplines with which Descartes is most often identified—is said to "depend on those parts of it which can only be learnt last of all"; namely, medicine, mechanics, and morals. In the tree of knowledge from the same preface, Descartes informs us that the roots of knowledge are metaphysics, the trunk physics, and the three branches, where the fruit is to be found, are medicine, mechanics, and morals.[36] Thus, in spite of our image of Descartes as primarily a physicist and metaphysician, there is ample evidence that he located the value of his project neither in metaphysics

36 For earlier uses of "the tree of knowledge," see R. Ariew, "Descartes and the Tree of Knowledge," *Synthèse* 92 (1992), 101–16.

nor even physics, but in philosophy's ability to deliver what is good for human beings, whether the instrumental goods of mechanics, the natural good of health, or the highest good of morals. In other words, Descartes understood the value of philosophy in terms of the life it could deliver and, specifically, as the means to a better life that included greater health than previously possible.

In providing guidance on how to live, including how to improve our knowledge, Descartes often resorted to the language of therapeutics and medicine, just as we saw Bacon doing in the previous section. Among his earliest extent writings, for example, likely from 1619 to 1622, Descartes declared, "I use the term 'vice' to refer to the diseases of the mind, which are not so easy to recognize as diseases of the body. This is because we have frequently experienced sound bodily health, but have never known true health of the mind" (*Writings*, 1:3). If this is not a full embrace of the traditions discussed in the early chapters of this volume, Descartes is certainly helping himself to the compelling rhetoric of health to describe the imperfect state of our knowledge and our minds.

Twenty years later, in what today is his most famous work, the *Meditations on First Philosophy* (1641), Descartes would again resort to the language of health and disease to defend his project. Against Hobbes's criticism that he had said nothing new when raising skeptical doubts in Meditation One—for example, perhaps I am dreaming? perhaps there is an evil deceiver?—Descartes replied: "I was not looking for praise when I set out these arguments; but I think I could not have left them out, any more than a medical writer can leave out the description of a disease when he wants to explain how it can be cured" (*Writings*, 2:121). Descartes's philosophy was a cure, but the question that remains unanswered in the reply to Hobbes is, a cure for what? In the *Meditations*, the answer would seem to be skeptical doubt, but this would confuse just one symptom for the true disease. What we must cure are our false beliefs more generally, and these, Descartes tells us, result from the prejudice of relying on the senses as a guide to truth, itself a by-product of being born into the world as children who must rely on our senses to survive.[37]

Conceiving Descartes's philosophy as a therapy tied to overcoming our false beliefs and the overreliance on the senses is also encouraged by the topic of his earliest book-length philosophical manuscript: the *Rules for the Direction of the Mind.* This is a difficult work—made more difficult by a newly discovered manuscript in 2011—but the *Rules* of the late 1620s has as its goal "to direct the mind with a view to forming true and sound judgments about whatever comes before it" (*Writings*, 1:9). Among the benefits of the "universal wisdom" Descartes sought to cultivate, he included an "intellect" that could "show his will what decision it ought to make in each of life's contingencies" (*Writings*, 1:10). From judgment, to universal wisdom, to the cultivation of the intellect, in the *Rules* philosophy equipped us to attain "the comforts of life . . . [and] the pleasure to be gained from contemplating the truth, which is practically the only happiness in this life that is complete and untroubled by pain" (*Writings*, 1:10, see also 1:179).

As we saw in the case of the later *Discourse* and *Principles*, the theme of guiding the mind to knowledge and virtue, and ultimately happiness, is not just to be found in Descartes's unpublished work. It is hinted at on many occasions over a lifetime of writing. Its fullest expression comes late, however, when he is called upon to address the concerns of Princess Elisabeth of Bohemia and, subsequently, in his *Passions of the Soul* (1649). To Elisabeth he wrote: "True philosophy . . . teaches that even amidst the saddest disasters and most bitter pains we can always be content, provided that we know how to use our reason" (*Writings*, 3:272). The cultivation of the mind, in other words,

37 There are, of course, many different interpretations of the *Meditations*, but for an elaboration of the interpretation I follow here, including one in which childhood is presented as the source of our errors, see D. Garber, "*Semel in vita*: The Scientific Background to Descartes' *Meditations*," in A. O. Rorty (ed.), *Essays on Descartes' Meditations* (Berkeley: University of California Press, 1986), 81–116; S. Menn, *Descartes and Augustine* (Cambridge: Cambridge University Press, 2002); and G. Hatfield, *Routledge Philosophy Guidebook to Descartes and the Meditations* (London: Routledge, 2002).

is the key to happiness, and with adequate effort virtue and happiness are attainable in this life. Though Descartes would go on to write in the final article of the *Passions*, "it is on the passions alone that all the good and evil of this life depends," what the medicine of the mind can offer, consistent with his letters to Elisabeth, is a contentment immune to the passions and our bodily condition.

Medicine for the Body

We will return to Descartes's medical advice, but before this we should document another of the currents that flows into Descartes's correspondence. It is his interest in theoretical medicine—anatomy and what we would now call physiology. These are the capstones of Descartes's natural philosophy or physics and the basis for his scientific medicine.

Roughly at the time when he abandoned the *Rules*, in a letter from December 1629 to his friend and intellectual ally in Paris, Marin Mersenne, Descartes announced that he had begun "to study anatomy." Early the next year, Descartes made his interest in both theoretical and practical medicine more explicit, and at the same time declared it more central to his ongoing work. Writing again to Mersenne, Descartes expressed regret in January 1630 that Mersenne was experiencing an outbreak of an acute skin disease. "Please look after yourself," Descartes wrote his friend, "at least until I know whether it is possible to discover a system of medicine which is founded on infallible demonstrations, which is what I am investigating at present." Shortly afterward, in April 1630, Descartes similarly indicated how medicine, and by implication medicine's central notion of health, consumed his time:

> I am now studying chemistry and anatomy simultaneously; every day I learn something that I cannot find in any book. I wish I had already started to research into diseases and their remedies, so that

> I could find some cure for your erysipelas, which I am sorry has troubled you for such a long time. (*Writings*, 3:21)

For Descartes these are among the earliest indications he was taking an active interest in the medicine of the body, and one of the few instances in which he cites chemical medicine among his interests.

Regrettably, any reconstruction of Descartes's activities during this period, due to gaps in the historical record, must be inferred from a limited number of surviving letters along with the ultimate fruit born from his research later in the 1630s. We do know that *The World*, a work occupying Descartes from 1630 to 1633, was initially conceived as a contribution to meteorology, portions of which appeared among the essays accompanying the *Discourse* in 1637. But we also know that Descartes repeatedly expanded the scope of *The World* between 1630 and 1633, so much so that what survives today as *The World* and the *Treatise on Man* were meant to be two parts of a single work of physics. And finally, we know that the content of the *Treatise on Man* presupposes and incorporates extensive knowledge of anatomy and physiology, preliminary studies to a medicine of the body. Evidence for Descartes's interest in the medicine of the body appears repeatedly throughout his subsequent publications and correspondence, the latter of which includes exchanges with physicians, such as Vopiscus Fortunatus Plempius, a professor of medicine in Amsterdam and later Leuven, and those, like Princess Elisabeth of Bohemia, who welcomed Descartes's medical advice.

To recall the terrain we have mapped so far, we have discovered in Descartes a diverse medicine of the mind meant to cure us of errors, aid us in gaining true beliefs, equip us to make true judgments, and even provide for happiness and contentment. We have also found a medicine of the body in Descartes, one aimed at curing diseases and preserving health through greater knowledge in physics, especially through anatomy and physiology. We do not yet know the relationship between these two medicines. It would seem that the medicine of the mind as a cure to error and a method of invention is a prerequisite to a medicine for the body for Descartes. That is, the medicine of the mind cultivates the intellect and it is through the intellect that we attain a better physics and, ultimately, a better physiology and medicine. Yet lurking here is a tension within Descartes's project. Do we need a medicine of the mind for an effective medicine of the body, or can the latter develop on its own, in which case the art of medicine might not need the science of physics. And how are we to understand the health of the body? This is not merely a question about how physicians might define health but, more fundamentally, whether the physics that comes on the heels of the medicine of the mind has the resources to differentiate between something that is healthy as opposed to diseased. In the next subsection, I consider the prospects for the concept of health, after which, I present Descartes's three therapeutic strategies. Although Descartes did not see any obvious tension between his medicine of the mind and his medicine of the body, we will have a chance to consider whether there is a tension between the biomechanics that develops from his medicine of the mind and his other therapeutic strategies in the conclusion to the paper, when we look at the reception of Descartes's medicine.

Medicine and Metaphysics

The pivotal passage relating Descartes's most fundamental metaphysical commitments about the natural world to health occurs in Meditation Six, where he emphasizes the natures we must cite when referring to artificial and natural kinds. He specifically tells us what is involved in imagining bodies suffering from, in his words, "errors of nature" (*Writings*, 2:58–59). The passage is long, but it is too important not to quote at length:

> Yet it is not unusual to go wrong even in cases where nature does urge us toward something. Those who are ill, for example, may

desire food or drink that will shortly afterwards turn out to be bad for them. Perhaps it may be said that they go wrong because their nature is disordered, but this does not remove the difficulty. . . . A clock constructed with wheels and weights observes all the laws of its nature just as closely when it is badly made and tells the wrong time as when it completely fulfils the wishes of the clockmaker. In the same way, I might consider the body of a man as a kind of machine equipped with and made up of bones, nerves, muscles, veins, blood and skin in such a way that, even if there were no mind in it, it would still perform all the same movements as it now does in those cases where movement is not under the control of the will or, consequently, of the mind. I can easily see that if such a body suffers from dropsy, for example, and is affected by the dryness of the throat which normally provides in the mind the sensation of thirst, the resulting condition of the nerves and other parts will dispose the body to take a drink, with the result that the disease will be aggravated. Yet this is just as natural as the body's being stimulated by a similar dryness of the throat to take a drink when there is no such illness and the drink is beneficial. Admittedly, when I consider the purpose of the clock, I may say that it is departing from its nature when it does not tell the right time; and similarly when I consider the mechanism of the human body, I may think that, in relation to the movements which normally occur in it, it too is deviating from its nature if the throat is dry at a time when drinking is not beneficial to its continued health. . . . As I have just used it, "nature" is simply a label which depends on my thought; it is quite extraneous to the things to which it is applied, and depends simply on my comparison between the idea of a sick man and a badly-made clock, and the idea of a healthy man and a well-made clock. But by "nature" in the other sense I understand something which is really to be found in the things themselves; in this sense, therefore, the term contains something of the truth. When we say, then, with respect to the body suffering from dropsy, that is has a disordered nature because it has

> a dry throat and yet does not need drink, the term "nature" is here used merely as an extrinsic denomination [*denominatio extrinseca*]. However, with respect to the composite, that is, the mind united with this body, what is involved is not a mere label, but a true error of nature, namely that it is thirsty at a time when drink is going to cause it harm. (*Writings*, 2:58–59; modified)

One way in which to understand Descartes here would begin by noting that there are cases of illness in which we desire something that will, in fact, lead to great harm. If we genuinely desire something when it is not good for us, the background worry for Descartes is that God would be deceiving us by giving us a "nature" susceptible to such errors; God would be responsible for our misperception. But, Descartes asks, what is it for us to have a nature in this sense? Consistent with his remarks from the *Treatise on Man* and *Discourse,* in Meditation Six Descartes initially uses "nature" to signify the bodies that are defined by their size, shape, and state of motion. To this extent our nature is no different than the nature of those things studied in physics or natural philosophy. In other words, our bodies, like the rest of the bodies in the world, are extended and necessarily obey the laws of nature along with the other general constraints appropriate to extended things.

In Meditation Six, Descartes extrapolates from this insight to the rest of the physical world. Thus, from the point of view of Descartes's physics, the human body and the animal body, and really all bodies, are alike just machines with dispositions to act. In terms of mechanical analysis, the desire for harmful food or drink is the result of a disposition or state of motion coupled with the laws of nature, just as any other physical phenomenon will be. There simply is no normative or evaluative judgment that can be made which calls the process of a human body or machine causing its own demise unnatural. As Descartes concludes, there is no room for straightforward characterizations of health and illness given the resources of dispositions and laws of nature alone; there is nothing that can be disordered in violation of its nature.

Even so, it seems undeniable that living things get sick, age, and eventually die. It is as true of the world we experience as anything we might claim to know.[38] Put another way, pain hurts, and so too, in a slightly different way, does the loss of vitality. If these are not caused by deviations from our healthy states or subjective experiences of such deviations, then what are they, and how should they be understood? Contrary to how it may at first appear from Meditation Six, Descartes believes that the language of function and malfunction appropriate to his mechanical physiology and pathology, especially where medicine is concerned, is more than either a figment of our imagination or, in contemporary language, a sophisticated account of biological function plus species-specific statistical normality.[39]

Scholars disagree about how to interpret what Descartes says next in Meditation Six, but it seems clear that for animals, the human body, and even clocks, assigning them a nature that justifies talking as though they are *supposed* to be like something, or *supposed* to do something, is to assign them a nature that they do not have from the standpoint of physics. It is to use "nature," Descartes says, as an "extrinsic denomination."[40]

---

38 G. Hatfield, "Animals," in J. Broughton and J. Carriero (eds), *Companion to Descartes* (Oxford: Blackwell, 2008), 404–25, emphasizes this fact and makes good use of it when discussing the ontological status of our bodies in Descartes' physics. It is also a fact that speaks strongly against eliminative readings of Descartes's conception of life.

39 See C. Boorse, "Health as a Theoretical Concept," *Philosophy of Science* 44 (1977), 542–73 for a defense of the view that health involves a value-free statement of biological function plus statistical normality. As I suggest above, Boorse's position is close to the view Descartes presents but ultimately rejects in Meditation Six. Boorse has defended his view, although understandably not against Descartes, in Boorse, "A Rebuttal on Health," in J. M. Humber and R. F. Almeder (eds), *What Is Disease?* (Totowa: Humana Press, 1997), 1–134; and Boorse, "A Second Rebuttal on Health," *The Journal of Medicine and Philosophy* 39 (2014), 683–724. For more on Boorse's view and its critics see Elselijn Kingma's contribution in the present volume.

40 Alternatively, in adopting this technical terminology Descartes may be indicating that the right way to understand our judgment that some objects are broken and others are in working order, or that animals are healthy or unhealthy, would follow accounts of the so-called secondary qualities. This claim does not enter my discussion here, but it may signal a neglected approach to

I have argued elsewhere that the proper interpretation of Meditation Six, at this critical juncture in the text, hinges on this technical and antiquated term.[41] In citing an "extrinsic denomination," Descartes is acknowledging that it is not a mistake to speak about a clock, an animal, or the human body as malfunctioning or being in error; our talk of a "nature" or of "health" is not arbitrary or simply imposed without justification. Instead, we are simply relying on a relation or analogy that exists between the clock, animal, or human body, and something else. A very different example will help illustrate these two points. When we refer to a healthy salad, "healthy" would be an extrinsic denomination. This does not mean we are incorrect to claim that a salad is healthy, it is just that salads are healthy because they are a cause of (related to) health in human beings.

The point we have reached in our survey of Descartes is this: according to the privileged reality described by his physics, the human body and the animal body are extrinsically denominated with a nature because, taken by themselves, they lack an individual nature; in the mechanical world there are no normative ideals but only necessities controlling the way a body's dispositions produce effects. Nevertheless, as in the case of a clock, when the clock is related to our intentions, this relation justifies our saying that the clock is for telling the time, and relative to our purposes is a good or a bad clock. In itself, of course, the clock is a complex bit of matter disposed to move in fixed patterns.[42] Following Descartes's account, the next step is to realize that what has

---

the issues of teleology and function in nonhuman living things that have become a focus of recent scholarship (see references in note 42).

41 G. Manning, "Descartes' Health Machines and the Human Exception," in D. Garber and S. Roux (eds), *The Mechanization of Natural Philosophy* (Dordrecht: Springer, 2013), 237–62.

42 There is a question of whether even this is so according to the strictures of Descartes's metaphysics and, if it is, whether one might accept teleology and derive an account of function from the dispositional unities that exist in the world. For discussion with a negative conclusion, see D. Des Chene, *Spirits and Clocks: Machine and Organism in Descartes* (Ithaca, NY: Cornell University Press, 2001), 125ff. For a more optimistic conclusion, see L. Shapiro, "The Health of the Body Machine? Or Seventeenth-Century Mechanism and the Concept of Health," *Perspectives on Science* 11 (2003), 421–42; D. J. Brown, "Cartesian Functional Analysis," *Australiasian Journal of Philosophy* 90 (2012), 75–92; and K. Detlefson, "Descartes on the Theory of Life and Methodology

just been said about a clock is true of animals and the human body. In relation to our purposes, or perhaps some other touchstone, as I explain below, we can make normative judgments about a thing's nature and our healthy states.

Beyond the purposes we might have for our bodies when thought of as instruments, in our case—when we understand ourselves to be more than our mechanical bodies, but a union between mind and body—there exists an individual nature not otherwise found in the Cartesian world. Descartes clearly believed that there are things about ourselves we do not choose. When the union's natural and intrinsic tendencies are frustrated, we have a case of what he calls in Medication Six a "true error of nature." In still other words, our nature is intrinsically denominated, and our embodied existence as human beings, as combinations of minds and bodies, serves as the touchstone for normative judgments about our healthy states. It specifically introduces the need for a body that can support union with the mind while minimally interfering with the mind's operations. The idea is that health and illness can be defined relative to the purpose of extending the existence of the union and the mind's productive activity while the union exists.[43] In all of Descartes's writing, Meditation Six appears to be the only place where he explicitly and thoughtfully endorsed attaching normative claims to bodies without the need to utilize an extrinsic denomination.[44]

Indeed, elsewhere in the *Meditations* and his correspondence, Descartes emphasizes how our experience of pain and pleasure mark our relation to our bodies and how it is bodily sensation, and not disembodied reflection on our minds or bodies alone, that is our route

---

in the Life Sciences," in P. Distelzweig et al. (eds), *Early Modern Medicine and Natural Philosophy*, 141–72.

43 As I understand her, this is also the view presented A. Simmons, "Sensible Ends: Latent Teleology in Descartes' Account of Sensation," *Journal of the History of Philosophy* 39 (2001), 49–75.

44 Scholars have not reached a consensus on this issue. Though my sympathies lie closer to Des Chene, *Spirits and Clocks*, alternatives can be found in Voss, "Descartes: Heart and Soul"; Shapiro, "The Health of the Body Machine"; Hatfield, "Animals"; Brown, "Cartesian Functional Analysis"; and Detlefson, "Descartes on the Theory of Life."

to understanding ourselves as a compound of mind and body. This does likely imply for Descartes that there is something about our earthly existence that cannot be fully known, and the sorts of "clear and distinct" cognitions he otherwise prized in knowledge of mind and of body are not available in the case of the compound of the two. Nevertheless, the point is that in spite of a long tradition portraying Descartes as denying the lived embodied experience of a human being, his position seems to be quite the opposite. For him, the human being, the composite of mind and body, plays a fundamental role in grounding even the health of the human body. Descartes is so invested in the role bodily sensations play in guiding us to preserve our health that he writes: "The best system [of the body] that could be devised [by God] is that it should produce the one sensation which, of all possible sensations, is most especially and most frequently conducive to the preservation of the healthy man. And experience shows that the sensations which nature has given us are all of this kind" (*Writings*, 2:60–61).

The initial challenge Descartes identifies in Meditation Six, of characterizing health in terms of errors or malfunctions—deviations from our nature—is still with us today, and there is not yet agreement on how to meet it. For his part, however, once having staked out a position in Meditation Six, Descartes does not appear to look back (and it is interesting to note that the long passage from Meditation Six does not figure prominently among his disciples or later reception). To the contrary, Descartes provided medical advice to his friends and correspondents as though the concept of health was easily or obviously defined, in spite of the apparent tension between his metaphysical depiction of bodies as merely extended and the fundamental place of the human being in his medical philosophy.[45]

45 Voss, "Descartes: Heart and Soul," reads the evidence slightly differently here. "To define the body's health in terms of the human being's welfare is to go against [the] grain" of Descartes's

Descartes's Medical Therapies

Descartes's interests in theoretical and practical medicine come together, as well as his interests in the medicine of the mind, in his correspondence with the exiled Princess Elisabeth.[46] Descartes has his hands full in these letters on nearly every subject that Elisabeth raises, but our purposes can be served if we focus on Elisabeth's "indisposition in the stomach," which is a topic of their correspondence in July 1644. Assuming the role of a corresponding physician, he writes "the remedies which Your Highness has chosen, diet and exercise, are in my opinion the best of all." With this, Descartes endorses a very traditional view of medical intervention: you cure the body by controlling a subgroup of the non-naturals, in this case what you eat and how much you move.

Yet, tellingly, Descartes also qualified this endorsement. Intervening through direct control of the body was best, "leaving aside those [remedies] pertaining to the soul." The medicine of the mind or soul, by which Descartes means the same thing here, was relevant because he had

> no doubt that the soul has great power over the body, as is shown by the great bodily changes produced by anger, fear and the other passions . . . I know no thought more proper for preserving health than a strong conviction and firm belief that the architecture of our bodies is so thoroughly sound that when we are well we cannot easily fall ill except through extraordinary excess or infectious air or some other external cause, while when we are ill we can easily

---

metaphysical commitments and "the evidence is that Descartes abandons the attempt" (189; see also Shapiro, "The Health of the Body" for the same conclusion). The evidence Voss cites is largely Descartes's silence on the issue, but I hesitate to equate silence with outright rejection.

46 For more extensive accounts of the medical content of Descartes's correspondence with Elisabeth see S. Mills, "The Challenging Patient: Descartes and Princess Elisabeth on the Preservation of Health," *Journal of Early Modern Studies* 2 (2013), 101–22; and the extensive editorial notes in L. Shapiro (ed. and trans.), *The Correspondence between Princess Elisabeth of Bohemia and René Descartes* (Chicago: University of Chicago Press, 2007).

> recover by the unaided force of nature, especially when we are still young. (*Writings*, 3:237)

All three of Descartes's therapeutic strategies for regaining physical health are on display and work in tandem with one another in the 1644 correspondence with Elisabeth. First, Descartes advises Elisabeth to attend to what she eats and how much she exercises. Though it is not explicit in the letter, this unremarkable advice is based on an entirely different etiology for Descartes than for other physicians. It reduces to a form of mechanical therapy or, as I labeled it earlier, biomechanics. In the example above, Descartes does not advise exercise because of an imbalance in the traditional four humors, but because exercise will facilitate circulation; thinning of the blood; and possibly even modulating the heat in the heart, which plays a central role in all physiological functions.[47] Biomechanics is also implicated in the two further therapeutic strategies Descartes endorses, as we are about to see, though it is not itself the most effective therapy on Descartes's view.[48]

---

47 In the *Description of the Human Body* (1649) Descartes writes, "it is so important to know the true cause of the heart's movement that without such knowledge it is impossible to know anything which relates to the theory of medicine. For all the other functions of the animal are dependent on this, as will be clearly seen in what follows" (*Writings*, 1:319).

48 Tad Schmaltz notes that though "Descartes sometimes suggests that therapeutic treatments require only our ordinary experience, and thus need not involve an investigation of the physiological details, for the most part his medicine presupposes a special form of 'biomechanics' ": T. Schmaltz, *Early Modern Cartesianisms: Dutch and French Constructions* (Oxford: Oxford University Press, 2016), 228–29. Schmaltz also finds that the art of medicine is "distinct in kind" from Descartes's biomechanical medicine (265). The relation between the art of medicine and the science of medicine on which the art is supposedly based according to both scholastics and Descartes is a complicated one. So far as I understand Schmaltz, however, I disagree with his assessment. It is worth noting that the medical advice Descartes offered was quite traditional and, as we are about to see, his psychosomatic medicine also presupposes the mind can affect the biomechanics of the body, something it admittedly does without knowledge of the physiological details. To me this suggests that psychosomatic medicine exploits biomechanics and is not wholly distinct from it. Further, the fact that Descartes's naturopathic medicine offers a direct route to an individualized therapeutic intervention (as such it allows for an informed rejection of a physician's recommendation) does not mean that the biomechanical science Descartes envisions would (ultimately) recommend a different medicine for the body than the one recommended by his naturopathic medicine. These two medicines would ultimately agree about the proper intervention; it is simply that the patient's

The additional complementary therapies recommended in the correspondence with Elisabeth are psychosomatic and naturopathic medicine. Let us take each of these in turn. We saw above that Elisabeth is advised by Descartes to think about how persistent and self-sustaining her body is, and this thought, this reassuring and happy thought, will serve as a kind of hygiene preventing illness except in "extraordinary" cases. Later in the correspondence to Elisabeth, from May or June of 1645, Descartes further elaborates the power of the mind and especially the imagination on our bodies. He cites two hypothetical men. The first man's thoughts make him physically sick, while the other's preserve and heal his body. Specifically, the first man has "every reason to be happy" but spends all his time "in the consideration of sad and pitiful objects." This "by itself," writes Descartes, "would be enough gradually to constrict his heart and make him sigh in such a way that the circulation of his blood would be delayed and slowed down" (*Writings*, 3:250). Linking the mind's thoughts to the body, as Descartes unquestionably does here, his biomechanical view of the body reemerges. For the "grosser parts of his blood, sticking together, could easily block the spleen, by getting caught and stopping in its pores." Alternatively, "the more rarefied parts, being continually agitated, could affect his lungs and cause a cough which in time could be very dangerous" (*Writings*, 3:250). The real message to Elisabeth is not about biomechanics and physiology, however, but about psychosomatic medicine: the wrong thoughts can make us physically ill.

In the second hypothetical case, Descartes asks Elisabeth to consider a man with little reason to be happy due to life's misfortunes. But this man, as opposed to the one before, spends his "time in the consideration of objects which could furnish contentment and joy," including, presumably, that earlier mentioned "conviction and firm belief that the architecture of [his body] . . . is sound." Descartes maintains that this second man's happier thoughts "would be capable of restoring him to health,

---

individual body is best known to the patient herself. This is precisely why Descartes can recommend that the patient be her own physician.

even if his spleen and lungs were already in a poor condition because of the bad condition of the blood caused by sadness" (*Writings*, 3:250). Descartes only adds that coupling psychosomatic therapy with "medical remedies to thin out the part of the blood causing the obstructions," which is to say traditional medicine informed by his biomechanical view of the body, would be the most effective approach. Once again, if we look beyond the proximate physical causes Descartes cites, the lesson for Elisabeth is about psychosomatic medicine: the right thoughts can preserve our bodies and even regain our health.[49]

While acting as a corresponding physician for Elisabeth, Descartes advises her to direct her mind toward ideas that will bring her joy. Initially citing the advantages of a medicine of the body, he acknowledges "the waters of Spa are very good," but "above all if Your Highness while taking them observes the customary recommendation of doctors, and frees her mind from all sad thoughts" she could maximize chances of recovery.[50] Indeed, in this way, using the biomechanical therapy indicated by her condition plus psychosomatic medicine, Descartes believes she "will recover perfect health." In addition, and repeating a sentiment from the *Discourse* some eight years earlier, he reminds her that bodily health "is the foundation of all the other goods of this life" (*Writings*, 3:250).

Descartes's medical views presented in the 1644 and 1645 correspondence with Elisabeth are repeated the following year, in October or November 1646. Again writing to Elisabeth about the advantages of psychosomatic medicine, Descartes points out that "bodily health

49 In presenting this Stoic-inspired idea of resisting life's circumstances, whatever they may be, Descartes's second hypothetical man turns out to be Descartes himself: "I take the further liberty of adding that I found by experience in my own case that the remedy I have just suggested cured an illness almost exactly similar, and perhaps even more dangerous" than Elisabeth's own (*Writings*, 3:250).

50 Descartes finishes his recommendation by advising Elisabeth to "even [refrain] from all serious meditations on scientific subjects." This added suggestion is not simply a remark informed by Elisabeth's gender. It is also something to which Descartes appeared committed in his own life (see the discussion in Garber, "*Semel in vita*" and the references to Descartes's work there).

and the presence of agreeable objects greatly aid the mind by chasing from it all the passions which partake of sadness and making way for those which partake of joy." Moreover, "when the mind is full of joy, this helps greatly to cause the body to enjoy better health and to make the objects which are present appear more agreeable" (*Writings*, 3:296). There are many additional passages in the correspondence with Elisabeth, but also in the later *Passions*, where Descartes reiterates these and similar claims.[51] We will see in the next section, however, that Descartes's psychosomatic medicine was not uniformly embraced by his followers. Indeed, the perceived tension between his biomechanical and psychosomatic medicine during the years after his death has contributed to the image of Descartes as an enemy to mind-body medicine. But before moving to his reception, we must discuss his third therapeutic strategy.

Descartes's naturopathic medicine also surfaces in his July 1644 letter to Elisabeth with which this subsection began. Recall that he told Elisabeth the "unaided force of nature" recovers our health when it is lost. Descartes's claim here is not obviously consistent with his well-known rejection of teleological explanation in physics (*Writings*, 2:38–39). But even so, Descartes unquestionably supports the idea that bodies can resist external causes that would otherwise change

51 The topic of the passions assumes a more prominent role in the final years of Descartes's life and deserves much fuller treatment than I have provided here. For entry into the discussion, see the essays collected in B. Williston and A. Gombay (eds), *Passion and Virtue in Descartes* (Amherst, MA: Humanity Books, 2003). Still, at least two things are worth noting. First, Descartes does not pursue the topic of the passions as a moral philosopher but only as a "natural philosopher" (*Writings*, 1:327). Second, Descartes's view of the passions includes a partial or qualified endorsement. Specifically, the passions are good insofar as they promote the body's health, understood, I believe, in terms of the preservation of the body, and they can even provide us with happiness understood in terms of our bodily states. However, even this notion derives from a prior commitment to the human being as a compound of mind and body, which (on my account) is the real measure of the body's health. The relevant questions for Descartes always become (1) is the body fit to support union with the mind, and (2) is it obstructing the mind's activities to the least possible extent? These two goals are not always realized together, but the key idea is that the body's preservation, though a point Descartes often emphasizes and though a precondition for the preservation of the union and the freeing of the intellect from our embodied distractions, is merely a means to answering yes to questions (1) and (2) and is valued accordingly.

their present state and that bodies even return to a state of health of their own accord. As he explained in a conversation recorded by Frans Burman: even "when we are ill, nature still remains the same . . . and makes light of any obstacles in her way, provided we obey her." Accordingly, physicians ought to "allow people the food and drink they frequently desire when they are ill . . . [because in] such cases nature herself works to effect her own recovery; with her perfect internal awareness of herself, she knows better than the doctor who is on the outside" (*Writings*, 3:354). In other words, physicians should be less intrusive or, better, should help their patients do as their patient's nature dictates because the patient knows the particularities of her case best. Indeed, the privileged insight of a patient into her own care is emphasized by Descartes when he cites Tiberius Caesar: "No one who has reached the age of thirty should need a doctor, since at that age he is quite able to know for himself through experience what is good or bad for him, and so be his own doctor" (*Writings*, 3:354).[52]

What we find in the correspondence with Elisabeth is that Descartes's advice fell into three categories: manipulate the traditional non-naturals, entertain happy thoughts, and let our natures guide us. In the 1644 correspondence, Descartes advises we pursue all three therapeutic strategies together, intervening in the body and in the mind, but also encouraging Elisabeth to use her knowledge of herself to take the lead in how to proceed. The potential tension between these three forms of medical advice will be a topic below, as will the lasting impression that only the biomechanical strategy is strictly Cartesian.

52 This idea that the patient was best suited to understand her own complaints and guide her own care was unique neither to Descartes nor Tiberius, nor a principled stand against trained physicians. For a brief discussion of self-management of medicine in the early modern period, see R. Porter, *Disease, Medicine and Society in England, 1550–1860* (Cambridge: Cambridge University Press, 1992), 17–26. This provides yet another reason to be cautious before concluding that Descartes's threefold strategy in therapeutics is internally inconsistent.

Cartesian Physicians

We have already seen that it is incorrect to identify Descartes as just a metaphysician or a mechanical scientist, and so it will be little surprise that his early influence was strongest in medical circles, where his views were taught and his ideas disseminated to younger generations who would themselves go on to enliven Europe's salons and scientific societies throughout the remainder of the seventeenth century.[53] Instrumental in shaping the medical reception of Descartes were physicians such as Henricus Regius, Johannes De Raey, Tobias Andreae, and Friedrich Gottfried Barbeck, as well as textbook writers like Johannes Clauberg and Jacques Rohault. Like us, however, they had to confront the ambiguities and potential inconsistencies among Descartes's texts.[54] In fact, many of the initial reactions to Descartes accepted one or more features of his system while rejecting others.[55] As we are about to see, this last observation applies equally to Descartes's

53 See P. Mouy, *Le développement de la physique cartésienne* (Paris: Vrin, 1934), 73–85; and, more recently, A. H. Munt, "The Impact of Dutch Cartesian Medical Reformers in Early Enlightenment German Culture (1680–1720)" (PhD diss., University College London, London, 2005); and Schmaltz, *Early Modern Cartesianisms*. Although his focus is on the history of physics, John Heilbron identifies the significant role of a "cadre of physicians" early in Descartes' reception: "Was There a Scientific Revolution?," in J. Z. Buchwald and R. Fox (eds), *The Oxford Handbook of the History of Physics* (Oxford: Oxford University Press, 2013), 7–24 (15ff). These works notwithstanding, historians of philosophy have by and large ignored the insight, explicitly provided by Gary Hatfield, that Descartes's physics includes what we think of today as biology and aspects of medicine. See, e.g., G. Hatfield, "*Descartes' Metaphysical Physics* by Daniel Garber; *Kant and the Exact Sciences* by Michael Friedman," *Synthese* 106 (1996), 113–38.

54 The choices made by Descartes's followers erase some of the complexity in Descartes's position and often introduce distortions. This is not to suggest that a "pure Descartes" can be discovered; we should not seek such a thing as our goal. As Geneviève Rodis-Lewis puts it, we "cannot of course go back in time," but we should try to avoid "the distortions of the Cartesian system which arise when one examines it from the perspective of the systems that succeeded it—systems which diverged very considerably from the source of their initial inspiration": "Descartes and the Unity of the Human Being," in J. Cottingham (ed.), *Descartes* (Oxford: Oxford University Press, 1998), 197–210, at 198.

55 Des Chene claims the "ambiguous legacy of Descartes is reflected in the attitudes of his successors" ("Life and Health," 733). However true this may be, it should not reflect on Descartes, who saw no tension between his three therapeutic strategies. He believed they complemented one another, as the correspondence with Elisabeth suggests. The idea that the three strategies point in different directions, and so contribute to an ambiguous legacy, is (I believe) itself a by-product of the interpretative choices of his successors, along with political and institutional constraints unknown to Descartes.

medicine and his therapeutic strategies, not all of which survived together among seventeenth-century Cartesians.

Recall that Descartes allows a "true error of nature" only in the human case, as when we are thirsty at a time that "drink is going to cause us harm." This links the concept of bodily health to what is intrinsically good for the union of mind and body, but Descartes's definition of health remains mostly "programmatic," linking what is good for the body to what is good for the union but without specifying what is good for the union beside bare survival and the body's minimal interference with the mind.[56] To be sure, Descartes does offer specific medical advice but his advice easily fits with established medical traditions. The exception is the etiological justification for medical intervention deriving from Descartes's mechanical view of the body and the scientific medicine his physiology is meant to support. So, while his biomechanical therapy is distinctive, its originality lies in the supporting theory and not the theoretically informed practice itself. The same combination of tradition and innovation, and the same gaps between innovative physiology and innovative practice, appear among Descartes's followers when they express views about the concept of health. In some cases, they emphasize one of Descartes's therapeutic strategies as opposed to the others, and, in at least one case, they separate medicine from Descartes' philosophy, emphasizing that medicine is an art, not a science, and so only a naturopathic therapy will be viable.

To appreciate the medical landscape of the early Cartesians, take, as a first example, advocates of Descartes's medical science and his biomechanical therapeutics, the pair of Henricus Regius and Jacques

56 Voss, "Descartes: Heart and Soul," 188–89. Efforts to fill in some of the details in Descartes's position beyond what I say here include L. Shapiro, "Descartes on Human Nature and the Human Good," in C. Fraenkel, D. Perinetti, and J. E. H. Smith (eds), *The Rationalists: Between Tradition and Innovation* (Dordrecht: Springer, 2011), 13–26; and Smith, "Early Modern Medical Eudaimonism."

Rohault.[57] Rohault, the younger of the two, includes an entire chapter titled "Of Sickness and Health" in his Cartesian textbook, the *Treatise on Physics* (1671).[58] In this work, Rohault defined health as "a particular Disposition of the Body whereby it is enabled readily to perform all the Duties belonging to it." What exactly these "Duties" are remains unspecified. But interestingly, this definition is all about the body, not the mind or the union of mind and body, and the very same definition was offered thirty years earlier by Descartes's most prominent medical follower, Regius. In Regius's *Physiology or the Knowledge of Health* (1641), a collection of disputations written with Descartes's help, health is defined as the "disposition of the parts of the human body such that it is able to perform its proper actions."[59] In Regius's later *Fundamentals of Physics* (1646) the earlier definition extends beyond human beings to animals in general: "health is the disposition of the parts of the animal body such that it is able to perform its proper actions."[60] His emphasis on dispositions continued the following year in the *Fundamentals of Medicine* (1647) where he explains that the "true form of illness . . . is a depraved disposition of the parts of the body."[61] Although it is hardly unexpected that Descartes's physiological views led some of his followers to characterize health in terms of

57 There are similarities between the two beyond their treatment of health. For example, Rohault choose to present Descartes's physics without his metaphysics, a perspective consistent with Regius's earlier efforts; and Rohault advocated an explicit melding of Descartes's physics with experimentation, which was a hallmark of Regius's earlier work as well. To my knowledge these similarities are not well explored.

58 All translations of the *Traité de physique* come from J. Clarke (trans.), *Rohault's System of Natural Philosophy, Illustrated with Dr S. Clarke's Notes, Taken Mostly out of Sir Isaac Newton's Philosophy. With Additions*, 2 vols. (London: Printed for James, John and Paul Knapton, 1723), pt. 4, chap. 25.

59 *Sanitas est dispositio partium humani corporis actionibus recte perficiendis apta*: H. Regius, *Physiologia, sive Cognitio sanitatis* (Vltraiecti: Ex Officinâ AEgidii Roman, Academiae Typographi, 1641), I.I.I; reproduced in E.-J. Bos, "The Correspondence between Descartes and Henricus Regius" (Ph.D. diss., Utrecht University, Utrecht, 2002), 199.

60 *Sanitas est dispositio partium corporis animalis, actionibus rectè perficiendis apta*: H. Regius, *Fundamenta physices* (Amstelodami: Ludovicum Elzevirium, 1646), 154).

61 *Cum enim morbus sit partium dispositio, nullam materiam habet. Forma vero morbis, quae, est dispositionis partium pravitas, est ipse morbus*: H. Regius, *Fundamenta medica*

the body's actions and dispositions—Regius lists among the body's actions: motion, sensation, sleep, appetites, pulse, and respiration—neither Regius nor Rohault follow Descartes in relating the body's "Duties" or "proper actions" to what is good for the union, and so by reference to something over and above the body itself.

Still, Regius and Rohault do elaborate on Descartes's medical science and his biomechanical therapy. As Rohault explains, "Two Things generally go to this [healthy] Disposition; namely, a fit Construction of the Parts, and a just Temperature of them." This is entirely consistent with viewing the body as a machine as Descartes's physiology surely does. The emphasis on the temperature of the parts is also noteworthy, though entirely traditional in associating the question of temperature with health. But here too it seems likely Regius served as Rohault's model, and not Descartes himself. In Regius's 1641 disputations, he also separates two elements of health: "the right temperature (*bona temperies*)" and "proper conformation of the parts (*apta partium conformatio*)."[62] He goes on to explain, "the right temperature" is not a question of the body's primitive qualities but the "position, figure, quantity and motion or rest of insensible particles, causing harmony of actions in the sensible parts."[63]

These physiological views are extended by Rohault to account for the diseases affecting the parts and the temperature of the body. "All Distempers are generally owning to the ill Regulation of our Lives, either from too much or too little Sleep, too much or too little Exercise, &c. Sometimes they are caused by Things without, and very often by

(Ultrajecti: Theodorum Ackersdycium, 1647), 27. For discussion of this definition of illness, see T. P. Gariepy, *Mechanism without Metaphysics: Henricus Regius and the Establishment of Cartesian Medicine* (PhD. diss., Yale University, 1990), 216ff.

62 *Sanitatis partes duae sunt, bona temperies et apta partium conformatio: Physiologia*, I.I.14; reproduced in Bos, *The Correspondence*, 201. Regius would use the same language in the *Fundamentals of Physics*, 159, the *Fundamentals of medicine*, 3, and again in the subsequent expanded edition of the *Fundamentals of physics:* H. Regius, *Philosophia naturalis* (Amstelodami: Ludovicum Elzevirium, 1654), 236.

63 *Idcirco bona temperies à nobis definitur: situs, figura, quantitas, et motus vel quies particularum insensibilium partes sensibiles constituentium, actionibus perficiendis conveniens: Physiologia*, I.I.14; reproduced in Bos, *The Correspondence*, 202.

Abuse of Food; that is, by our intemperance in eating and drinking; which is so much the more injurious to us, because it affects us inwardly."[64] Remarkably, just as Rohault embraces the biomechanical implications of Descartes's physics for medical theory, he offers very traditional advice tied to the regulation of the non-naturals. This is yet another instance of the theme I mentioned earlier in connection with Descartes himself, where we found a melding of philosophical innovation with traditional forms of therapy.

Among Descartes's followers, this melding of innovation and tradition sometimes favored one side of the synthesis over the other, as in the case of Johannes De Raey (1622–1702), the third Cartesian we will consider. De Raey had been Regius's student at Utrecht and later became a professor of philosophy at Leiden, where he also lectured on medicine, before becoming a member of the Amsterdam Athenaeum. His evolution from a strong believer in Descartes's scientific medicine and the biomechanics it engendered to a skeptic about the very idea of a scientific medicine, Cartesian or otherwise, is well documented.[65] But it would be wrong to suggest De Raey broke entirely with Descartes. While De Raey certainly rejected biomechanics as the basis for medical practice, it is also the case that he endorsed naturopathic therapy. The shift in De Raey's thinking appears to have occurred after an encounter with the Leiden physician Franciscus de la Boë (Sylvius), who had suggested, "in medicine as in physics what is known truly is known only by experience."[66] De Raey did not approve of the extension of such a claim to physics, which he assigned to philosophy. Instead, De Raey believed medicine had its own "foundation and subject" and this

---

64 Clarke, *Rohault's System of Natural Philosophy*, 187–8.

65 See especially Verbeek, "Tradition and Novelty," 167–96, and the recent discussion in Schmaltz, *Early Modern Cartesianisms*, 264 ff.

66 ... *tam in Physica quam in Medicina id omne quod vere scitur, sola experientia sciri*: J. De Raey, *Cogitata de interpretatione* (Amstelædami: Henricum Wetstenium, 1692), 659; cited and translated in Schmalz, *Early Modern Cartesianisms*, 266. This claim appeared in an October 1680 letter De Raey wrote to another Cartesian, Christopher Wittichius (mentioned below), later reprinted in the *Cogitata*.

insulated philosophy from any overlap with medical epistemology.[67] While it is certainly not a view we might expect to find in a Cartesian, according to De Raey it was essential that "one should not philosophize outside of philosophy" and accordingly "medicine . . . [has] never been or can ever be a part of philosophy."[68]

De Raey's shift reorients him away from Descartes's biomechanics and the dispositional accounts of health he would have learned from Regius and toward un-theorized tenets of medical practice, especially faith in the body's ability to recover and the individuals' capacity to heal themselves. No other possibility presents itself if, following De Raey, we must accept that philosophy does not affect medicine. As Tad Schmaltz explains, De Raey's "conception of medicine as an art can be understood in terms of the notion in Descartes of a *médicin de soi-méme*."[69] In other words, De Raey excises all but Descartes's naturopathic medicine, embracing the implication that each of us has a privileged knowledge of our bodies and that we need not look beyond ourselves to philosophy for effective medicine. We each know when we are healthy, in other words, and we know what to do about it when we are not.

Along with representatives of Descartes's biomechanical and naturopathic medicine, there were other Cartesians in the late seventeenth century who emphasized the virtues of psychosomatic medicine. If we move east from Paris, where Rohault was active, passing through the Netherlands, which was the home of Regius and De Raey, and into early modern Germany, we discover a diverse Cartesianism with

---

67 *Ita longius progressus, imprimis perspicue intellexi, atque non uno loco & temper Claris verbis dixi & scripsi olim, Medicinam, Jurisprudentiam, Theologiam, in communi omnium hominum intellectu habere fundamentum & subjectum suum, verum hoc partier dici non debere de Philosophia, cuius pars Physica est, quam à Medicina distinguimus*: De Raey, *Cogitata*, 660; cited and translated in Schmatlz, *Early Modern Cartesianisms*, 265.

68 *Praecipuus Philosophiae fructus est ac debet esse quod nos doceat, haec duo potissumum: Unum, Extra phiosophiam philosophandum non esse; Alterum, quod hinc sequitur, Medicinam artesque mechanisas, neque suisse hactenus, neque unquam posse esse, philosophiae huc usque promotae partem*: De Raey, *Cogitata*, 654; cited and translated in Verbeek, "Tradition and Novelty," 194.

69 Schmaltz, *Early Modern Cartesianisms*, 265.

both biomechanical and psychosomatic medicine clearly represented. Cartesianism was taught at Jena, Marburg, Leipzig, Wittenberg, and Frankfurt on the Oder, though even among these universities it was the gymnasium at Duisberg that stands out. Founded in 1654, the Cartesians Johann Clauberg and Christopher Wittichius were active at Duisberg from the start. Its medical faculty would later include Theodor Craanen, Tobias Andreae, and Friedrich Gottfried Barbeck, making it one of the only medical institutions in Europe where Cartesianism was continuously and openly taught for five decades. Yet in spite of this continuity, Duisberg's Cartesians did not always agree with one another and were often willing to point out the limitations of the other therapeutic strategies they did not prefer. For example, Clauberg, who had been a student of De Raey's, was a proponent of Descartes's medicine of the mind. For Clauberg, this medicine was associated with logic and curing prejudices through, first and foremost, an appreciation of the distinction between mind and body.[70] Clauberg was also committed to Descartes's vision of a mechanical physiology and the need to perform experimental and observational work in order to discover medicines for the body.[71] Barbeck, like Clauberg, was a believer in biomechanics. And he would explicitly discuss the physician's need to understand the body's actions and the internal (mechanical) sources of its motion.[72]

---

70 *Et praecipua causa infinitarum confusionum & densissimarum tenebrarum in Physica & Medicina fuit hactenus, quod neque solius mentis proprietates sola mente seorsim considerant, sed quia perpetuo adhaerent sensibus & mentem cum corpore confundunt*: J. Clauberg, *Notae in Cartesii Principia,* in J. T. Schalbruch (ed.), *J. Clauberg: Opera Omnia Philosophica*, 2 vols. (Amstelodami: Ianssonio-Waesbergii, Boom, à Someren, and Goethals, 1691), 1:lxvi; cited in Trevisani, *Descartes in Deutschland*, 94. For discussion of Clauberg's logic in English, see A. Strazzoni, "A Logic to End Controversies: The Genesis of Clauberg's *Logica Vetus et Nova*," *Journal of Early Modern Studies* 2 (2013), 123–49.

71 *Medicina quodcunque boni habuit non ex illis, quae in Scholis Physicis frequentabantur . . . ; sed potius ab experientia & observatione*: Clauberg, *Disputationes*, I, 11; cited in Trevisani, *Descartes in Deutschland*, 83.

72 *[F]acilius construi potest definitio generalis vitis corporeis omnibus conveniens, nimirum quod sit actio proveniens ex certa corporis organisatione seu structura & transfluxu alicujus fluidi ex interna & propria causa moti proveniens*: Barbeck, *Disputatio medica de vita*, sec. 7; cited in Trevisani,

Different from Clauberg and Barbeck, Andreae saw the special value of psychosomatic medicine. Not to be confused with his cousin of the same name, who had been a teacher of Clauberg's, Andreae's student years at Duisberg included a medical dissertation touching on the causes of epilepsy. In this work, he showed awareness of Descartes's prioritization of the union between mind and body. He seemed especially taken by a passage from Meditation Six where Descartes explains that "the part of the brain that immediately affects the mind produces just one corresponding sensation" and that this correspondence, owing to the union, is part of the "best system that could be devised" because it produces "the one sensation which, of all possible sensations, is most especially and most frequently conducive to the preservation of the healthy man" (*Writings*, 2:60). For Andreae, this meant that when we are healthy similar thoughts are associated with similar bodily motions, and similar bodily motions correspond to similar thoughts.[73] In the case of epilepsy, this has the unfortunate implication that fear of an epileptic seizure can cause an attack. The solution, which is not foolproof, is to seek to control our passions by, among other things, focusing on intellectual pleasures.[74] This advice could have come straight from Descartes's letters to Elisabeth.

Indeed, part of the message to extract from this account of Descartes's reception is that it was his medical advice that his followers pursued, even if they were willing to follow one strategy more than others. Germany would continue to serve as the home for some of the most interesting responses to Descartes's medicine

---

Descartes in Deutschland, 146. For discussion of Barbeck in English, see Des Chene, "Health and Life," 733–34 and Smith, "Heat, Action, Perception," 118–19.

73 *Ita enim Natura nos formavit, ut similibus cogitationibus similes motus corporei, similibusque spirituum motibus similes cogitationes respondeant*: T. Andreae, *Disputatio philosophica inauguralis explicans naturam & phaenomena cometarum* (Duisburg, 1659), sec. 20; cited in Trevisani, *Descartes in Deutschland*, 102.

74 See Trevisani, *Descartes in Deutschland*, 103, on whose analysis I rely, as well as the more general discussion in Des Chene, "Health and Life," 733; and Smith, "Heat, Action, Perception," 116.

and his preferred medical advice, where the ever more explicit sides of Descartes's medicine gained lives of their own both inside and outside the universities. Eventually, the links to Descartes began to fray, though never entirely in the case of biomechanics and its materialist implications. Due to the traditional character of so much of Descartes's medical advice, however, it would be hard to claim that the psychosomatic and naturopathic medicine he championed was as distinctively his own. We might speculate that it is for this reason that Descartes is so often vilified for the excesses of a scientific and impersonal medicine but rarely credited with the other therapeutic strategies he endorsed and led others to support. Be that as it may, it must also be remembered that the Cartesian story continued through other equally complex and innovative figures whose medical interests have only recently begun to receive the attention they deserve. Leibniz, for example, who would offer his own renovation of tradition, and Enrenfried Walther von Tschirnhaus, the only early modern figure to have written, so far as I am aware, a work with the title *Medicine of the Mind and Body* (1686),[75] were both students of German Cartesianism and deserve a place on the map this chapter has attempted to reveal.[76]

75 For Leibniz, see J. E. H. Smith, *Divine Machines: Leibniz and the Life Sciences* (Princeton, NJ: Princeton University Press, 2011); and especially Smith, "Heat, Action, Perception." Tschirnhaus remains a figure relatively unknown outside of his correspondence with Spinoza. A valuable exception in English is J. Adler, "The Education of Ehrenfried Walther von Tschirnhaus (1651–1708)," *Journal of Medical Biography* 23 (2015), 27–35.

76 I am deeply indebted to Cynthia Klestinec and especially Peter Adamson who read and commented on many drafts of this paper with great generosity, wisdom, and patience.

CHAPTER SEVEN

Health in the Eighteenth Century

Tom Broman

In many respects, to speak of "health" is to speak of an absence, to place a label on something for which there is no clear concept. Take, for example, the essay "Concepts of Health and Disease" in the recently published volume on philosophy of medicine in the *Handbook of the Philosophy of Science*. The essay begins, unsurprisingly, by linking "health" with "normality," and then using normality as a standard for determining whether a given physical or behavioral condition warrants designation as an illness or disorder that requires therapeutic intervention.[1] No doubt there is good reason to approach health from this point of view. After all, if there is no condition that can be determined, either subjectively or instrumentally, as lying beyond the boundaries of normal function—as a pain, for example, or an unusual

1 C. Boorse, "Concepts of Health and Disease," in F. Gifford (ed.), *Philosophy of Medicine* (Amsterdam: Elsevier, 2010), 1–52. Boorse has long been a champion of the idea that normality is best understood as a statistical concept, although the essay considers a broad range of concepts of health, some of which, at least, recognize potential drawbacks to treating it from a negative point of view. See further Elselijn Kingma's chapter in the present volume.

reading for some metabolic product—then it makes sense to call the person "healthy."

The problem of how to describe health other than as an absence of pathological indicators does not just arise with statistically inflected measures of normality. To the contrary, our uncertain grasp on the concept of health is the product of three entrenched ways of conceiving the human body, all of which pertain to our understanding of health, but none of which address it directly. The first tradition can be called the "anatomical-morphological" or "physiological" tradition and has its origins in writings such as Aristotle's *Parts of Animals* and Galen's *Usefulness of the Parts*, as well as the heaps of works that have followed their template. This is a view of the body that assesses the functions of living beings, with an eye toward cataloguing and explaining those functions. The gradual transfer of attention from gross anatomy and organs to cells to molecular biology has not fundamentally altered the basic explanatory thrust of this tradition. However, to describe the body's functions from this point of view is not the same as describing what it means to be healthy.

A second originally independent "pathological" tradition dealt directly with disease as a phenomenon to be interpreted semiotically and countered therapeutically. One can observe this orientation in many Hippocratic writings, such as the *Epidemics* and *Regimen in Acute Diseases*. Conspicuously, this view of the body did not depend directly on comprehension of physiological function, although it would be incorrect to say that ancient writers overlooked any possible relationship between function and illness as a deviation from normal function. The Hippocratic *Places in Man* clearly suggests otherwise, although the way it links a discussion of anatomy to an account of various fluxes in the body is not very consistent. In short, while various ancient writers did link pathology to physiology, such efforts did not express a belief in a necessary connection between the two.

Over time, of course, a more intimate association between physiology and pathology did develop, as Georges Canguilhem described in his famous 1943 doctoral thesis, *Essay on the Normal and the Pathological*. For Canguilhem, the redefinition of pathological states as quantifiable deviations of "more or less" from physiological norms, which he attributed to the work of Auguste Comte and, above all, the nineteenth-century physiologist Claude Bernard, was not an entirely happy development, because it downplayed the subjective experience of illness. If physiological knowledge can be extracted from the pathological states of one organ or another, he insisted, that can only be because of the prior clinical reality of someone having suffered a sickness.[2] Thus illness as a condition that befalls individuals necessarily precedes any possibility of analyzing the relationship between physiology and pathology.

Even for Canguilhem, an insistence on illness as a distinct state of being did not translate into a similar emphasis on what it means to be healthy, although he did not overlook it entirely. In Canguilhem's language, life could be described as a "normative" process, a dynamic interplay between organism and environment by means of which the norms that define healthy function are constantly defined and redefined.[3] Put another way, when healthy, living things exhibit what could be called an adaptive responsiveness whereby they adjust their body's functions to their variable surroundings. A hibernating bear displays far lower levels of metabolic activity than an active one, but the bear plainly is not ill. Instead it exhibits an altered state of "normativity" and thus is healthy by Canguilhem's definition.

The third tradition of writing to emerge from the ancient world, dietetics, or hygiene, is probably the one that brings us closest to a clearly articulated concept of health. In its rich evocation of what came later to be called the "non-naturals," dietetic advice on food, drink, exercise, and living environment presented a glimpse into what Greek writers,

2 C. R. Fawcett and R. S. Cohen (trans.), *Georges Canguilhem: The Normal and the Pathological* (New York: Zone Books, 1991), 88.

3 Fawcett and Cohen, *Canguilhem: The Normal and the Pathological*, 126–27.

ranging from the Pythagoreans to Galen, considered the good life. Yet in this respect too, "health" was characterized neutrally as a balance between humors that was appropriate for a given person, without further specification of what is meant by the term.[4]

If we now turn to some eighteenth-century writings, we can see how the meaning of "health" easily slips between the three different senses of the body just presented. In Herman Boerhaave's *Institutiones medicae*, the most widely used medical textbook of the era, the body's overall physiology is introduced by means of mechanical metaphors, as in the following well-known passage:

> The solid parts are either membranous pipes, or vessels including the fluids; or else instruments made up of these, and more solid fibers, so formed and connected, that each of them is capable of performing a particular action by the structure, whenever they shall be put into motion we find some of them resemble pillars, props, cross-beams, fences, coverings, some like axes, wedges, levers, and pulleys; others like cords, presses, or bellows; and others again like sieves, strainers, pipes, conduits, and receivers; and the faculty of performing various motions by these instruments, is called their function; which are all performed by mechanical laws, and by them only are intelligible.[5]

Similarly, the body's fluids perform their functions "agreeable to the laws or principles of Hygrostatics, Hygraulics, and Mechanics."[6]

---

4 L. Edelstein, "The Dietetics of Antiquity," in O. Temkin and C. L. Temkin (eds), *Ancient Medicine: Selected Papers of Ludwig Edelstein* (Baltimore, MD: Johns Hopkins University Press, 1967), 303–16; F. Steger, "Antike Diätetik—Lebensweise und Medizin," *NTM Zeitschrift für Geschichte der Wissenschaften, Technik und Medizin* 12 (2004), 146–60.

5 H. Boerhaave, *Institutiones medicae* (Leiden: Severinus, 1730), sec. 40, 12–13. The translation is taken from *Dr. Boerhaave's Academical Lectures on the Theory of Physick* 6 vols. (London: W. Innys, 1742–46), 1:81.

6 Boerhaave, *Institutiones medicae,* sec. 41, 85. In choosing the terms "hygrostatics" and "hygraulics" in preference to the more familiar "hydrostatics" and "hydraulics," Boerhaave

As useful as they might have been for understanding, or at least analogizing, the body's physiology, Boerhaave's mechanical metaphors were not well suited for accounting for the specific qualities of living beings, such as "life" itself. In a section describing the causes of life and health that are present in a body—and let us note that Boerhaave separated the two conditions—he remarked with reference to knowledge of causes that "so long as the heart continues its motion, so long does life remain; but whenever the organ ceases to move, life itself also ceases to be; the motion of the heart is therefore the cause of life."[7] Meanwhile, Boerhaave described life as "the condition of the several solid and fluid parts of the body, which is absolutely necessary to maintain that mutual commerce between that and the mind to a certain degree, so as to be not perfectly removed beyond the power of being restored again." Recognizing perhaps that this was not the most pellucid definition imaginable, he added that "life cannot be defined well till its physiology, or [the] nature and principles of action have been first considered; for it is the sum and aggregate of all the actions performed in the human body."[8]

The meaning of "health" in the *Institutiones* was similarly complicated. The text opens by presenting a basic definition of health as the performance of "the several actions of the human body with ease, pleasure, and a certain constancy."[9] We should take note of the subjective criterion in this statement. *Being* healthy is *feeling* healthy—at least, in part. From that point, the *Institutiones* made repeated and numerous references to "health," but offered no further specification of its meaning until Boerhaave turned to the section on "Pathologia."

---

expressed his intention to speak not just about water but also about fluids more generally. See sec. 41, notes 9 and 10, 87–88.

7 Boerhaave, *Institutiones medicae*, sec. 43, note 2, 92.

8 Boerhaave, *Institutiones medicae*, sec. 42, note 1, 90–91.

9 Boerhaave, *Institutiones medicae*, sec. 1, 2. The subjective component of health is reinforced in sec. 33, note 1, 77, where the "sum or aggregate of all the actions resulting from the structure of the several parts" is designated "life" and the ease and comfort of those actions is called "health."

There he presented a summary of the main points previously set out in the lengthy section on physiology, reiterating a threefold division of living functions into the vital, the natural, and the animal. This categorical division of functions was conventional, indeed ancient, in its roots. Vital functions refer to those that are essential to the maintenance of life, such as breathing and the heartbeat, along with the "secretory action of the cerebellum," while natural functions pertain to processes such as nutrition, growth, and reproduction. Finally, animal functions are designated as "those by which the human understanding conceives Ideas agreeable to the corporeal action with which they are united."[10] At this point, Boerhaave reintroduced the idea of health, using much the same language as before, but this time without any suggestion of feelings of ease or comfort. As described in the section on pathology, health was "that faculty of the body, in which all parts are duly enabled to perform their respective offices with perfection."[11]

Considering that the *Institutiones* and other textbooks were written for aspiring physicians whose work would largely consist of the diagnosis and treatment of illness, perhaps it is only to be expected they devoted little attention to exploring heath, a condition that leaves them with little to do, apart from offering hygienic advice. Even if we grant this objection, however, other more or less contemporary definitions of health displayed similar points of view. The *Cyclopedia* published by Edward Chambers in 1728 defined health in a brief entry in two ways, first in terms of the body's constituent parts as "a due temperament, or constitution, of the several parts whereof an animal is composed, both in respect of quantity and quality." Second, it addressed health functionally as "that state of the body, wherein it is fitted to discharge the natural functions perfectly, easily, and durably."[12] Another

---

10 Boerhaave, *Institutiones medicae*, sec. 695, 255–56.

11 Boerhaave, *Institutiones medicae*, sec. 695, 256.

12 E. Chambers, *Cyclopædia: or, an Universal Dictionary of Arts and Sciences*, 2 vols. (London: W. Strahan et al., 1728); the entries "Health" and "Healthiness" are in volume 1.

encyclopedia from the early eighteenth century, Johann Heinrich Zedler's *Grosses vollständiges Universal-Lexicon aller Wissenschaften und Künste*, which began appearing in 1732 and eventually reached sixty-four volumes plus four supplements, offered a more expansive and moralistic view. In the tenth volume (1735), "Gesundheit" was presented with two meanings. The first one offered the conventional definition as "that condition of the human body, in which all its uninjured parts can carry out their natural functions without hindrance."[13] But then the *Universal-Lexikon* added a second, less typical definition: "One attributes health to human understanding (*Verstand*) if it finds itself able to distinguish the true from the false, and moves the will to dispose its actions accordingly." The reason for considering health as pertaining both to mind and body immediately became clear as the article added that "natural health consists of a fitting balance between the mutually interactive forces of the soul and the body."[14] The article then described health as a gift bestowed upon humans by God and condemned as sinful the failure to preserve it through attentive regulation of the non-naturals.[15]

One last example from the encyclopedic literature can be cited here, the article on "Santé" in the *Encyclopédie* of Denis Diderot, authored by Arnulphe d'Aumont, a graduate of the medical faculty

---

13 "Einmahl ist es ein solcher Zustand des menschlichen Leibes, in welchem derselbe an allen seinen Theilen unverletzt seine natürlichen Verrichtungen ungehindert ausüben kann." The syntactical positioning of "unverletzt" makes its precise meaning unclear, and the translation offered here is merely one possible rendering. J. H. Zedler, *Grosses vollständiges Universal-Lexicon der Wissenschafften und Künste*, 64 vols. (Halle: J. H. Zedler, 1732–50), 10: col. 1334.

14 Zedler, *Grosses vollständiges Universal-lexicon der Wissenschafften und Künste.*

15 Zedler, *Grosses vollständiges Universal-Lexicon der Wissenschafften und Künste*, 10: cols. 1335–36. Articles in Zedler's *Universal-Lexikon* were not attributed to individual authors, and citations in the article offer no clear clue as to the author's intellectual allegiances. Only two recent authors are referred to: Friedrich Hoffmann and Michael Alberti, both professors of medicine at the University of Halle. But since Hoffmann is usually thought of as a mechanist, while Alberti was one of the most prominent German followers of Georg Ernst Stahl's animist physiology, their pairing in this article is not very indicative, apart perhaps from suggesting the author's connection with Halle.

at the University of Montpellier.[16] D'Aumont opened by declaring that health was "the most perfect state of life," and then specified health as a "natural accord, the reasonable arrangement of the living body's parts from which it follows that the exercise of all its functions happens in a lasting manner, with all the ease and freedom and in all of the extension of which each organ is susceptible, according to its purpose, and relative to the actual situation, to the different needs of age, sex, the temperament of the individual person, and the climate in which the person lives."[17] The "ease and freedom" in this description of healthy organs refers not only to their physiological function but also to the feelings of well-being enjoyed by the healthy person: "Thus it is by the ease with which one feels the exercise of the body's and soul's functions; by the satisfaction that one has for his physical and moral existence; by the fitness [*convenance*] and constancy of this exercise; by the testimony which one renders of this feeling, and the connection of these effects that one knows that one is living as healthy and perfect a life as possible."[18]

The individuality of health received repeated emphasis in the article. "Health does not consist of a precise point of perfection common to everyone," d'Aumont declared at one point, and then he delivered an exhortation on the centrality of regulating the non-naturals as the best means to maintain health. Just as with health, the proper balance to be maintained with the non-naturals could only be comprehended

16 The article was unsigned and its attribution is somewhat contested. According to Victoria Meyer, the author was Louis de Jaucourt, a student of Boerhaave's who authored a considerable number of articles for the *Encyclopédie*, including several on medical topics. See "Health," available in the online *The Encyclopedia of Diderot & d'Alembert Collaborative Translation Project*. However, W. Coleman attributed the article to d'Aumont in his "Health and Hygiene in the *Encyclopédie*: A Medical Doctrine for the Bourgeoisie," *Journal of the History of Medicine and Allied Sciences* 29 (1974), 399–421, and this attribution has been supported by Elizabeth Williams in personal communication with the author. On d'Aumont's role in the *Encyclopédie*, see E. A. Williams, *A Cultural History of Medical Vitalism in Enlightenment Montpellier* (Burlington, VT: Ashgate, 2003), esp. 121–23.

17 Jaucourt, "Health."

18 Jaucourt, "Health."

individually, with all the complexities that differences in age, sex, occupation, temperament, and other factors brought into consideration. The article did not go into specifics; rather it referred the reader to other articles in the *Encyclopédie* on "Régime" and "Non-naturelles, choses," both of which were signed articles by d'Aumont, along with the entry on "Hygiene," which was unsigned, but which he may also have written.[19]

So much for direct definitions of health. Other examples could be added to these, but the range of meanings would not be broadened appreciably. So instead of pursuing this path further, it will be more fruitful to approach the idea of health indirectly by examining other important concepts that were closely associated with it during the 1700s. In the rest of this chapter I will focus on two such concepts: sensibility and public health. As we shall see, each of these introduced significant novelties to the understanding of health during the period.

Sensibility and Health

To talk about "sensibility" in the eighteenth century as if it were a single concept would restrict the range of its applicability far too narrowly. Instead, it would be far more accurate to describe it as a set of associated concepts, or, as Henry Martyn Lloyd suggests, as a discursive formation—arguably the most characteristic such formation of the period.[20] At the most general level, sensibility denotes the ability

---

19 As claimed by Coleman, "Health and Hygiene in the *Encyclopédie*," 402.

20 If ever something deserved to be labeled a discursive formation, according to Foucault's definition in *The Archaeology of Knowledge*, then sensibility surely qualifies. See H. M. Lloyd, "The Discourse of Sensibility: The Knowing Body in the Enlightenment," in H. M. Lloyd (ed.), *The Discourse of Sensibility: The Knowing Body in the Enlightenment* (New York: Springer, 2013), 1–23, at 2–3. There are an enormous number of studies of sensibility, many of which focus more or less exclusively on its use in literature. For two excellent studies that range widely over literature, science, and moral philosophy, see G. J. Barker-Benfield, *The Culture of Sensibility: Sex and Society in Eighteenth-Century Britain* (Chicago: University of Chicago Press, 1992); and J. Mullan, *Sentiment and Sociability: The Language of Feeling in the Eighteenth Century* (Oxford: Clarendon Press, 1988). For a similarly wide-ranging and well-crafted study that does the same for France but

of an individual to perceive and respond to stimuli, and the study of these responses became the preoccupation of novelists and dramatists, as well as moral philosophers and physicians. Its use in medical writing was usually accompanied by a more or less explicit renunciation of mechanistic models of the body, such as Boerhaave's. Sensibility, by contrast, was offered as a uniquely vital function that could not be redeployed in mechanistic language.[21]

Sensibility possessed three distinct yet closely related meanings. First, it referred in an epistemological frame to the sources of human knowledge in sensory experience. As is well known, eighteenth-century philosophy was dominated by empiricist theories of knowledge in the writings of John Locke, David Hume, and the Abbé Condillac, and for them sensibility denoted the mind's receptiveness to external stimuli.[22] The empirical sources of knowledge also connected sensibility to its second meaning—the basis for our judgments of right and wrong. In contrast to systems of ethics that depended on particular doctrines of revealed religion or on a deduction from a consideration of human nature in the abstract, eighteenth-century moralists such as Hume, Adam Smith, Denis Diderot, and Jean-Jacques Rousseau claimed that moral sensibility is a product of our day-to-day intercourse with other people. This experience, coupled with our ability to

---

with greater emphasis on medicine, see A. Vila, *Enlightenment and Pathology: Sensibility in the Literature and Medicine of Eighteenth-Century France* (Baltimore, MD: Johns Hopkins University Press, 1998).

21 See P. H. Reill, *Vitalizing Nature in the Enlightenment* (Berkeley: University of California Press, 2005). More narrowly focused on Scottish developments but more wide-ranging in its consideration of the intellectual currents that informed sensibility is C. Packham, *Eighteenth-Century Vitalism: Bodies, Culture, Politics* (Houndsmills: Palgrave Macmillan, 2012); on vitalism at the University of Montpellier, which was one of its leading centers, see Williams, *Medical Vitalism*. Although from its title one might suppose that sensibility plays a major role in Stephen Gaukroger's recent synthetic historical account, S. Gaukroger, *The Collapse of Mechanism and the Rise of Sensibility* (Oxford: Clarendon Press, 2010), it is subordinate to the author's stated aim of charting the rise of natural science as a model for all cognitive claims in the modern world. To this end sensibility is mobilized as evidence of the spreading influence of natural science.

22 Lloyd, "The Discourse of Sensibility," 7–8.

share the feelings of others via sympathy educates the sensibility, and by means of this this process we formulate our ideas of morality.

For these writers, to claim that moral judgments should be tempered by social intercourse represented an effort to place ethics on a solid scientific basis. This theme was sounded by Hume in his *Treatise of Human Nature* (1739), where he rejected as "metaphysical reasonings" any philosophy that attempted to derive human nature deductively from first principles. Instead, Hume asserted that "careful and exact experiments" would furnish the foundation of his examination of human nature.[23]

An explicit experimental basis (in our contemporary meaning) for sensibility emerged from the research of the Göttingen medical professor Albrecht von Haller, and this provides us with the third meaning of sensibility. The aim of Haller's experiments was to determine which vital forces are present in different parts of the body. He identified two such forces, the first of which was irritability, associated anatomically with muscle tissue. In his experiments, Haller could observe irritability directly, by poking a tissue with a needle or applying alcohol or a caustic chemical and then watching it contract. However sensibility, the other vital force identified by Haller, proved a more difficult phenomenon to observe. "I call that a sensible part of the human body," Haller wrote in his *Dissertation on the Sensible and Irritable Parts of Animals* (1752), "which upon being touched transmits the impression of it to the soul; and in brutes, in whom the existence of a soul is not so clear, I call those parts sensible, the irritation of which occasions evident signs of pain and disquiet in the animal." Haller localized this effect in nervous tissues while denying that sensibility occurred

23 Hume, *Treatise of Human Nature* (1739), xvii. For his purposes, Hume's use of "experiment" here is nicely ambiguous. In the 1730s, Hume's readers would have understood the word in the context of the experimental approach to natural philosophy promoted by Royal Society luminaries from the seventeenth century such as Robert Boyle and Robert Hooke, and, of course, Isaac Newton. On the other hand, Hume plainly did not perform experiments of that kind in his study of human nature, and thus his use of the term also called upon its older classical meaning as *experimenta*, the fruits of experience. On this point, see Gaukroger, *The Collapse of Mechanism*, 450–51.

anywhere else. But the method of his experiments in the two cases was strikingly different: whereas Haller could observe irritability in contractions that were the immediate result of his experimental intervention, sensibility could only be observed indirectly, requiring that the effects of his interventions come to the animal's (or his own, or his students') consciousness for their expression.

In the context of how sensibility served as a link between the organic and the mental and moral spheres of human life, such an indirect description of its effects would make sense to Haller and his contemporaries. But the lack of equivalence between irritability and sensibility as empirically demonstrable properties of living matter also exposed Haller to criticism regarding their supposed independence from one another. His most committed opponent was the Edinburgh physician Robert Whytt, whose own experimental work had suggested the presence of what he described as a "sentient principle" that resided in the nerves and was distributed throughout the body. It was this principle, Whytt maintained, that perceived external stimuli and then prompted the muscles to respond. Thus what appeared to Haller as an independent force of irritability in muscles was, according to Whytt, actually a force that was secondary and subordinate to sensibility.[24]

These interpretive difficulties notwithstanding, what made Haller's experiments compelling to his contemporaries was that they illustrated how sensibility sits at the intersection of three key domains of human life. We think, we move, we act: sensibility negotiated the transition between these domains and knit them together. An illustrative example of how sensibility functioned in this capacity can be seen in Denis Diderot's *Lettre sur les aveugles* (1749). Diderot claimed that blind people, by virtue of their lack of sensible contact with the outside world, were given to a highly abstract form of reasoning that

---

24 R. K. French, *Robert Whytt, the Soul, and Medicine* (London: Wellcome Institute for the History of Medicine, 1969), esp. 63–76. On the larger ramifications of the controversy over irritability and sensibility, see Reill, *Vitalizing Nature*, 128–31.

made them excel at endeavors such as mathematics. At the same time, however, their sensory deprivation prevented blind people from developing their sensibility to an adequate degree, and this would handicap their attempts to forge social bonds with others. To be sure, a blind person could communicate via language. But lacking the richness of sensory experience that most people have, so Diderot claimed, the blind person would lack those feelings that attend human social interactions, specifically those feelings of sympathy mediated by our sensibility.[25]

Although the cultivation of sensibility was linked to a healthy human life, it also posed a significant risk because excessive levels of sensibility made one susceptible to moral decadence, as Rousseau acerbically and notoriously pointed out in his first and second *Discourses*. Moreover, the dangers arising from excessive sensibility also made individuals susceptible to the vapors and other kinds of nervous exhaustion. Nowhere was this connection made more explicit than in the writings of Samuel Auguste Tissot, the most popular writer on moral and physical hygiene during the entire eighteenth century. Like Rousseau a native of Switzerland, Tissot shared his compatriot's alarm at the decadence and self-indulgence of the wealthier classes, and in his writings he delivered a caustic scolding that, if the numbers of translations and reprints of his writings are any indication, captured the anxieties of his readers, even if it did little to force them to change their lifestyles.

An example of Tissot's attitude toward moral decay in concert with physical illness can be found in the *Essay on the Illnesses of Fashionable People* (1770). Tissot opened the essay with some general considerations

25 D. Diderot, "Lettre sur les aveugles, à l'usage de ceux qui voient," in *Oeuvres de Denis Diderot*, tome I (Paris: J. L. J. Brière, 1821), 283–382. See 286, where Diderot claims that the blind lack a sense for beauty. On the lack of sensibility more generally, see 297–98. Evidently, Diderot discounted the role of the other senses in educating the sensibility. For a discussion of Diderot's *Lettre*, see J. Riskin, *Science in the Age of Sensibility: The Sentimental Empiricists of the French Enlightenment* (Chicago: University of Chicago Press, 2002), 52–59.

of what makes someone healthy. Alongside the standard criteria of regularity and ease of vital functions, he also noted that the healthy person was not likely to be affected by routine and unavoidable alterations in the state of the non-naturals.[26] Against this condition of perfect health Tissot contrasted that of someone with a "delicate" constitution, whose health may be good at any one moment, but whose grasp on health was ever threatened by changing circumstances. Such people, he continued, are hardly ever well, without being able to determine a precise cause of their suffering.

Tissot next turned his attention to the causes of good health, of which he named three. The first was strong fibers, which gives the proper tone to muscles and blood vessels and maintains the regularity of "animal functions." The second cause was good perspiration, which, because it is the most general of the body's evacuations is also the most important. The third cause was "firm nerves" that are not too sensible to impressions and "don't disorder the whole frame for a trifling cause."[27] Such disorders, he added, are common in people who have weak nerves and too great a degree of sensibility. With respect to the question of who enjoys the best health, Tissot declared that rural laborers were in the best position to do so. Not surprisingly, the counterpoint to the healthy routines of the sturdy peasant came in the form of the dissipative pursuits of the idle rich, "who, to defeat the insupportable tediousness of a life disagreeably inactive, attempt to kill time by pleasure."[28]

This stereotypical contrast between the hearty rural laborer and the overly delicate *bourgeois* allowed Tissot to specify the different ways that each group interacted with the non-naturals. With respect to food and drink, for example, the peasant eats "the coarsest

26 Note the correspondence with Canguilhem's discussion of health as adaptive responsiveness, discussed earlier in this chapter.

27 S. A. Tissot, "Essai sur les maladies des gens du monde," in *Oeuvres de Monsieur Tissot*, vol. 4 (Lausanne: François Grasset, 1784), 22–23. Translations mine.

28 Tissot, "Essai sur les maladies des gens du monde," 26.

bread," buttermilk, "vegetables, and those commonly the least savory," and very little butchered meat. By contrast, the wealthy bourgeois consumes "the juiciest meats, the most highly flavored game, the most delicate fishes stewed in the richest wines, and rendered still more inflammatory by the addition of aromatic spices," and so on.[29] Similar contrasts were offered throughout the rest of Tissot's discussion of the non-naturals, making it plain that what he was aiming for was not merely a reform of the hygienic practices of the wealthier and more educated classes, rather a more thorough social critique.

The attack on the over-refinement of sensibility in modern society achieved its apotheosis in *L'onanisme*, Tissot's notorious discourse on masturbation, which unsurprisingly became his most famous work as well—translated, reissued, and pirated for decades after its appearance in 1760. In the subject of masturbation, Tissot seemingly found the perfect vehicle for his analysis. Accordingly, he filled the book with case studies of nervous exhaustion and other physical ailments to which those who practice "self-abuse" are subject, the stories so arranged as to lead the reader on a titillating path toward a crescendo of moral depravity. This summit was represented by the story of a young clockmaker who began indulging himself as a boy and who, by his early twenties, had so exhausted himself that he was unable to control his vital functions and could do nothing other than lie in his own excrement.[30] With the aid of such lurid narratives, Tissot could tap a rich vein of moral outrage, diagnosing an epidemic of self-abuse that other writers fell over each other to exploit. One of the earliest such efforts, the article on "Mansturpation" for the *Encyclopédie* in 1765, was in large measure plagiarized from Tissot; many others would follow.[31]

Although from these examples it might seem that the hygiene and pathology of sensibility was almost exclusively a French obsession,

29 Tissot, "Essai sur les maladies des gens du monde," 28.

30 S. A. Tissot, *L'onanisme*, in *Oeuvres*, vol. 1, 33–6.

31 For discussion of the article from the *Encyclopédie*, see Williams, *Medical Vitalism*, 228.

such was not the case. To be sure, Rousseau's powerful presence in French letters as both social critic and epitome of the *homme sensitive* made French writers especially responsive to its significance. But English-language writers could claim a share of the discussion as well, and many of them published their work decades before Rousseau and Tissot appeared on the scene. We have already mentioned Hume's writing in the context of linking epistemology and ethics via sentiment. For their part, English physicians preceded Tissot in diagnosing an epidemic of diseases stemming from overindulgence and nervous exhaustion. Richard Blackmore, the personal physician to Queen Anne, published his *Treatise on the Spleen and Vapors* in 1726, and it was followed by an even more successful guide to health and ill health, *The English Malady*, by George Cheyne, a physician whose practice was located in Bath, but who was consulted by the learned and the powerful far and wide.

Finally, it deserves mentioning that the language of sensibility was highly gendered. Men certainly possessed sensibility, but women overflowed with it, to such an extent that women's heightened capacity for sensibility made them moral exemplars and beacons of sympathy. Best-sellers such as Samuel Richardon's *Pamela* (1740) played on such notions in telling the story of a servant girl whose steadfast virtue in resisting the advances of her rakish aristocratic master and suitor succeeds in converting him into a good man. But alongside these tributes to feminine virtue and sympathy came a view of feminine bodies as vessels far too weak to contain the torrents of sensitive energy coursing through them. And of course contemporaries, usually males, criticized women's self-indulgent pursuit of leisure for making things worse. Already in 1711, the Earl of Shaftsbury found himself called upon to denounce one such leisure activity, reading novels, for its affect on girls, writing how "tender virgins, losing their natural softness, assume the tragic passion, of which they are highly susceptible."[32]

32 Quoted in Barker-Benfield, *The Culture of Sensibility*, 119.

Shaftesbury made this prescient observation, it should be noted, years before the novel-reading craze had properly begun. By century's end, political reformers, such as Mary Wollstonecraft, could turn such sentimental wallowing into a call for the reform of women's education in her *Vindication of the Rights of Woman* (1792), and shortly afterward Jane Austen satirized the attitude in her novel *Sense and Sensibility* (1811).

Heightened sensibility manifested itself not only in women's weepy attachment to novels and other behaviors but also in their bodies. The age-old women's disease of hysteria was reinscribed as a disorder of nervous excess, while the French physician François Boissier de Sauvages accounted for chlorosis as a distinctly female affliction because of women's heightened emotional sensitivity. When men succumbed to various forms of burning passion, according to Tissot, they began to resemble women physiologically, succumbing to "melancholic anorexia" that manifested in symptoms such as erotomania. The pathological feminization of men through overly excited sensibility, as Elizabeth Williams points out, was a dominant theme of the French medical literature in the last third of the century, but it was certainly not uniquely French.[33]

By the latter decades of the century, the gendering of women's bodies in terms of their heightened sensibility had hardened into a model of physiological uniqueness. Pierre Roussel's *Système physique et moral de la femme* (1775) described women's bodies as entirely configured around the essential task of reproduction. Roussel described the female reproductive organs—womb, Fallopian tubes, and ovaries—as centers of extraordinary sensibility that subjugate the rest of the body, becoming "the dominant center of movement and action." This greater share of sensibility made women unsuitable for physically demanding

---

33 Williams, *Medical Vitalism*, 234–35. On the longer history of chlorosis, see H. King, *The Disease of Virgins: Green Sickness, Chlorosis and the Problems of Puberty* (London: Routledge, 2004).

tasks such as brisk walking or horseback riding, as well as for the intellectually demanding pursuit of scholarship and research.[34]

Thus sensibility was framed in the eighteenth-century as a highly sensitive register of the intersection between health and morality. While providing the indispensable foundation for social life in general and the cultivation of an appropriate ethos of behavior in society, when developed too far sensibility also opened the door to degradations and abuses that threatened both physical health and moral standing. Moreover, when inscribed physiologically onto women's bodies, sensibility served as a flexible device for describing the contrasting visions of women's nature promoted by their male counterparts. On the one hand, their supposed possession of heightened sensibility made women exemplars of virtue, a point exploited repeatedly in literature. On the other hand, the same wellsprings of sensibility placed women in danger of becoming slaves to their passions, thirsting after self-indulgence in a deepening spiral of degradation and ultimately in ill health.

Health as a Public Good

The other important concept that gives us purchase on concepts of health in the eighteenth century was the idea of health as a collective social good, a direct antecedent of our contemporary concept of "public health." In some respects, to identify the origins of public health in the 1700s might appear doubly dubious. The roots of public health arguably go back well before 1700, at least as far back as late medieval efforts to combat the bubonic plague. Already in the fifteenth century, the responses undertaken by municipal and princely governments to repeated onslaughts of the bubonic plague suggest a clear concern for health in a collective sense and an attempt to institute measures to

34 Williams, *Medical Vitalism*, 244–45. Vila, *Enlightenment and Pathology*, 243–55, describes the broader literary and cultural context for Roussel's ideas.

preserve it.[35] Although such initiatives were far from insignificant in the longer history of public health, those medieval and early modern efforts tended to be episodic in nature, relapsing into quiescence when the latest outbreak of plague receded. The most significant applications and developments of what we recognize as public health occurred in the nineteenth century to combat the appalling problems of urbanization and epidemic disease that became so prominent at that time.

Yet the eighteenth century did introduce some novelties to public health that would figure significantly in later developments. One such novelty was based on the concept of political arithmetic—the new science of political arithmetic formulated by John Graunt and William Petty in the late seventeenth century. Graunt, a London tradesman with little formal education and no official standing in England's scientific elite, published his *Natural and Political Observations Made upon the Bills of Mortality* in 1662. The timing of the publication was significant. In the aftermath of the Civil War and the subsequent restoration of the English monarchy in 1660, Graunt sought to use tables of mathematical data about London's population to document the city's stability as a vehicle to promote the prosperity of the country. The source used by Graunt for constructing his tables were the London bills of mortality first published in the fifteenth century as a way of providing an early warning of plague outbreaks. Graunt's enthusiasm for the quantification of population for administrative and economic ends was shared by William Petty, a physician and charter member of the Royal Society when it was founded in 1660. A member of the medical staff in Oliver Cromwell's army in Ireland in 1652, Petty was charged with conducting a survey of the island, the first ever. Thus when Graunt's *Natural and Political Observations* appeared in 1662,

35 See A. G. Carmichael, *Plague and the Poor in Renaissance Florence* (Cambridge: Cambridge University Press, 1986); and J. Henderson, "The Black Death in Florence: Medical and Communal Responses," in S. Bassett (ed.), *Death in Towns: Urban Responses to the Dying and the Dead, 100–1600* (Leicester: Leicester University Press, 1992), 136–50.

Petty held a favorable position to see its potential, and he applied similar techniques to an analysis of Dublin bills of mortality.[36]

While the inspiration for assessing the national population lay in the seventeenth century, the economic benefits of having a growing population became a favorite topic in the eighteenth century; namely, in the political economic writings of David Hume and Adam Smith in the United Kingdom; Anne-Robert-Jacques Turgot (1727–1781) and Pierre Samuel du Pont de Nemours (1739–1817) in France; and Johann Heinrich Gottlob von Justi (1717–1771) and Joseph von Sonnenfels (1733–1817) in Prussia and Austria, respectively. Insofar as these writers prompted rulers and government ministers to look to trade and to the vitality of their domestic economies, and not merely to their treasuries, as a measure of wealth, the health of the productive forces in those economies became a matter of national concern.[37]

The responses to this interest in population and the support of economic vitality took different forms in different places. In Great Britain, it manifested itself in part as attempts to understand the spread of epidemics and the promotion of inoculation against smallpox. Inoculation—the introduction of pus from smallpox pustules into healthy people as a preventive measure against the disease—had long been practiced in areas as diverse as China, the Ottoman Empire, even in Wales, but its acceptance in learned medicine was virtually nonexistent before the early 1700s. Two accounts of the practice by Italian physicians living in Ottoman Constantinople were presented to the Royal Society in 1713 and 1714 but failed to ignite much interest in the technique. A more sustained interest came from the accounts

---

36 On the origins of political arithmetic, see J. C. Riley, *Population Thought in the Age of the Demographic Revolution* (Durham: Carolina Academic Press, 1985); and A. A. Rusnock, *Vital Accounts: Quantifying Health and Population in Eighteenth-Century England and France* (Cambridge: Cambridge University Press, 2002).

37 As is well known, Michel Foucault famously coined the now-ubiquitous term "biopower" to describe this intersection of governance with the disciplining of bodies as a cornerstone of modern public health. On the reception of the concept of biopower, see R. Cooter and C. Stein, "Cracking Biopower," *History of the Human Sciences* 23/2 (2010), 109–28.

of inoculation communicated by Mary Wortley Montagu, the wife of the British ambassador to the Ottoman court in 1717 while resident in Istanbul. Upon her return home, Montagu, who had been a victim of smallpox as a child, had her daughter inoculated in 1721, and the next year she was one among several advocates who succeeded in persuading Caroline, the wife of the future George II, to have their two daughters inoculated against the disease.[38]

The publicity attending the successful treatment of the two young princesses gave inoculation a considerable boost among the British public, but it remained controversial for a number of reasons. In the first place, physicians encountered ethical problems in deliberately giving their patients a disease, a clear violation of the Hippocratic dictum of "first do no harm." Moreover, the purported benefits of inoculation ran counter to the individualistic thrust of humoralist thinking, which emphasized adaptation by individuals to local environments. Even granting that inoculation might have some protective value in foreign lands such as Turkey, skeptics asked what proof could be offered that the same would hold in the United Kingdom, with its different climate acting on temperamentally different British bodies. The marshaling of such environmentalist perspectives in medical thinking, which were clearly derived from *Airs, Water, Places*, offers powerful evidence of their enduring persuasiveness.[39]

Such objections were met by the mathematically inclined physicians John Arbuthnot and James Jurin, who used quantitative measures of the population similar to those pioneered by John Graunt to argue

---

38 The reception of Montagu's news about inoculation and subsequent royal patronage of the technique is described in G. Miller, *The Adoption of Inoculation for Smallpox in England and France* (Philadelphia: University of Pennsylvania Press, 1957), 70–100.

39 On this line of thinking, see most recently J. Golinski, *British Weather and the Climate of Enlightenment* (Chicago: University of Chicago Press, 2010), 140–50. Like Riley, Golinski believes that the eighteenth century experienced a "Hippocratic revival," Golinski, *British Weather*, 140, although it is not evident what supposedly was undergoing revival. Clearly many of the same doctrines about bodies and climate were present in sixteenth- and seventeenth-century writing as well. See A. Wear, "Place, Health, and Disease: The Airs, Waters, Places Tradition in Early Modern England and North America," *Journal of Medieval and Early Modern Studies* 38 (2008), 443–65.

for the overall value of inoculation. Arbuthnot satisfied himself with attempting to measure the overall cost to the population of mortality resulting from smallpox. Jurin attempted to argue a different and more complicated point, an assessment of the comparative risks to an individual from inoculation versus a naturally occurring case of smallpox. Jurin's number made the case that inoculation was the safer alternative, and his results were widely appealed to throughout the eighteenth century by advocates of the practice. Yet resistance continued, in part because although Jurin's calculations quantified a measure of risk to the individual, they could not adequately account for the risk to the population from the artificial inducement of cases of smallpox resulting from inoculation. In the absence of a more widespread appreciation by the reading public of what it meant to talk about risk as a collective phenomenon, Jurin's arguments could only succeed to a limited extent.

In France, meanwhile, the organization of public health to promote population growth took a number of different forms. One such initiative involved the organization of defenses and quarantine measures against epidemic diseases. The outbreak of bubonic plague in the vicinity of Marseilles in 1720—the last such appearance of the disease in western Europe—saw the government mobilize the army to erect a massive quarantine operation that effectively prevented the epidemic from spreading beyond Provence. The utility of a well-administered quarantine was not lost on France's neighbors, nor, indeed, on the French government itself. The British Parliament replaced the Quarantine Act of 1710 with a more stringent version in 1721, while in France the lessons learned in Provence provided a blueprint for combatting an economically serious wave of cattle epidemics during the century.[40]

40 L. Brockliss and C. Jones, *The Medical World of Early Modern France* (Oxford: Oxford University Press, 1997), 350–54 and 744–45. For Great Britain, see M. DeLacy, *The Germ of an*

Population growth also underlay initiatives begun in France and elsewhere to improve delivery of health care by physicians and other healers. In 1759, the French government awarded a royal warrant to Angélique-Marguerite Le Boursier du Coudray, a midwife who had practiced for a number of years in Paris. Coudray undertook a lengthy tour of various regions in France, offering courses in midwifery to local women. To judge by contemporary accounts, the courses were both well attended and successful in raising the standards of practice. By the time of Coudray's retirement from teaching in 1783, she was said to have trained some 5,000 women as midwives. The efforts to improve the practice of midwifery in France were matched by similar efforts to increase the skills of surgeons, male *accoucheurs*, and others who might attend women during birth.[41] In the Holy Roman Empire, meanwhile, a number of principalities passed medical ordinances that, among other things, subjected midwives, apothecaries, and other healers to more stringent training requirements and recurrent inspections of their practices. Even if in many cases those enhanced requirements did little more than steer additional income in the direction of town doctors (*Stadtphysici*) and medical faculties tasked with enforcing the ordinances and receiving fees for inspections and training courses, at a minimum they expressed a desire to improve the level of care provided to the public.[42]

The cameralist impulses that drove these health reforms, supported by the conviction that well-designed administrative regimes could promote the commonweal and enrich the ruler's treasury,

Idea: Contagionism, Religion, and Society in Britain, 1660–1730 (Houndmills, UK: Palgrave Macmillan, 2016), 157–58.

41 Brockliss and Jones, *Medical World of Early Modern France*, 740–42. In England, man-midwives largely replaced female midwives during the eighteenth century, a move that was partially justified on the basis of the men's supposedly superior training and skill. See A. Wilson, *The Making of Man-Midwifery: Childbirth in England, 1660–1770* (Cambridge: Harvard University Press, 1995).

42 T. H. Broman, *The Transformation of German Academic Medicine, 1750–1820* (Cambridge: Cambridge University Press, 1996), 42–72.

received their fullest expression in Johann Peter Frank's *System einer vollständigen medicinischen Polizey* (*A Complete System of Medical Policy*), a sprawling six-volume treatise that began appearing in 1779 and was eventually completed in 1817. In Frank's expansive view of cameralist administration, nearly everything was fair game: He urged that marriages between younger men and older women be legally discouraged, because the man's seed would be wasted on a barren woman; he called on Catholic principalities to discourage young men and young women from entering convents and monasteries, again on the grounds that their valuable reproductive potential was going to waste; and he advocated dramatic improvements in orphanages and birthing facilities available to pregnant women. Beyond policies aimed at increasing the population directly, Frank also presented a host of suggestions for regulating the quality of food, building safer and more salubrious housing, cleaning streets, and much, much more.

Unsurprisingly, no government stepped forward to embrace Frank's vision for medical police in its gargantuan totality, although much of what he proposed represented initiatives undertaken piecemeal in a number of places.[43] Instead, his work merits our attention for another reason. The *System* and related writings present clear evidence of what was fast becoming a widely accepted realization that the population's heath had become an ongoing concern for governments at all levels. Whereas in previous centuries, health magistracies and other regulatory apparatuses had been periodically resuscitated during outbreaks of bubonic plague and other epidemics, by the eighteenth century, governments thought about public health in the context of economic vitality. That connection would provide the foundation for many of the major developments in public health after 1800.

43 For example, R. Porter describes a number of initiatives undertaken in eighteenth century London in "Cleaning Up the Great Wen: Public Health in Eighteenth-Century London," *Medical History*, suppl. no. 11 (1991), 61–75.

Reflection

PICTURES OF HEALTH?

Ludmilla Jordanova

It is common enough to hear someone described as a "picture of health," a compliment, possibly tinged with gendered assumptions, that hints at contentment and bonny appearance. The phrase has been used as a title for books on subjects such as sickness and inequality, advice for young girls, and health propaganda. "Picture of health" sounds straightforward, but once probed, it reveals many of the complexities of the concept "health." What does "health" look like, especially if it is an ideal, a notion that tends to be future directed? Is it more than the absence of obvious illness? How does it vary from one person to another? How is it related to changing aesthetic ideals?

We have difficulty picturing "health," perhaps precisely because the notion expresses some kind of ideal, a state to be desired and aspired to—an absence more than a presence. Arguably, pictorial traditions have been more eloquent, vivid, and confident when it comes to negatives compared to positives. "Health" is, however, rendered visible in the context of other personified abstractions, which often take the form of a female figure. In the present day, it is alluded to in many forms of visual culture, such as images of fitness that derive equally from the sport and beauty industries. "Health," we might say, is a particularly treacherous keyword; the visual expressions of which invite close scrutiny. Both its cognates, such as cleanliness, goodness, and purity, and its opposites, including disfigurement and monstrosity, help us understand its elusive richness.

Simple gestures and expressions can effectively convey extreme pain and suffering. Distorted bodies can be used to ridicule and criticize, as in cartoons. Although there is considerable cultural and historical variation in these matters, untutored audiences can, through their somatic understanding, recognize another person who is not well or experiencing anguish. In Christian traditions there have been many opportunities for exploring how the absence of "health" can be rendered visible, in images of martyrdom and crucifixion, for instance. We currently possess a vast visual repertoire, deeply indebted to photojournalism, that represents sickness, injury, and suffering, both physical and mental. However, it is more difficult to depict the presence of health, which is an elusive abstraction. What would this look like—a newborn infant, a young child, or a breastfeeding mother? These examples suggest some generic attributes of health, which are associated not just with life but also with future promise. In the case of older people, since the manifestations of health vary markedly from one individual to another, it is difficult to imagine what a picture of health would look like. Then at particular times and places there are ideals, which may lead to a general preference for a plump baby, or a skinnier one, a tanned body or perhaps a pale one, women with flat stomachs or protruding ones, and so on. Although there is rarely consensus on such matters, there are dominating norms, which can be traced in a range of artifacts from paintings to advertising, and from book illustrations to toys. Evidently, then, health never stands alone but is bound up with other concepts, values, and trends. The visual arts have portrayed ideals, but the attractions of representing the absence of health seem to have been greater. While my evidence is impressionistic, this is a claim that would be worth pursuing. My hypothesis is that for artists, and for their audiences, the visual relish of illness, disability, disfigurement has been considerable, allowing a range of moralizing commentaries to be constructed that are relatively easy to decipher. Traditions of the nude, of depicting physical beauty have certainly been significant, but they did not necessarily invite viewers to contemplate "health" in particular. By contrast, depictions of its absence, especially in settings where there is an extensive written culture promoting healthful practices, could prompt such contemplation. The image of a sickly child could prompt reflection on the parenting practices most likely to produce a healthy one, for example. William Hogarth's well-known series *Marriage à la Mode* (ca.1743; plate 6) provides a graphic account of descent into ill health. Indeed the deterioration of health or death is the end point of a number of his moral series. There are no comparable suites of pictures that tell a wholly positive story—that would have been far too bland, I suspect, to command interest.

It seems clear that notions of health, hygiene, fitness, and cleanliness are variable, even if they are widely desired. It can be no coincidence that, over many hundreds of years, people have turned repeatedly to idealized figures to express them, to presiding gods and goddesses, and to stylized representations. Personifications are useful because they make explicit notions that people at the time took for granted. In reflecting on the visualization of abstract ideas, I have often turned to George Richardson's 1779 *Iconologia* (plate 7).[1] Relatively small prints depict a vast array on concepts, including health—a short commentary is provided for each idea. Richardson's account of "health," which is shown as an attractive, mature woman, is fuller than many others. "She"

1 G. Richardson, *Iconology: or, a Collection of Emblematical Figures, Containing Four Hundred and Twenty-Four Remarkable Subjects, Moral and Instructive; in Which Are Displayed the Beauty of Virtue and Deformity of Vice* (London: Printed for the author by G. Scott, 1779).

also comes up in relation to "Goodness," who "is represented by the figure of a very beautiful and comely woman, dressed in white robes. . . . She is represented beautiful and comely to express health and contentment." The image itself is, predictably enough, rather bland, while the connections among physical and moral goodness, virtue, health, and happiness are perfectly explicit in the text. Richardson treatment of health drew upon centuries of personifying and worshipping health. The images are "classical" in somewhat unspecific ways, in showing simple, flowing gowns, for example. Another example is the golden globe held by Health, which "signifies that health is the most precious treasure of human life." This idea that health is necessary for the other parts of life to be enjoyed was familiar in ancient Greece, with the worship of Hygieia, the daughter and assistant of Asklepios, who represented the preservation of health and the curing of sickness. So Hygieia helps her father, the son of Apollo, and presides over the continuing state of good health. Images of her on vases and in stone are indeed pictures of health.

The existence of representations of Hygieia suggests the wish to have something graspable to represent an idea that is so elusive. Images of deities facilitate worship. Worship invites mediating objects of veneration. What the classical evidence suggests is the force of an idea—health or hygiene—can be visualized as a woman, whose figure can carry a multitude of ideas about the desirability of human well-being. There is plenty of evidence to suggest that personifications of health have also had strategic value in more recent times.

These examples of ideas of health taking simple feminine forms support my point about the difficulties of visualizing "health." Visual analysis of health propaganda hints at similar conclusions. Words and images work together to reinforce simplified notions, which link exercise, or good food, or routine cleanliness with health. The promise of health if only certain practices are routinely performed indicates a future directed, even utopian, orientation. Onto such basic imagery we can project a deep desire for health, which may be imagined with vivid intensity but remains in the realms of fantasy. Visual representations suggest ways of thinking about health and its history that respect its complex characteristics.

CHAPTER EIGHT

Freud and the Concept of Mental Health

Jim Hopkins

Freud and his successors have often described the aim of psychoanalysis in terms of three goals: resolving or mitigating patients' emotional conflicts; strengthening patients' sense of or relation to reality; and enabling patients better to love and to work. Freud's therapy thus linked mental health with the ability to form affectionate, nurturing, and cooperative relationships and to gain satisfaction from purposeful efforts to alter conditions in reality; and these with a realistic and unconflicted attitude toward oneself, toward one's own motives, and toward other people and the world generally.

These aspects of health are connected by Freud's overarching hypotheses about the "alienation from reality" that he took to be constitutive of mental disorder. This alienation, Freud held, was to be seen both in psychiatric symptoms, like hallucinations and delusions, and in the more ordinary, but nonetheless debilitating, failures in understanding and feeling—self-defeating misconstructions of their own motives, projects, and relationships with others—that created the difficulties in loving and working (or more generally the difficulties in living) that led people to seek psychoanalysis. In this sense, his therapy can be seen as an extension of the Delphic injunction to know oneself.[1]

Alienation, Phantasy, and Memory

Freud regarded this alienation as a form of natural shielding, or *defense*, by means of which the brain (or mind) protected itself against harsh internal realities of emotional frustration, *psychic conflict*, and pain, or causes of *unpleasure* more generally.[2] These are internally related: conflict produces frustration, frustration produces anger, and both conflicts and frustrations are inevitable in our attempts to satisfy needs and desires. He formulated preliminary ideas about conflict and defense early in his psychiatric career. Finding contemporary physical treatments for mental disorder useless, he followed the example of his senior colleague Joseph Breuer, who told him of a patient, now called Anna O., who had been diagnosed with hysteria. She and Breuer had so thoroughly investigated the occurrence of her symptoms in the

---

1 This essay builds on previous work of mine, including "Psychoanalysis Representation and Neuroscience: The Freudian Unconscious and the Bayesian Brain," in A. Fotopolu, D. Pfaff, and M. A. Conway (eds), *From the Couch to the Lab: Psychoanalysis, Neuroscience and Cognitive Psychology in Dialogue* (Oxford: Oxford University Press, 2012), 230–65, on psychoanalysis and Bayesian neuroscience; "Understanding and Healing: Psychiatry and Psychoanalysis in the Era of Neuroscience," in K. W. M. Fulford et al. (eds), *The Oxford Handbook of the Philosophy of Psychiatry* (Oxford: Oxford University Press, 2013), 1264–92, on psychoanalysis and psychiatry; "The Significance of Consilience: Psychoanalysis, Attachment, Neuroscience, and Evolution," in S. Boag et al. (eds), *Psychoanalysis and Philosophy of Mind: Unconscious Mentality in the 21st Century* (London: Karnac Books, 2015), 47–136, on psychoanalysis, evolution, attachment, and neuroscience; "Free Energy and Virtual Reality in Neuroscience and Psychoanalysis," *Frontiers in Psychology* 7 (2016), art. 922, which coordinates some important recent work in neuroscience on dreaming and mental disorder with the discussion here.

2 Freud regarded *psychical conflict* as a pivotal concept for understanding mental disorder, and he used *unpleasure* for a range of *homeostatic* and emotional sufferings and discomforts, as exemplified by those attending the sensory initialization of birth. For information on italicized psychoanalytic terms, see J. Laplanche and J.-B. Pontalis, *The Language of Psychoanalysis*, trans. D. Nicholson-Smith (London: W. W. Norton, 1973); for neuroscientific terms such as *homeostasis* below see neuroscience.uth.tmc.edu.

context of their occurrence that, as Breuer said, "her life had became known to me to an extent to which one person's life is seldom known to another" (*Works* 2: 21–2).[3]

In this relationship of full disclosure doctor and patient found together that her symptoms expressed forms of fictive experience or belief—now psychoanalytically described as *phantasy*—related to emotionally significant events she had apparently forgotten. Moreover, her symptoms eased when she remembered these events and expressed the emotions connected with them.[4] For example, for a time, and despite "tormenting thirst," she refused to drink. She pushed away water like "someone suffering from hydrophobia" (*Works* 2:34), as if drinking itself had become disgusting or dangerous. Under hypnosis she remembered remaining silent despite great disgust, when a companion let a little dog ("horrid creature") drink from her glass. After expressing her outrage and disgust, she was able again to drink again without difficulty.

In this case the patient's phantasy—a delusory conviction that there was something about drinking plain water that made it disgusting or dangerous—had apparently stemmed from a memory she was *unconscious* of, in the sense that that it was inaccessible to awareness, understanding, and reason. Accordingly, Breuer and Freud focused on the *memories* in which Anna O.'s phantasies were rooted, as opposed to the phantasies themselves. Hence, they hypothesized that their hysterical patients suffered, not from engrossment in phantasy, but from memories of events they have found *traumatic*, and so had banished them from consciousness and thought by *repression*. Insofar as these memories could be made conscious, the symptoms could be relieved.

---

3 References to Freud are to J. Strachey (ed.), *The Standard Edition of the Complete Psychological Works of Sigmund Freud*, 24 vols. (London: Hogarth Press, 1957). Cited by volume and page number.

4 Toward the end of the last century, Breuer's treatment of Anna O., like much else to do with Freud, was the subject of much uninformed historical discussion. This is clarified in R. Skues, *Sigmund Freud and the History of Anna O.* (Houndmills, UK: Palgrave Macmillan, 2006), in which Breuer's modest, but apparently genuine, therapeutic success can be discerned.

Freud was soon to revise the assumptions about memory and phantasy embodied in this theory. Before doing so, however, he pressed his patients for memories connected with their symptoms, and he was able to trace a range of disorders back through childhood to origins that apparently lay in infancy. Although this risked conflating the roles of memory and phantasy, these researches were nonetheless pioneering and valuable. They can now be seen to accord with more recent work in psychopathology indicating that many disorders do indeed have an early developmental history, and with research in attachment and developmental psychology indicating that basic patterns in emotion and conflict—for example, those that give rise to secure as opposed to disorganized attachment—are initially achieved in the first months of postnatal life.[5]

The Interpretation of Dreams and the Conception of Free Energy Neuroscience

Freud's radical revisions began when he started to analyze his own and his patients' dreams. This required engaging in the process of *free association*, in which patients described the rapidly changing contents of their own conscious states of mind in as much detail as possible and without omission or censorship. This enabled Freud to learn as much about his patients' experiences, memories, thoughts, and feelings as they were able to put into words and to extend this understanding by observing how their expressions in analysis were related to one another,

5 Current journals related to this topic include *Child Development, Attachment and Development*; *Development and Psychopathology*; and *The Journal of Child Psychology and Psychiatry*. On early attachment see B. Beebe and F. Lachman, *The Origins of Attachment: Infant Research and Adult Treatment* (New York: Routledge, 2014). For a psychoanalytic case history focusing on the intergenerational transmission of trauma see V. Volkan, *A Nazi Legacy: Depositing, Transgenerational Transmission, Dissociation, and Remembering Through Action* (London: Karnac Books, 2015); and for related empirical hypotheses as to the mechanisms involved see C. Cohen et al. "The Lasting Impact of Early-Life Adversity on Individuals and Their Descendants: Potential Mechanisms and Hope for Intervention," *Genes, Brains, and Behavior* 15/1 (2016), 155–68.

to their dreams and symptoms, and to their actions in life outside their sessions. This radical mode of self-disclosure—and the range of data it provided—was unprecedented in previous psychological research and even now remains without parallel in any other discipline. Hence, although other forms of investigation of the emotional functions of dreaming increasingly acknowledge the evidential importance of the dreamer's associations,[6] none so thoroughly takes them into account.

Freud's first love in research had been the study of the nervous system, in which he had shown unusual distinction.[7] As his new sources of data led to new hypotheses, he initially tried to formulate them in a new approach to neuroscience, in a "psychology for neurologists," later published as the *Project for a Scientific Psychology*. Here he observed "the pathological mechanisms which are revealed in the most careful analysis in the psychoneuroses bear the greatest similarity to dream-processes" (*Works* 1:336). Both dreams and symptoms, as he now saw, served the common function of protecting the conscious self or "ego" from otherwise distressing or disruptive arousals of conflicting emotions. They did so, moreover, in the same way: via the creation of fictive experiences and beliefs, forms of phantasy that entailed an "alienation from reality" that masked and pacified (or in neuroscientific terms *inhibited*) aspects of emotion and conflict.

Freud's new approach to neuroscience envisaged the nervous system as operating to minimize *free energy*, a conception he apparently developed on the basis of Helmholtz's account of the thermodynamic energy in a system that was available for conversion into work. Freud thought of this energy as introduced into the nervous system by sensory impingement, both external (as in visual, auditory, or tactile

---

6 J. Malinowski and C. Horton, "Metaphor and Hyperassociativity: The Imagination Mechanisms behind Emotion Assimilation in Sleep and Dreaming," *Frontiers in Psychology: Psychopathology* 6 (2015), art. 1132.

7 While Freud was a medical student, he was invited by the celebrated physiologist Ernest Bruke to conduct research in his laboratory. Prior to practicing as a psychiatrist, Freud had published well over 100 papers in neurology, as well as monographs on disorders of movement and childhood cerebral palsy that established him as a leading expert in these fields.

perception) and internal (as in *proprioception* and *interoception*). The most important source was the constant and inescapable flow of "endogenous (interoceptive) stimuli" (*Works* 1:297) that reflected the "peremptory demands of the internal needs." These inputs created a "demand for work" on the part of the nervous system, to produce the "specific actions" in the external world that would satisfy the needs and thereby minimize the energy introduced by their arousal.

Thus the "scream" of the helpless infant was also her first communication of urgent need, expressing the demand for work that others had to perform on her behalf if she was to survive in order to learn to do so herself (*Works* 1:318). The first mental process by which the infant accomplished this was the innate *primary process* that generated phantasies—fictive dreamlike beliefs and experiences—that helped mitigate frustration and pain while promoting learning. Just as a hungry and distressed infant finds comfort and relief in the experience of being held and nursed by her mother, so an infant could also (temporarily) secure a degree of comfort by engaging in a dreamlike phantasy of such experience. Such phantasies could shield the infant from "the first great anxiety state" generated by the sensory assault that occurred with birth, and later from "the infantile anxiety of longing—the anxiety due to separation from the protecting mother" (*Works* 19:58).

This could in turn promote learning, as Freud illustrated by an example in which a nursing infant used the pacifying "wishful cathexis" provided by phantasy to "experiment" by turning her head in such a way as to secure the nipple. But this mode of learning required the infant to forgo the immediate relief provided by wishful phantasy to focus on the "indications of reality" available in perception (*Works* 1:328–29, 356). Thus against the background of the regulation of pleasure and unpleasure by the primary process, the secondary processes introduced the infant—cry by cry, feed by feed, excretion by excretion—to the harsh realities of the world into which she has been born, as well as to her resources for coping with them.

Whereas the fictive satisfactions provided by phantasy are instantaneous, those obtained in accord with the reality principle require tolerance of frustration and delay. But these real satisfactions are essential for the maintenance of life, are deeper and more lasting, and are therefore more effective in minimizing free energy, than those provided by phantasy. Hence, the infant's waking image of the world, as constantly registered and re-registered in memory, steadily diverges from its dreamlike beginnings. As development proceeds the secondary processes increasingly inhibit, overlay, and supplant the primary process in waking life, so that finally it operates without realistic constraint only in the processing of memory and emotion during sleep.

Waking and dreaming consciousness contribute to the minimization of free energy in distinct ways. In waking life, sense-perception arouses emotions and desires, and these drive thought and action so that the desires are satisfied and the emotions pacified and the free energy introduced with their arousal is minimized. In sleep, by contrast, arousals of emotion are prompted by memory rather than perception, and these, and the associated free energy, are pacified by the fictive satisfactions of dreaming. Still, if memories of particularly traumatic events or conflicts are aroused during sleep, this may cause a dream experience of anxiety or terror for which no pacifying solution can be devised, as in the nightmares that are part of post-traumatic stress disorder. In such cases, the minimization of free energy requires inhibition of the primary process and a return to waking consciousness and thought. Alternatively, powerful conflicts and frustrations rooted in early experience can be aroused in waking life. In this case, particularly in individuals who already rely excessively on engrossment in phantasy for the regulation of their emotions, the primary process may again become dominant in waking consciousness, producing the fictive beliefs and experiences constitutive of mental disorders.

Thus on Freud's account, a main part of mental health consists in coordinating the primary and secondary processes, or again in coordinating phantasy with reality-oriented thought. As we will briefly consider below, phantasy and imagination are instrumental in imbuing real activities with emotional significance derived from the past; but the balance between mental health and disorder turns on the achievement of real as opposed to phantasied satisfactions. Health is shown in the appropriate ordering of the emotional significance of relationships and projects, and this requires bearing frustration sufficiently over the course of development to be able to replace phantasy by thought.

The Priority of Phantasy

In Freud's account, the primary process is "in the apparatus from the first" (*Works* 5;603) and so prior to memory. As Freud saw, this entailed that the possibility of tracing the history of his patients' conflicts via memory was severely limited. As well as being established to replace phantasy, memory was apparently the product of a continual reconstruction, in which memories were re-formed as they were re-aroused. Thus as Freud described in "Screen Memories" (*Works* 3), what might seem to be emotionally significant memories from childhood may turn out on analysis to be interwoven with distorting phantasies. Something similar held within the therapeutic process itself, in which Freud's patients could clearly be seen to develop phantasies about him as their therapist that replicated those about their parents in their memories, dreams, and symptoms.

Freud described this as *transference*. It enabled him to triangulate among his patients' dreams, symptoms, apparent memories, and experiences of transference in therapy to gain a fuller sense both of their phantasies and of the remembered events that had given rise to them. Nonetheless, the relation between apparent or seeming memory and genuinely veridical memory remained to be determined in each individual case. Hence as Freud later said, "the phantasies possess *psychical* in contrast to material reality, and we gradually learn to understand that in the world of the neuroses it is psychical reality that is of the decisive kind" (*Works* 16:368).

Identification and Projection

In addition, Freud and his successors came to understand phantasy as including two families of mechanisms that we can describe as *identification* and *projection*. These are mechanisms of development as well as defense, and they are related to individual and group cooperation and conflict. Freud described identification as "the assimilation of one ego to another" (*Works* 22:63), and he took this to include at least three developments. The first was the basic assimilation of attitudes and other dispositions from those around us by which we initially articulate our selves and characters. An example might be the way some 60–70 percent of infants come to share basic emotional dispositions (as shown in categories of attachment) with their mothers by the end of the first year. The second was the assimilation of abilities and skills in personal learning, such as infants' acquisition of language. The third, facilitated by the first two, was the assimilation of attitudes and stances in empathy, sympathy, and other socially coordinating emotions that distinguish members of cooperating ingroups, of which the family is the first example.

Projection, by contrast, represents and creates *difference*. Whereas identification often involves the assimilation of admired or desired characteristics, projection often involves the imaginary relocation of morally condemned or unwanted characteristics from the self onto others. Thus where identification often creates an image of a good self in relation to a good other or others (a *good us*), projection often creates an image of a good self in relation to a bad other or others (a *bad them*). Identification thus mediates relations of amity in cooperative ingroups, while projection mediates relations of competition and hostility with members of outgroups. The two mechanisms coordinate in creating and maintaining the many forms of *ingroup cooperation for outgroup competition and conflict* characteristic of human sociality.[8]

The development of identification and projection thus begins in the family. In early infancy emotions such as rage, fear, and distress at separation are directed at the mother, so that these "negative" emotions serve as honest signals of urgent need. The first *good us* is that of the mother nursing the child; and this entails the exclusion of others (father and siblings) from this relationship. Early projection creates images of both parents as liable to punish aggression within the family with retaliatory moralistic cruelty, so that infants learn to inhibit aggression partly by identifying with these "earliest parental *imagos*" (*Works* 22:50). [9] In this they create the harshly self-critical part of the self that Freud called the *superego* or *ego ideal*, which comes to the fore in depression and suicide, and some of whose manifestations we will consider below. Projection then shifts the locus of anger and fear away from the parents and outside the family as the infant comes to represent the mother as enduring, unique, and irreplaceable.

This is shown by the angry protests at separation from their mothers that infants begin to show at about eight months, and the fear of strangers (particularly those with beards) that arises at the same time. This indicates the establishing of the overall structure of ingroup-outgroup relations, as captured by the well-known proverb:

Myself against my brother
My brother and I against the family
My family against the clan
All of us against the foreigner.

---

8 See: J. Hopkins, "Evolution, Emotion, and Conflict," in M. Chung and C. Feltham (eds), *Psychoanalytic Knowledge* (Houndmills, UK: Palgrave Macmillan, 2003), 132–56, and J. Hopkins, "Conscience and Conflict: Darwin, Freud, and the Origins of Human Aggression," in D. Evans and P. Cruse (eds), *Emotion, Evolution, and Rationality* (Oxford: Oxford University Press, 2004), 225–48.

9 Freud uses the Latin term *imago* (image) for mental representations.

Such structures place each individual in series of groups, groups of groups, and so forth, up to the level of competing coalitions of nations. The cohesion of cooperating ingroups is based on identification, often with an idealized leader or creed that acquires the role of superego to the group, and so facilitates the projection of bad qualities into rival groups (*Works* 18:69–143). For this reason, although the mechanisms of identification and projection are continually in operation, some of their clearest manifestations are visible in group competition and conflict. Obvious examples would be the idealization of Hitler and the denigrating projections onto the Jews that were part of the unifying ideology of the Nazis, or again the denigrating projections onto Muslims that foster current anti-Muslim hostility, the idealization of particular versions of Islam and denigrating projections onto "unbelievers" that foster Islamic terrorism, the denigrating projections into available immigrant minorities that have become part of the electoral politics of most democracies, and so on *ad finem nostrum*. Members of groups with such common good and bad objects feel unified by identification and purified by projection and are consequently able to show hostility to their rivals that is justified by common ideals.

Conflict, Dreams, and Symptoms

In what follows I will illustrate some of Freud's basic ideas with simple examples, which I have elsewhere discussed in more neuroscientific terms.[10] Let us start with core case of the secondary processes (perception, desire, belief, action) in reducing free energy by satisfying a desire based on sensory signals of need. In general, we can represent desires in sentential form, that is, as desires that **P**, where "**P**" is replaced by a sentence specifying the content of the desire. Thus an agent's desire to drink can be more specified as her desire *that she get a drink*. When an

10 J. Hopkins, "Free Energy and Virtual Reality in Neuroscience and Psychoanalysis," *Frontiers in Psychology: Cognitive Science* 7 (2016), art. 922.

agent acts successfully on this desire this will bring about the situation **P** (that she gets a drink) that *satisfies* the desire; and this in turn will cause a perceptual experience of desire-satisfaction (here, the experience of drinking water and slaking thirst).

Such an experience serves to *pacify* the desire whose satisfaction it registers, freeing the agent for work on further desires and the biological imperatives that underlie them. This is the general pattern of the secondary processes, operating in accord with the reality principle to minimize free energy by ensuring the satisfaction of desire and need. This is schematized in the following diagram:

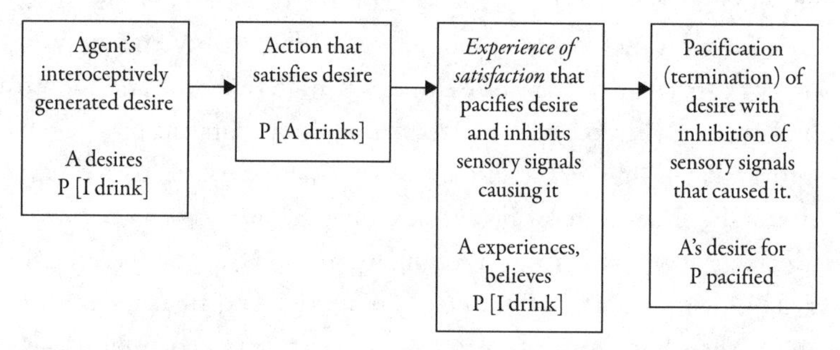

This also illustrates how the operation of the neural mechanisms related to need and desire are reflected in commonsense psychology. We describe desires in terms of the actions they will produce if acted on, so that the description of a desire tacitly predicts its effect on the agent's behavior. Also, we tacitly understand how emotions and desires are related to perceptual experiences that will pacify them. Such experiences of satisfaction go beyond individual desires to inhibit, at least temporarily, the drives or emotions in which they originate. In the right circumstances an experience of eating or drinking (or nursing) can terminate not only hunger and thirst but also the fear of starving or dying, as that of a single breath can end the panic of apparently impending suffocation, or a yearned-for look or touch can end the pain of separation, exclusion, or social isolation. Even a minimum of satisfaction can go a long way. Such satisfactions are particularly important to the newborn, for whom obtaining physical *and emotional* care is a continual matter of thriving or failing.

The discovery that prompted the developments in Freud's thinking discussed above was that dreams *also* used experiences of satisfaction to inhibit sensory signals and pacify drives, emotions, and desires. Thus—continuing the simple example above—he reported that when he had eaten anchovies or some other salty food, he was liable to dream that he was drinking cool, delicious water. After having this dream, he would wake, find himself thirsty, and get a drink. In waking from such a dream, we intuitively regard them as caused by, and representing the satisfaction of, desires familiar from waking life.

This natural conclusion is based on a tacit comparison of the desire we feel on waking with the experience of the dream. The desire is to drink, and the dream is of drinking; so the dream is the *dreamt fictive experience of satisfaction* of the desire. In light of this, the dream-experience seems best understood as caused by the desire and—since it has apparently been doing so prior to waking—as serving to pacify the desire, if only temporarily, so that sleep and dreaming can continue. These fictive pacifying experiences both mask and mitigate conflict—in this case as between nocturnal arousals of drive, emotion, desire or feeling on the one hand, and the wish to sleep, or again the mechanisms that operate to continue the biological functions of sleep and dreaming on the other. We can represent their role as follows:

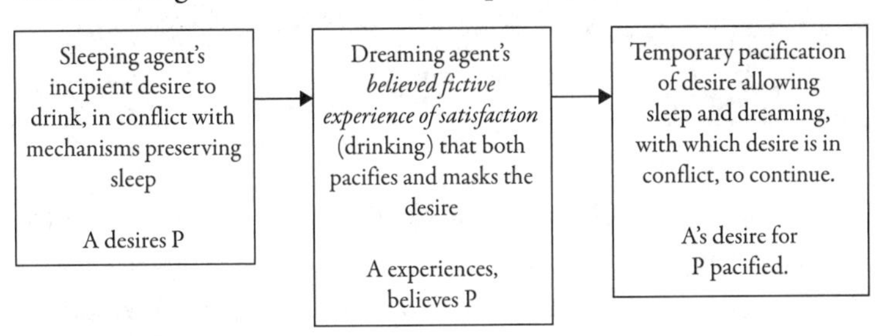

This is a simple example of the structure Freud discovered in dreams, which he thought played a particularly important role in easing the infant into the sensory initialization of birth. Two things in particular are worth noting. First, even this simple dream is similar to the kind of hallucination or delusion characteristic of mental disorder. In responding to the onset of thirst with a dream of drinking, a dreamer instantly obscures both internal and external reality: the reality of her own state of mind, and her engagement with the extra-mental world. Her real state of mind is that she has an incipient and growing desire (need) for water; but this is overlaid and masked—and so, as in repression, rendered unavailable to consciousness—by the fictive experience of satisfying a desire that in fact remains unfulfilled. The dreamer is so divorced from her own needs, wants, and real situation that if she remained shielded in this way she would die. This, in miniature, is the situation of someone suffering from a psychotic hallucination or delusion. So even this simple dream illustrates what Freud means by speaking of "alienation from reality."

Second, it seems that the dream process, like that of the pacification achieved in real action, operates by temporarily eliminating or reducing such parameters of motivation as desire and emotion. Just as the experience of a real drink eliminates the thirst and desire to drink, so (albeit temporarily) does the fictive experience of drinking in a dream. In the case of such emotions as anger and guilt, this reduction or elimination of desire or emotion is particularly important. For this let us consider another relatively simple dream, reported by the neuroscience blogger Neurocritic, who had recently suffered an injury to his leg, in a recent discussion of the suppression of pain in rapid-eye-movement (REM) dreaming.[11]

> Yesterday morning, I had a terrible nightmare in which my real life leg pain was projected onto someone else in an exceptionally gruesome way. I was driving along an unknown neighborhood street

---

11 neurocritic.blogspot.co.uk/2011/09/neurophysiology-of-pain-during-rem.html.

> when suddenly a man . . . had fallen under my car and had both his legs amputated from being run over. . . . The gravely injured man was still alive . . . I was absolutely horrified. All I could do was say "oh my god oh my god oh my god" over and over. . . . It was an awful nightmare, and in the dream I was quite traumatized by the entire experience.

This dream has the same wish-fulfilling form as diagrammed above, except that in this case the phantasy has further consequences for the dreamer. As Neurocritic says, this dream illustrates projection, involving the imaginary relocation of some aspect of the self in another person. While often the imaginary relocation is used to increase self-esteem or diminish guilt, in this case the projection was of pain and was realized by imagining causing terrible pain in someone else, which in turn made the dreamer himself feel anxious and guilty. So this dream could also exemplify the mechanism that Freud's successor Melanie Klein described as *projective identification*, in which the psychic relocation in the projection is imagined as occurring via bodily activities.

Examples from One of Freud's Case Histories

The emotional conflicts with which Freud was concerned were of two related kinds. The first was conflict between love and hate (or positive and negative emotion more generally) for a single person. The second, based on this, was conflict between parts or aspects of the self, for example those involving the self-critical and self-punishing superego. The relations between such conflict, the primary process, and mental disorder are displayed in a case history that predated Freud's formulation of his ideas about the superego, and from which some of his session-by-session notes survive. The patient, known as the Rat Man (hereafter R) suffered from what Freud called an obsessional neurosis, nowadays described as obsessive-compulsive disorder (OCD).

As R told Freud in the consultation before his analysis began, his main compulsive ideas concerned "a fear that something bad might happen to two people he loved very much," his father and a lady he venerated and hoped to marry. Also, he suffered impulses "to do some injury to the lady" and to harm himself. The impulses to harm his lady did not occur when she was with him, but when she went away. Thus when she went to visit her grandmother, and would not agree for him to join them, he had felt "commanded" to cut his own throat. As he went to fetch his razor as if to do so, however, it occurred to him that things were not so simple: he must kill the old woman instead. At this he had fallen down in horror (*Works* 10:260).

Freud understood this episode in a way that was close to common sense but also made use of his distinctive hypotheses. In commonsense terms, R was frustrated and pained by his lady leaving him alone. He first directed his anger at himself and then at the old woman the lady had left him to visit. But also—and here is the characteristically Freudian part of the account—R was so alienated from the reality of his own emotions he was unaware of his separation distress, the anger it caused, and of their suddenness and force. Thus, he had "an unconscious fit of rage" (188), which he experienced, in accordance with the mechanism of projection, as something coming from outside, in the form of a command to cut his own throat.

This provides a clear example of the second kind of conflict described above, the direction of anger or hatred by a part of the self against the self. In this case, the directing part was the agency internal to R that commanded him, as if from outside, to cut his own throat in response to the aversive emotions he was feeling. The first kind of conflict, that of love and hate directed at the same individual, was also clear in R's main compulsive symptom, the fear that "something bad might happen" to the lady and father he loved.

This was expressed in his repeated compulsive imagining—as in a waking nightmare—that his beloved father or venerated lady were being subjected to a terrible torture, in which hungry rats ate their way into the anus of the victim, causing an agonizing death. Imagining this, particularly in the case of his father, made R guilty, anxious, and depressed, even though he knew that his father was beyond harm since he had been dead for years. He constantly sought to forestall or prevent it by performing the actions or rituals that were the symptoms of his OCD. These included undertaking meaningless but onerous tasks or uttering various preventive formulae, sometimes with the insertion of "without rats" (291).

Despite R's genuine affection for both his father and his lady, one might reasonably think that his repeatedly imagining them being tortured in this way was an expression of hostility toward them. Freud understood these imaginings—like the "commands" he received to kill the old lady or to cut his own throat—as stemming from unconscious rage, harbored, in the case of his father, from early childhood. This was confirmed in his analysis both by transformations of his unconscious rage into conscious expressions of rage, and by his recovering both his anger and his fear toward his father in early childhood, and in subsequent episodes in which his anger had been expressed. He realized that when he was angry with people he often "wished the rats on them" and that he had long harbored such rat-wishing rage against his father.

Thus R's symptom had a pattern that we can illustrate as similar to those above:

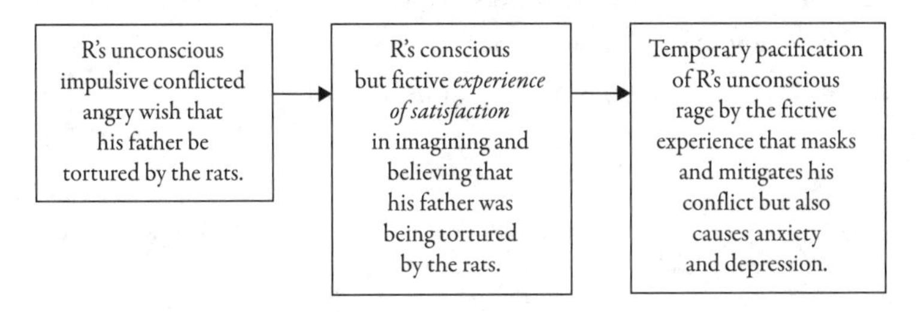

An Example from Post-Freudian Psychoanalysis

Finally, let us consider some delusions described by Elyn Saks in her account of her descent into schizophrenia.[12] She entered a psychiatric hospital in a savagely self-critical depression, in which she kept repeating, "I am a piece of shit and I deserve to die." Such self-hating internal conflict is a mark of the rage against the self that characterizes "introjective" depression,[13] which Freud described in terms of the superego, as seen in the examples from R above. When antidepressants gave Saks some relief, she told her doctor that she felt less angry and remarked on "how much rage I had felt, directed mostly at myself."

Later, however, her self-reproaches returned in force, and her increasingly unbearable depression altered only when she began to imagine herself "receiving commands" from "shapeless powerful beings that controlled me with thoughts (not voices) that had been placed in my head." These commanded, for example, "Walk through the tunnels and repent. Now lie down and don't move. You are evil." She was also commanded to injure herself, which she did by burning herself with cigarette lighters, electric heaters, or boiling water, so that finally she spent most of her time "alone in the music room or in the bathroom, burning my body, or moaning and rocking, holding myself as protection from unseen forces that might harm me."[14]

These delusions also fit the generalizations diagrammed in the figures so far. They served to mitigate the conflict of Saks's self-punishing depression, by replacing it with an imaginary relationship in which the self-punishing part of herself was projected, as in Neurocritic's dream, into imaginary punishing others. Thus, while greatly simplified, we can represent Saks's development of paranoid symptoms parallel to the simple dream and symptom diagrammed earlier.

---

12 E. Saks, *The Center Cannot Hold: My Journey Through Madness* (New York: Hyperion, 2008).

13 S. Blatt, *Experiences of Depression* (Washington, DC: APA Press, 2004).

14 Saks, *The Center Cannot Hold*, 84–86.

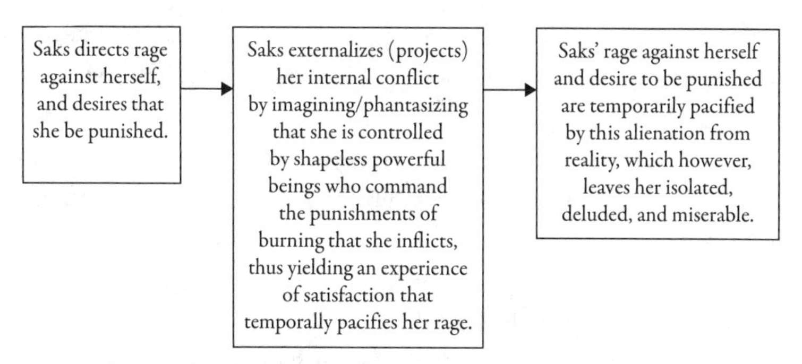

This kind of transition from depressive to paranoid functioning has often been discussed in psychoanalysis,[15] and it is usually said to involve the projection and fragmentation of the superego. This goes back to Freud's account of how, in schizophrenia, "the voices, as well as the undefined multitude [of critical presences embodied in the superego] are brought into the foreground again by the disease" so that the sufferer's harshly critical conscience "confronts him in a regressive form as a hostile influence from without" (*Works* 14:96). Thus once her delusions set in, as Saks says, "the commanding influence" responsible for her self-directed moralistic cruelty "came from within my own head, but was not mine. It was someone else commanding me."[16]

In such a case, while the reduction in internal conflict may relieve the unbearable internal hostility and depressive pain that can cause suicide, it also marks a deeper alienation from reality, and a deeper failure of self-regulation than the depression it relieves. Saks's self-critical faculty was already punishing her dysfunctionally, for some imagined or phantasied transgression.[17] Projecting this part of herself into the "shapeless powerful beings" of her delusions was a *further* step from

15 See for example D. Bell, "Who Is Killing What or Whom? Some Notes on the Internal Phenomenology of Suicide," in S. Briggs, A. Lemma, and W. Crouch (eds), *Relating to Self Harm and Suicide: Psychoanalytic Perspectives on Practice, Theory and Prevention* (London: Routledge, 2008), 38–45, and in the same volume J. T. Maltsberger, "Self Break-up and the Descent into Suicide," 38–44.

16 Saks, *The Center Cannot Hold*, 85.

17 Cf. her phantasies of killing babies, as discussed in Hopkins, "Understanding and Healing."

reality. This is why the projective externalization constitutes a deeper regulatory failure; why Freud describes it as regressive in the quotation above; and why such paranoia involves a deeper alienation from reality than depression, even though a main risk in schizophrenia is suicide in the depressive phases in which the subject is attempting to re-establish internal regulatory control.

Freud remarked that R's relief from his OCD was related to his recognizing *himself* in the invading rats of his phantasy; and we can now see that this symptom, like Saks's delusions or Neurocritic's dream, effected an imaginary relocation (projection) of an enraged (and biting) part of himself in another. The oscillation between depression and mania in bipolar disorder (BPD) shows a related mechanism, in which attempts to maintain internal but overly depressive regulatory control give way to delusions in which, for example, the subject imagines himself accomplishing or expecting wonderful things. (An important difference between Saks's paranoid solution to depressive conflict and the manic one is that the superego is fragmented and projected by the delusions constituting the former but only massively inhibited by the delusions of excellence constituting the latter.) A third conflict-reducing response is the inhibition of the generation of purposive action itself, as seen in the response to separation distress in infant animals.[18]

Symbolism and the Cognitive Regulation of Emotion

Since the body-invading rats of the phantasy that was R's main symptom also represented R's own biting rage, we can also see this as

18 D. F. Watt and J. Panksepp, "Depression: An Evolutionarily Conserved Mechanism to Terminate Separation Distress? A Review of Aminergic, Peptidergic, and Neural Network Perspectives," *Neuropsychoanalysis* 11/1 (2009), 7–51.

an example of Freudian *symbolism*.[19] The forming of this particular symbol seemed traceable to R seeing what he took to be a large rat in the graveyard where his father was buried and imagining the creature had been eating his father's corpse (*Works* 10:297). A comparable example of symbol-formation could be seen in R's transference phantasy of Freud's mother dead, with R's two Japanese swords stuck through her breast, and the lower part of her body, and especially her genitals, eaten up by Freud and his children (282–83).

The swords were a military souvenir that hung in R's bedroom, and he had previously named them "marriage" and "copulation" (267). Here he used this established symbolic relation to create a phantasy of a phallic and oral attack on the nurturing and reproductive capacities of the maternal breast and genitals, as well as the projection onto Freud and his children of the primitive oral rage and greed that he often symbolized in his phantasies about the rats. We can now see such symbolism as instantiating psychodynamic forms of what cognitive scientists have described as *conceptual metaphor*.[20] Thus as R's symptoms and phantasies indicate, such symbol formation, clearly rooted in bodily activities and processes, is a form of *enactive embodied cognition*, and one that plays a significant role in the processing of emotion.

Freud described this symbol formation as enabling the *sublimation* of the drives, that is, the diversion of affect from basic bodily processes (e.g., those of biting and oral incorporation, or again genital aggression, as seen in R's case) to those to which the symbolization imparted emotional significance (e.g., activities with swords). Later analysts, such as Melanie Klein, came to regard this as central to the development of the ego, so that, for example, the lack of symbolic play and inhibition

---

19 A. Petocz, *Freud, Psychoanalysis, and Symbolism* (Cambridge: Cambridge University Press, 1999).

20 See J. Hopkins "Psychoanalysis, Metaphor, and the Concept of Mind," in M. P. Levine (ed.), *The Analytic Freud: Philosophy and Psychoanalysis* (London: Routledge, 2000), 11–35.

of interests found in some autistic individuals were explainable by failures in this basic process. Hence, as discussed by Malinowski and Horton,[21] such symbolic but cognitive work seems a particularly important part of processing emotion in dreams.

Psychoanalysis and Recent Work in Development and Neuroscience

We saw that Freud initially cast the basic ideas of psychoanalysis in terms of an account of neuroscience in which the brain operated to minimize free energy. This bears a striking resemblance to the new paradigm for neuroscience advanced over the last decade by Karl Friston and his colleagues under the *free energy principle*.[22] They represent the brain as operating to minimize variational free energy, an information-theoretic analogue of Helmholtz's conception. This implies, consistently with both Helmholtz and Freud, that the brain naturally embodies a model of the causes of the impingements on its own sensory receptors and uses this to minimize its own errors in predicting them.

This conception of free energy as prediction error makes it possible to solve one of the great epistemic problems of neuroscience (and arguably of philosophy more generally); namely, how brains (or persons) can ensure the correctness of their representations or model of the world on the basis of information internal to the nervous system. It has attracted widespread interest in the sciences of the mind, including recent philosophically informed books.[23] Owing to the use of the concept of free energy, the account of the brain that emerges has the same overall structure as that given by Freud.

21 Malinowski and Horton, "Metaphor and Hyperassociativity."

22 For more detail, see Hopkins, "Free Energy and Virtual Reality."

23 A. Clark, *Surfing Uncertainty: Prediction, Action, and the Embodied Mind* (Oxford: Oxford University Press, 2016); J. Howhy, *The Predictive Mind* (Oxford: Oxford University Press, 2013).

In both accounts free energy enters the nervous system via sensory impingement, with "endogenous [interoceptive] stimuli" as a particularly important source. As in Freud these inputs reflect "the major needs," or biological imperatives, so in Friston they predict departures from a continuously recalculated overall homeostatic (or *allostatic*) equilibrium, in which free energy is minimized.[24] As Freud speaks of these inputs as creating a "demand for work" to produce "specific actions," so Friston speaks of "an imperative to minimize prediction error" via the "kinematic trajectories" involved in action.[25] In both accounts the process of minimization requires the brain to embody a representation or model of the world, including the agent's body (in Freud the "bodily ego"); and for both this requirement is initially met by the innate generation of a prior virtual (or phantasy) version of reality, which will subsequently be modified by experience.

During the trimester prior to birth, infants spend most of their time in a state resembling later REM sleep and dreaming, and their facial expressions indicate both positive (laughter-related) and negative (cry-related) emotions.[26] Accordingly, Hobson, Hong, and Friston hypothesize that the brain is "genetically endowed with an innate virtual reality generator" whose working "is most clearly revealed in rapid eye movement sleep dreaming."[27] We are "born with a virtual reality model" of what we will subsequently discover to be the causes of sensory impingement, and this model is "entrained by sensory prediction errors" to become "a generative or predictive model of the world."

---

24 G. Puzzlio et al., "Active Inference, Homeosatis Regulation, and Adaptive Behavioural Control," *Progress in Neurobiology* 134 (2015), 17–35.

25 K. Friston, "Prediction, Perception and Agency," *International Journal of Psychophysiology* 83 (2012), 248–52.

26 V. Schöpf et al., "The Relationship between Eye Movement and Vision Develops before Birth," *Frontiers in Human Neuroscience* 8 (2014), 775; N. Reissland et al., "Do Facial Expressions Develop before Birth?" *PLoS ONE* 6/8 (2011), e24081; N. Reissland, B. Francis, and J. Mason, "Can Healthy Fetuses Show Facial Expressions of 'Pain' or 'Distress'?" *PLoS ONE* 8/6 (2013), e65530.

27 A. Hobson, C. Hong, and K. Friston, "Virtual Reality and Consciousness Inference in Dreaming," *Frontiers in Psychology: Cognitive Science* 5 (2014), art. 1133.

The operation of this "virtual reality generator" thus almost exactly parallels that of Freud's primary process, or post-Freudian accounts of innate infantile phantasy. Like the primary process/phantasy, it is an innate precursor of dreaming that serves for the minimization of free energy from birth and paves the way for perceptual learning, the establishing of memory, and free-energy minimizing action. Freud described these latter as the *secondary processes*, and Friston describes them in terms of *active inference*. Hence on both accounts, the role of innate phantasy/virtual reality at birth is supplanted during waking by the reality-oriented processes and the actions they inform, so that the primary process/virtual reality generator come to operate without sensory constraint only in the sensory attenuation and bodily paralysis of dreaming.

As Friston and his colleagues remark, "if the brain is a generative model of the world, then much of it must be occupied with modelling other people."[28] This suggests that the generative model of the infant should coincide with the "internal working models" of self and other as studied in the development of the emotional bonds that constitute attachment. Indeed, postnatal active inference—and hence the fundamental shift from virtual reality to generative model—seems to be driven by the subcortical "prototype emotion systems" limned by Panksepp.[29] These play the role in contemporary affective neuroscience that Freud assigned to the drives. Watt and Panksepp describe the systems as "sitting over homeostasis proper (hunger, thirst, temperature regulation, pain, etc.)" and "giv[ing] rise to attachment," which in turn serves as "the massive regulatory-lynchpin system of the

---

28 K. Friston et al., "Computational Psychiatry: The Brain as a Phantastic Organ," *The Lancet Psychiatry* 1/2 (2014), 148–58, at 151.

29 J. Panksepp, *Affective Neuroscience* (Oxford: Oxford University Press, 1998). See also L. Bivens and J. Panksepp, *The Archaeology of Mind: Neuroevolutionary Origins of Human Emotions* (New York: Norton, 2013); and A. Damasio and G. B. Carvalho, "The Nature of Feelings: Evolutionary and Neurobiological Origins," *Nature Reviews Neuroscience* 14 (2013), 143–52.

human brain" exercising "primary influence over the prototype systems below."[30]

As this indicates, from the turn of the century there has been in increasing consilience among the findings of psychoanalysis, attachment and other forms of developmental psychology, and affective and cognitive neuroscience.[31] The free energy paradigm seems ideally fitted to serve as a framework for consolidating these results. Hence, Alan Hobson, for four decades the most persistent and zealous neuroscientific critic of Freud, has produced two books applying the notion of a virtual reality generator/generative model in support of what he has recently described as Freud's "visionary" conclusions linking dreaming and psychopathology.[32] In this, as Hobson says, he "takes up the *Project for a Scientific Psychology* exactly where Freud left it in 1895."[33] But of course Freudian and post-Freudian developments of the ideas in Freud's *Project* should also be taken into account, as well as the specific notion of computational complexity that Friston relates to dreaming.

The Complexity Theory of Dreaming and Mental Disorder

In Friston's framework, the minimization of free energy is effected via the maximization of the predictive accuracy of the brain's model together with the minimization of its computational complexity. Complexity is measured by the number of hypotheses (parameters) engaged in predicting a range of data and the extent to which they require modification during this engagement. Hence in his accounts,

---

30 Watt and Panksepp, "Depression."

31 See Hopkins, "The Significance of Consilience."

32 A. Hobson, *Ego Damage and Repair: Towards a Psychodynamic Neurology* (London: Karnac, 2014); A. Hobson, *Psychodynamic Neurology: Dreams, Consciousness, and Virtual Reality* (New York: Taylor and Francis, 2015).

33 Hobson, *Psychodynamic Neurology*, 5.

the brain tends to maximize accuracy while also increasing complexity (by introducing and altering hypotheses or parameters) during waking, and then to reduce complexity during sleep, by means that include synaptic pruning during slow wave sleep (SWS) and dreaming/virtual reality in REM.

Friston's notion of complexity is conceptually related to the psychoanalytic and psychiatric notions of emotional conflict and trauma and hence to key notions in Freud's account of dreaming and mental disorder. Emotional conflict consists in the activation of emotions that produce inconsistent kinematic trajectories, and these perforce require reduction or elimination for coherent behavior. Again, the alterations that constitute complexity can be regarded as measuring the load of learning that sensory impingement places on the model in waking life.

The most important part of this, moreover, can be described as affective or emotional learning, as required for the computation of behavior ("kinematic trajectories") in relation to others. In this case, the load of learning indexed by complexity becomes the load of emotional learning or adjustment that the experience of relating to others places on the brain's model.[34] Experiences are rightly regarded as traumatic when, as in PTSD or BPD, the emotional adjustments (complexity, conflict) required for integrating them into thought and action are greater than the brain can manage.[35]

Taking Friston's conception of complexity in these Freudian terms enables us fully to integrate psychoanalytic findings about the role of conflict and trauma in dreaming and disorder into the new free energy paradigm. This yields what Chris Mathys has called "the complexity

34 Cf. the notion of affective load in R. Levin and T. Nielsen, "Nightmares, Bad Dreams, and Emotion Dysregulation," *Current Directions in Psychological Science* 18/2 (2009), 84–88.

35 M. Enlow et al., "Mother–Infant Attachment and the Intergenerational Transmission of Posttraumatic Stress Disorder," *Development and Psychopathology* 26/1 (2014), 41–65; D. Mosquera, A. Gonzalez, and A. Leeds, "Early Experience, Structural Dissociation, and Emotional Dysregulation in Borderline Personality Disorder: The Role of Insecure and Disorganized Attachment," *Borderline Personality Disorder and Emotion Dysregulation* 1/1 (2014), art. 15, 1–8.

theory" of dreaming and mental disorder: as I have argued elsewhere,[36] the theory that the conflicts and trauma that Freud thought expressed in the virtual realities of dreaming and mental disorder should be seen as forms of neurocomputational complexity, and that mental disorder is the product of such complexity, together with the mechanisms—phantasy/virtual reality at the level of the mind and synaptic pruning at the level of the brain—that have evolved to reduce it.

Such an account enables us to provide operational descriptions of the reduction of complexity in terms of psychoanalytic accounts of dreaming and mental disorder such as sketched above. Also it is nearly a consequence of the basic account of complexity that informs the work of Hobson et al.[37] If the accumulation of complexity in waking is so serious a problem for the brain that it requires processes of complexity reduction in sleep, then it would seem to follow that inadequacies or malfunctions in these processes—like prolonged sleep deprivation itself—might foster potentially damaging accretions of complexity that show as a waking mental disorder. In addition, just as the mechanisms of inflammation that have evolved to protect the body from injury can themselves cause bodily disorders, so the mechanisms that have evolved to reduce emotional complexity in sleep might cause disorders of the brain and mind in waking. This latter hypothesis enables us to integrate psychoanalysis and neuroscience with a range of observations about the correlation of disturbances in sleep with waking forms of disorder. If this updated and thoroughly testable integration of psychoanalytic and neuroscientific hypotheses proves correct, it may lead to advances in the understanding of the causes of mental disorder, both mental and physical, and as these operate both in waking and in sleep and dreaming.

---

36 Hopkins, "Free Energy and Virtual Reality."

37 Hobson, Hong, and Friston, "Virtual Reality and Consciousness."

This would fit with current approaches to mental disorder as involving "harmful dysfunction," explicable in evolutionary terms.[38] Here, although the dysfunction would concern the reduction of complexity, the reduction itself would be required by the complexities involved in the management of aggression in our uniquely group-cooperative and lethally group-competitive species, as well as emotions rooted in the triangle of parental investment, parent-offspring, and sexual conflict.[39] Our susceptibility to emotional conflict, trauma, and mental disorder, as well as our pervasive recourse to forms of conflict-relieving virtual reality, has deep evolutionary roots, particular to our uniquely complex and emotionally conflicted species.

Freud and the Philosophical and Scientific Traditions

Freud's investigations of symptoms and dreams yield an overall account of psychological development that encompasses normal life and work as well as mental health and disorder. In framing these ideas, Freud both reaffirmed an ancient understanding of dreaming and disorder and reformulated it in scientific terms. The Socrates of the Platonic dialogues had argued that human beings have "a wild beast within us," whose emotions and desires may be expressed in the committing in dreams of "every conceivable folly or crime—not excepting incest or any other unnatural union, or parricide." Like Freud, Socrates held that such desires can be "controlled by law and reason,

38 J. Wakefield, "Taking Disorder Seriously: A Critique of Psychiatric Criteria for Mental Disorders from the Harmful-Dysfunction Perspective," in T. Millon, R. F. Krueger, and E. Simonsen (eds), *Contemporary Directions in Psychopathology: Scientific Foundations of the DSM-V and ICD-11* (New York: Guilford Press, 2010), 275–302; and J. Wakefield and J. C. Baer, "The Cognitivization of Psychoanalysis: Toward an Integration of Psychodynamic and Cognitive Theories," in W. Bordon (ed.), *Reshaping Theory in Contemporary Social Work: Toward a Critical Pluralism in Clinical Practice* (New York: Columbia University Press, 2010), 51–80. For discussion, see Elselijn Kingma's chapter in the present volume.

39 Hopkins, "The Significance of Consilience."

and the better desires prevail over them." But in some persons—such as the "tyrannical man," whose development is described in the ninth book of Plato's *Republic*—they become the main principles of action. Later Descartes, among others, compared dreams with delusions, and Kant argued that "the deranged person" was "a dreamer in waking."[40]

Freud's work provides a culmination to this line of thought. Socrates's incestuous, parricidal, and potentially tyrannical "wild beast within us" is the unconscious but continually active residue of the powerful, impulsive, and unregulated emotions of early infancy, as these are expressed in dreams, psychoanalytic transference, and the symptoms of mental disorder. Thus Freud explained to R that the conflicts of his disorder had both arisen and been rendered unconscious in his infancy. The part of him that harbored these distressing emotions "was the infantile":

> That part of the self which had become separated off from it in infancy, which had not shared the later stages of its development, and which had in consequence become *repressed*. It was the derivatives of this repressed unconscious that were responsible for the involuntary thoughts which constituted his illness. (*Works* 10:177–78)

Freud also decisively advanced another of Plato's aims: that of understanding the relation between harmony and discord in the individual and in society. Plato's overall aim in the *Republic* was to understand the virtue of justice as involving a kind of psychological harmony in the individual that corresponded to, and could best flourish in, the social harmony of a just society. For him discord in the individual was both causally and conceptually linked to discord

40 H. Wilson (trans.), "Essay on the Maladies of the Head," in P. Guyer and H. Wood, *The Cambridge Edition of the Works of Immanuel Kant in Translation: Anthropology, History, and Education* (Cambridge: Cambridge University Press, 2007), 71.

in society. A similar understanding appears in Freudian group psychology, in which identification and projection link individuals and their frustrations and conflicts both to the ingroups with whom they cooperate and the outgroups with whom they engage in competition and conflict. Thus the same mechanisms that render us liable to group-on-group destructiveness in war, and helpless even in peace to curb the group competition for resources that threatens to destroy our environment, are also those that render the management of individual aggression so fraught with conflict to leave us vulnerable to mental disorder.

Also, and again like Freud, Plato's Socrates seems to have taken the aim of *knowing the self* to encompass knowing the nature and scope of such unconscious emotions and desires. In wanting to know himself, Socrates said, he wanted to know whether he was "a beast more complicated and savage than Typhon, or a tamer, simpler animal."[41] He was expressing his sense of the potentially radical nature of his inner divisions and the extremities of his own rage and other emotions, as well as the possible burial of these things beneath the part of himself that seemed gentle and simple. Typhon was "complicated" because he had a hundred different heads, and his fury led him to attack Zeus, the king of the gods. His punishment, like that of the Titans, was to be buried beneath Mount Etna so that his rage was hidden and expressed only in occasional volcanic eruptions.

This too approaches a Freudian view of the self, and a Freudian understanding of mental disorder as rooted, like that of R above, in rage and aggression that escapes self-knowledge. Freud was to use the same image as Socrates for the unconscious mind, comparing the wishes that give rise to dreams and symptoms to the Titans buried at Etna. Investigating and integrating these hidden but still furious aspects of the self, and thereby rendering them more rational and gentle, is

41 *Phaedrus* 230a, translation from J. M. Cooper (ed.), *Plato: Complete Works* (Indianapolis: Hackett, 1997).

the main task of psychoanalysis. Freud thus integrated this Platonic image of the self into a theory of the mind that radically extended the commonsense psychology of emotion, desire, and belief that we share with Socrates and fused this extension with a new form of neuroscience and developmental psychology. The scientific cogency of his results may only now be gaining full recognition.

Reflection

PORTRAIT OF THE HEALTHY ARTIST

Glenn Adamson

What does a healthy artist look like? Perhaps artists themselves are not the ones to tell us. Oftentimes, it's true, they have presented themselves as respectable and entirely well-adjusted, suited, or smocked, standing at the easel with palette in hand. But the more indelible images left to us by art history, particularly since the onset of Romanticism in the nineteenth century, are more vivid and less peaceful of mind. Take, for example, Theodore Géricault, an early nineteenth-century painter who defined the Romantic sensibility both in his work and life. In one of his earliest works, Géricault showed himself slouched in a chair with a skull perched on the shelf above him. The pictorial analogy between his own youthful, handsome visage and the death's head was clear: the artist was haunted by his own eventual demise. In 1824, shortly after completing a series of sensitive portrayals of "monomanics" (the inhabitants of asylums), which form a visual canon of mental disturbance, he would make good on the prediction of his early self-portrait, showing himself hollow-eyed and ghoulish, a dying man, as indeed he was. Géricault was soon to pass away at the age of thirty-two of tuberculosis.

Two decades later, Gustav Courbet painted himself in equally dire straits. Two self-portraits show him respectively as wounded and raving mad. In *L'Homme Blessé*, he lies back languidly, blood spilling over his shirt, regarding us with sensual calm as his life ebbs away. (The painting had originally featured a female companion and an unhurt Courbet, but when his affair with her ended, he adapted it into this more theatrical image.) A related work, *Le Déséspère* (*the Desperate Man*), shows the painter clawing at his own loosely flowing hair. Tightly cropped as if in a photograph, he stares out of the picture like a man possessed (plate 8).

Both Géricault and Courbet deployed the imagery of mental disturbance as an emblem of the Romantic temperament. They presented themselves as fired by an imagination that burned so intensely it hurt. From this metaphorical approach it is only a short step to the all-too-real mental illness of Vincent Van Gogh. The self-portraits he made shortly before he took his own life, alternately startled and mournful, stand for all time as the symbol of the artist as a disturbed genius. By this time in his life, he had become like one of those asylum dwellers shown by Géricault. Yet he was producing some of the greatest (and most self-aware) works ever painted. With his bandaged head and swirls of paint emerging from his pipe, he showed himself as both fragile and potent, the embodiment of the artist who cannot contain the force of his own creativity.

The uncomfortable question posed by Van Gogh's art is, ultimately, a philosophical one. If he had not been so troubled, it seems reasonable to conclude, he would also have been unable to create images of such originality. This makes his talent seem like a symptom, as much as a gift. Perhaps we should ask ourselves: is it ethically suspect to take aesthetic pleasure in the condition of a man like this? What do we make of a situation in which art and health seem at odds?

These questions relate in an interesting way to the concept of "moral luck," a term popularized by Thomas Nagel and Bernard Williams in the late 1970s. Their goal was to think in a new way about an old philosophical problem, that is, the dilemma that arises from the uncertain relationship between actions and consequences. The problem is at root a simple one. Imagine two people committing the same unwise action, for example, dropping a hammer off a high balcony. In one case, the hammer clatters harmlessly to the pavement. In the second case, however, it hits a child and kills him. The first person will simply have lost his hammer; the second will likely go to jail.

In practical terms, it is impossible for society to hold people accountable for their actions, regardless of their effects—who would keep track of all those falling hammers? In philosophical terms, matters are no less difficult. We must first consider the relation between action and intention. In our example, did persons A and B mean to drop the hammer, or did it get bumped off the ledge accidentally? And then there is the inherent messiness of intention itself, which may not be clear to someone committing a given action, much less those observing it. Taking these various factors into account, Nagel and Williams concluded, it seems that we do inhabit a world involving moral luck—"a significant aspect of what someone does depends on factors beyond his control, yet we continue to treat him in that respect as an object of moral judgment."[1]

This line of thinking has an interesting parallel in thinking about the case of Van Gogh, and art's relation to health in general. In his work on moral luck, Williams used the example of Paul Gauguin, Van Gogh's friend and rival. By all accounts, Gauguin was a nasty piece of work. He abandoned his wife and children to

---

1 T. Nagel, *Mortal Questions* (Cambridge: Cambridge University Press, 1979), 59. See also B. Williams, *Moral Luck* (Cambridge: Cambridge University Press, 1981).

pursue his art. His stay with Van Gogh in Arles, during which he proved to be selfish and abusive, helped push poor Vincent over the edge. His later relations with Tahitian women struck people at the time as sexually indecent, and many since as a cardinal instance of colonialist exploitation and exoticism. Despite this track record, Gauguin's name is venerated rather than despised because he created artworks that were both gorgeous and highly original. Moral luck was on his side.[2]

Van Gogh's case is the mirror image of Gauguin's, placing us as viewers in a position that involves the opposite quandary. If we might feel uneasy appreciating Gauguin's work because of his unpleasantness, we might feel equally uncertain about appreciating Van Gogh's expressive genius, because it derived in part from his ill health. Gauguin was a perpetrator, and Van Gogh a victim. But the result is to this extent the same: a disturbing dissonance between our ethical and aesthetic response. To enjoy the works they created without qualms, we feel that we need to sweep aside what we know of their biography, the lives that led to the work we see before us.

Nor is Van Gogh's a unique case. The stereotype that underlies the popular cult of the bohemian artist is of an individual driven by passion, often self-destructively so, and certainly unable to find a congenial place within the bourgeois social order. (Émile Zola created the classic text of the genre in his 1886 novel *The Masterpiece*, whose main character, the painter Claude Lantier, eventually hangs himself in his studio.) In more recent years, the canon of art history has been expanded to include individuals who are far more mentally ill than Van Gogh was. They have been embraced as masters of so-called outsider or visionary art, or in French *Art Brut*. The first artist to be celebrated in this way was Adolf Wölffli, a Swiss man who had been the victim of abuse as

2 See A. Gopnik, "Van Gogh's Ear," *The New Yorker*, January 4, 2010.

a child and spent his adult life in an asylum in Bern. Wölffli's artworks, and those of other artists whose biographies paralleled his (such as Henry Darger, Martín Ramírez, and Down syndrome sufferer Judith Scott), are appreciated for their intensity, often taking the form of repetitive marks that suggest a mental state of relentless obsession. Advocates of these artists argue that their value lies precisely in showing us the beauty and originality of supposedly "abnormal" minds.[3] The implication is that we should accept mental illness (whether it leads to great art or not) simply as a form of difference and not as a condition to be treated.

Matters are still more complex when we consider the fact that if artists are themselves aware of these dynamics, they can also manipulate them. That was the case with Courbet, who adopted the pose of a madman for dramatic effect. It was also true of Bada Shanren, an "unorthodox" painter and poet who lived in seventeenth-century China. One of the distinctive features of the history of Chinese ink painting is that unorthodoxy is itself a tradition. By Shanren's time, there had already been many *literati* (scholar-artists) who withdrew from society, living as hermits in natural surroundings. This brought them away from court life, and indeed, this role was often adopted purposefully to escape political conflict.

Shanren would have been aware of the long tradition of such supposed eccentrics, and he seems to have played on it as a way to build his own fascinating reputation, though period sources suggest that he had at least one genuine nervous breakdown. This makes his enthusiastic splashes and stabbing, cursory, sideways brushstrokes difficult to evaluate. They were certainly a dramatic departure from precedent. But was he innovating on purpose or because he could not help it? Either way, it seems, there is a

3 C. J. Morris, *Judith Scott: Bound and Unbound* (Brooklyn: Brooklyn Museum of Art, 2014).

quandary for the viewer of his work. If he was genuinely mentally unstable, then we are again venerating his symptoms. If not, then we might feel there is something inauthentic, even cynical, in his work.

The uncertainty is heightened still further if we shift from the mental to the physical domain. What if an artist cannot see very well? Surely that would reduce the quality of their work. Yet consider the Venetian master Titian or the impressionist Claude Monet, both of whom created their most revolutionary and effective works late in life. Both suffered from significant loss of eyesight in their older years; they even painted with extra-long brush handles, so that they could stand farther away from their canvases. In both Titian's and Monet's late works, there is an approximate quality, a lack of focus. The paintings were made by men with poor eyesight and trembling hands, but they are all the more evocative and lush as a result.

The conceptual artist Robert Morris has staged an even more extreme situation in a series of drawings entitled *Blind Time*. The first of these were executed in 1973 by the artist himself, blindfolded and carrying out a series of predetermined regular marks in graphite. Three years later he recruited a woman (identified only as A. A.), who had been blind from birth, to continue the series. Working to his verbal instructions, she made a set of drawings that she was of course unable to see. If we enjoy them—as well we might, given their atmospheric quality and the poignant story behind their making—does that mean that we are in some sense enjoying her disability? Or conversely, that we are rejecting the idea that her blindness *is* a disability, that she is simply "differently abled"? Morris's collaboration with A. A. returns us to the parallel with moral luck, in that her drawings cannot in any normal sense be considered the direct result of her own intention. If we appreciate them, can we even say that we are valuing her creativity? Or are we not rather appreciating blindness itself, and the compositional freedom that results from it?

I began this short Reflection by posing the question, "What does a healthy artist look like?" We have touched on artworks made by the mentally ill, the aged, and the blind. In none of these cases does the artistic quality of the work seem to have suffered as a consequence. Are we to conclude that health and aesthetics simply operate independently from one another—that there is no correlation between the two? No. In each case, the condition that would typically be considered a malady proves not only to be compatible with the realization of the work but also to be fundamental to its character. This points to a key trait of artists. They always work within constraints, and these boundaries of practice are not a negative factor in art but rather the source of creative friction. Think of the four sides of a painting, or the particular qualities of materials such as clay, stone, or plaster. Artists invariably encounter such limiting factors for their own creativity—format, materiality, scale, and many others—and make something of them. From the artistic perspective, health is just another of these constraints.

Viewed from this point of view, the health of artists takes on a different aspect. What is debilitating in other walks of life proves to be a spur to creativity. More than this, for if art is one of the purest expressions of our human experience, then perhaps its most important boundaries are those of human experience itself, the limits of the body and mind, which is exactly where illness carries us. When art shows what it is to suffer a condition that we conventionally label as madness, disease, or disability—and when we, in turn, take aesthetic pleasure in its doing so—we are recognizing that art and sickness are both a part of life. The one, considered properly, may help us accept the other.

CHAPTER NINE

Contemporary Accounts of Health

Elselijn Kingma

Much of the contemporary philosophical discussion about health and disease finds its origins in the 1960s and 1970s, which saw a period of great controversy surrounding psychiatry. So-called anti-psychiatrists argued that psychiatry was not a legitimate medical discipline but a "mere tool of social control": acting to affirm, enforce, and police certain social and evaluative norms.[1] The revision of the "bible of psychiatric classification" in the early 1970s, the *Diagnostic and Statistical Manual* (DSM)-II, formed the stage for a heated social and professional discussion about homosexuality:[2] Should this be included as a disease in the new version of the DSM—as it has been previously—or not? A very active gay lobby pushed the latter point, but psychiatrists were divided on the issue. In the discussion amongst them, two

1 See e.g. T. S. Szasz, "The Myth of Mental Illness," *American Psychologist* 15 (1960), 113–18; T. S. Szasz, *The Myth of Mental Illness* (London: Paladin, 1972); R. D. Laing, *The Divided Self* (London: Tavistock, 1959); and R. Howard (trans.), *Michel Foucault: Madness and Civilisation: A History of Insanity in the Age of Reason* (London: Tavistock, 1961).

2 R. Bayer, *Homosexuality and American Psychiatry: The Politics of Diagnosis* (Princeton, NJ: Princeton University Press, 1987).

competing positions on the definition of health and disease became apparent. The first proposed a value-free account of health and disease that embodied a vision of health as normal biological function. On this view—assuming that the sole biological function of sexual behavior is the production of direct biological offspring—homosexuality was thought to be a disease. The second and opposing position embodied a clinical focus: diseases are things that bother patients that prompt them to seek help and alleviation. On this value-laden view of health and disease, homosexuality would only be a disease if it was experienced as a problem by the person who "suffered" from it.

In the DSM debate, the gay protesters and the proponents of the second, clinical definition of health and disease won. In the final, 1974 revision of the DSM-II, homosexuality was replaced by the much weaker "sexual orientation disorder." In the DSM-III (1980), only "ego-dystonic homosexuality"—for example, homosexuality that was experienced by the "sufferer" as a problem—was listed. That edition of the DSM was also the first to provide an explicit definition of mental disorder (preceded by many caveats) as a "clinically significant behavioral or psychological syndrome." This clinical definition of disease has persisted unchanged into subsequent versions of the DSM. In the philosophical literature, however, the debate on defining health and disease would continue, and the clinical and value-free positions on defining mental disorder would crystallize into two opposing positions on health and disease: naturalism and normativism.[3]

---

3 D. Murphy, "Concepts of Disease and Health," in E. N. Zalta (ed.), *The Stanford Encyclopedia of Philosophy*, 2009, following P. Kitcher, *The Lives to Come: The Genetic Revolution and Human Possibilities* (New York: Touchstone, 1996), contrasts "objectivism" and "constructivism." Note that in this debate "disease" or "disorder" is meant to be inclusive, denoting not just what we ordinarily call disease but any condition that is a departure from health (including, e.g., trauma, disability, etc.). See, e.g., C. Boorse, "On the Distinction between Disease and Illness," *Philosophy and Public Affairs* 5 (1975), 49–68; C. Boorse, "A Rebuttal on Health," in J. M. Humber and R. F. Almeder (eds), *What Is Disease?* (Totowa, NJ: Humana Press, 1997), 1–134; and R. Cooper, "Disease," *Studies in History and Philosophy of Biological and Biomedical Sciences* 33 (2002), 263–82. I shall stick with that convention.

Naturalism

Naturalism is the view that health and disease are objective, empirical, or value-free concepts.[4] The central idea is that we could read the distinction between health and disease directly off the natural world, without having to appeal to the values that these states hold for us. Thus, proponents would hold, just as we can say that a dog, rabbit, or tree has a disease without having to appeal to the value that such a condition holds for either us, the dog, or the tree, so we should be able to say when a human has a disease without having to appeal to the value that condition holds for that human or for humanity.

The Descriptivity Problem

The central problem for any naturalistic account of health and disease is this: both working eyes and nonworking eyes are eyes. Both appear in nature. And both obey natural laws. How, then, does the naturalist about health and disease distinguish—as she should—healthy, working eyes from unhealthy nonworking ones? And specifically, how does she do that *without* leaving the descriptive realm?

4 Naturalists include M. Ananth, *In Defense of an Evolutionary Concept of Health: Nature, Norms and Human Biology* (Aldershot: Ashgate, 2008); Boorse, "On the Distinction between Disease and Illness," 49–68; C. Boorse, "What a Theory of Mental Health Should Be," *Journal for the Theory of Social Behaviour* 6 (1976), 61–84; C. Boorse, "Health as a Theoretical Concept," *Philosophy of Science* 44 (1977), 542–73; C. Boorse, "Concepts of Health," in D. van de Veer, and T. Regan (eds), *Health Care Ethics: An Introduction* (Philadelphia: Temple University Press, 1987), 359–93; Boorse, "A Rebuttal on Health"; C. Boorse, "Concepts of Health," in F. Gifford (ed.), *Philosophy of Medicine* (Oxford: Elsevier, 2011), 13–64; C. Boorse, "Replies to my Critics," *The Journal of Medicine and Philosophy* 39 (2014), 648–82; J. Garson and G. Piccinini, "Functions Must Be Performed at Appropriate Rates in Appropriate Situations," *British Journal for the Philosophy of Science* 65 (2014), 1–20; D. Hausman, "Is an Overdose of Paracetamol Bad for One's Health?" *British Journal for the Philosophy of Science* 62 (2011), 657–68; D. Hausman, "Health, Naturalism and Functional Efficiency," *Philosophy of Science* 79 (2012), 519–41; L. R. Kass, "Regarding the End of Medicine and the Pursuit of Health," *The Public Interest* 40 (1975), 11–42; R. Kendell, "The Concept of Disease and Its Implications for Psychiatry," *British Journal of Psychiatry* 127 (1975), 305–15; J. G. Scadding, "Health and Disease: What Can Medicine Do for Philosophy?" *Journal of Medical Ethics* 14 (1988); 118–24; J. G. Scadding, "The Semantic Problem of Psychiatry," *Psychological Medicine* 20 (1990), 243–48; T. Schramme, "A Qualified Defence of a Naturalist Theory of Health," *Medicine, Health Care, and Philosophy* 10 (2007), 11–17; and Szasz, "The Myth of Mental Illness," 113–18.

One could think of this as a problem of *natural normativity*,[5] for instance: How do we arrive at natural norms for eyes? But I am reluctant to use that terminology because it seems to carry a whiff of value-ladenness—of "ought"—about it. The central aim of the naturalistic project is precisely to avoid such value-ladenness. It is to state a *descriptive* norm, from which no ought follows, not even weakly. This kind of "norm" could thus never establish whether Fido *should* be able to wag his tail, only that there is a non-ought-carrying tail-wagging norm that Fido does or does not comply with. No "ought" follows. I shall therefore not call this the "normativity problem" but the "descriptivity problem."

As foreshadowed by early anti-psychiatrists and in the context of the DSM debate,[6] the naturalist's strategy for tackling the descriptivity problem has been to appeal to notions of biological function and dysfunction. In the literature on philosophy of biology there are two dominant proposals for analyzing biological function. Accordingly, two closely related but subtly different naturalistic accounts of disease have been proposed: a statistical account by Christopher Boorse and an etiological one by Jerome Wakefield.

Disease as Dysfunction

Christopher Boorse employs a so-called causal role account of function to define health and disease, where functions are the causal contributions made by traits to the organism's goals: survival and

5 P. S. Davies, *Norms of Nature* (Cambridge, MA: MIT Press, 2001), employs this terminology.

6 Szasz (see above, n. 1) defined a disorder as a dysfunction or lesion at a structural, cellular, or molecular level. He then submitted that no such lesion is present in so-called mental disorders; therefore these aren't disorders but mere problems in living. Later naturalists, by contrast, argued that such conditions *do* present such dysfunctions or lesions, though not at the level of the brain but the *mind* ("The Myth of Mental Illness," 113–18). E.g., Boorse, "What a Theory of Mental Health Should Be," 61–84; D. Papineau, "Mental Disorder, Illness and Biological Dysfunction," in A. Griffiths (ed.), *Philosophy, Psychology and Psychiatry* (Cambridge: Cambridge University Press, 1994), 73–82.

reproduction.[7] The heart, for example, makes a causal contribution to the organism's survival and reproduction by pumping blood. This is therefore its function. Boorse then offers a *statistical* solution to the descriptivity problem: normal functions—a.k.a. health—are those contributions to survival and reproduction that are statistically typical. This appeal to statistics is meant to distinguish normal functions, first, from *accidental* functions. An accidental function is a causal contribution to survival and reproduction that is not a biological function, as when my heart contributes to my survival by making noises or emitting a weak electrical current that allows a search and rescue team to find me after an earthquake. Second, it distinguishes normal, healthy function from *subnormal* function: a damaged heart that may still pump blood well enough to keep me alive but not well enough to be healthy or normal. Subnormal function and other departures from statistically normal function are dysfunctions, and hence departures from health on Boorse's account.

The alternative account of biological function and health, employed by Wakefield, solves the descriptivity problem in a historical fashion.[8]

---

7 C. Boorse, "On the Distinction between Disease and Illness," 49–68; C. Boorse, "Wright on Functions," *Philosophical Review* 85 (1976), 70–86; see further the references in note 4 above. Causal role accounts of function are also known as "Cummins Functions" (R. Cummins, "Functional Analysis," *The Journal of Philosophy* 72 (1975), 741–65).

8 J. C. Wakefield, "Disorder as Harmful Dysfunction: A Conceptual Critique of DSM-III-R's Definition of Mental Disorder," *Psychological Review* 99 (1992), 232–47; J. C. Wakefield, "The Concept of Medical Disorder: On the Boundary between Biological Facts and Social Values," *American Psychologist* 47 (1992), 373–88; J. C. Wakefield, "Dysfunction as a Value-Free Concept: A Reply to Sadler and Agich," *Philosophy, Psychiatry and Psychology* 2 (1995), 233–46; J. C. Wakefield, "Evolutionary versus Prototype Analyses of the Concept of Disorder," *Journal of Abnormal Psychology* 108 (1999), 374–99; J. C. Wakefield, "Mental Disorder as a Black Box Essentialist Concept," *Journal of Abnormal Psychology* 108 (1999), 465–72; J. C. Wakefield, "Spandrels, Vestigial Organs, and Such: Reply to Murphy and Woolfolk's 'The Harmful Dysfunction Analysis of Mental Disorder,'" *Philosophy, Psychiatry and Psychology* 7 (2000), 253–69. The etiological account of function is developed and defended in detail by K. Neander, "Functions as Selected Effects: The Conceptual Analyst's Defense," *Philosophy of Science* 58 (1991), 168–84; K. Neander, "The Teleological Notion of 'Function,'" *Australasian Journal of Philosophy* 69 (1991), 454–68; and R. G. Millikan, *Language, Truth and Other Biological Categories* (Cambridge, MA: MIT Press, 1984); R. Millikan, "In Defense of Proper Functions," *Philosophy of Science* 56 (1989), 288–302. Wakefield is often not considered a naturalist because he maintains that only *harmful* dysfunction is a disorder. That seems to me an unhelpful way of classifying the debate: he sufficiently resembles naturalists to be classified as such because he defines disease *in part* as biological dysfunction. See E. Kingma, "Health and Disease: Social Constructivism as a Combination of Naturalism and

On this view, functions of traits are the effects that explain why traits were naturally selected.[9] Thus the *pumping* effect of my heart is its function because the pumping effect of hearts explains the differential reproductive success of my ancestors and hence my present existence.

The difference between these accounts is subtle but not irrelevant. Consider the turtle's flippers.[10] These (or so the story goes) were selected for their current shape and form because of their ability to propel turtles in water. But turtles now also employ their flippers for the useful purposes of digging nest-holes and burying eggs. According to Boorse's account, the function of turtle flippers is swimming, digging, *and* burying—because all of these contribute to survival and reproduction of the turtle. Wakefield, by contrast, must argue that flippers' function is only their etiological function, that is, the effect for which they were selected: swimming. Whether they also dig and bury is neither here nor there. The upshot of this is that if we could imagine a flipper that retained its swimming ability but lost its burying and digging ability (for example because a mutation in its muscle fibers means it overheats when exercised outside the water), then this flipper would be disordered according to Boorse but not Wakefield.[11]

This example immediately illustrates what I and others have claimed is a key problem for Wakefield: it appears that our bodies, and especially our minds, perform many functions that they were never selected to perform. Right now, for example, I am writing and reading a foreign language, typing, sitting on a chair, manipulating abstract symbols,

---

Normativism," in H. Carel and R. Cooper (eds), *Health, Illness and Disease: Philosophical Essays* (Durham, NC: Acumen, 2012).

9 Neander, "Functions as Selected Effects," 168–84; and Neander, "The Teleological Notion of 'Function,'" 454–68; Millikan, *Language, Truth and Other Biological Categories*; and Millikan, "In Defense of Proper Functions," 288–302.

10 M. Perlman, "The Modern Philosophical Resurrection of Teleology," *The Monist* 87 (2004), 3–51.

11 Whether a Wakefield-type of account truly fails to account for digging being a function is a matter of dispute that hinges in part on the role of maintenance selection. See, e.g., P. Godfrey-Smith, "A Modern History Theory of Functions," *Noûs* 28 (1994), 344–62.

and so on. None of these functions, arguably, were ones that my mind or body was selected to perform, even though performing them must—like the flipper's digging ability—at least in part rely on other, selected, functional mechanisms. It thus appears that Wakefield's account can only ever single out a subset of the many things we do with our bodies and minds as subject to health and disease judgments. Other performances—such as the flipper's digging—simply fall outside the realm of health and disease altogether. This does not square well with modern medicine that, when push comes to shove, is interested in what our bodies do for us now, not in the performance of body parts that drove their selection. As the turtle's flipper illustrates, selection and useful function can come apart. But modern medicine (if it were to be interested in turtles) would surely want to consider the digging-disabled turtle ill. I therefore favor Boorse's account.[12]

The Biostatistical Theory and Its Critics

So, does Boorse's biostatistical theory (hereafter BST) of health map onto our intuitions and succeed in solving the descriptivity problem? Since its inception, nearly forty years ago, BST has been seen as the most viable naturalistic candidate and, as such, has been subjected to a staggering array of criticisms. Broadly speaking these criticisms are of three kinds. One set of criticisms contends that BST fails to map onto our intuitions about particular conditions and must thus be discarded.[13] A second set, which perhaps aren't really criticisms, but friendly improvements, propose amendments or alternative versions

---

12 E. Kingma, "Naturalist Accounts of Disorder," in K. W. M. Fulford et al. (eds), *Oxford Handbook of Philosophy and Psychiatry* (Oxford: Oxford University Press, 2013). See also D. Murphy and R. L. Woolfolk, "The Harmful Dysfunction Analysis of Mental Disorder," *Philosophy, Psychiatry and Psychology* 7 (2000), 241–52; and D. Murphy and R. L. Woolfolk, "Conceptual Analysis versus Scientific Understanding: An Assessment of Wakefield's Folk Psychiatry," *Philosophy, Psychiatry and Psychology* 7 (2000), 271–93.

13 Boorse, "A Rebuttal on Health," painstakingly documents and responds to these criticisms; and he does so again to more recent ones in Boorse, "Concepts of Health," 13–64, and Boorse, "Replies to my Critics," 648–82.

of Boorse's account, which are meant to preserve its spirit of naturalism.[14] This is often done in response to a third set, which contends that BST is not value-free.

One version of the latter criticism focuses on Boorse's use of reference classes. These reference classes—age, sex, and (perhaps) race—are groups that ground statistical abstraction; what is healthy is not what is normal per se, but what is statistically normal for a reference class. This move is necessary because statistical normality alone is too crude to define normal function: normal toddlers would be thoroughly dysfunctional if we compared them with what is normal in the whole population. But what justification can Boorse give for employing the reference classes he uses? If, as the criticism maintains, Boorse's is to be a value-free account of health, then his use of reference classes needs to be justified. Boorse needs to explain why it is acceptable to treat people with a Y chromosome as a separate reference class for the purposes of defining normal testosterone levels but not people who are blind as a separate reference class for the purposes of defining normal vision. The answer can't—on pain of circularity—be that being male is normal and healthy, but that being blind or deaf is not. But Boorse has not provided another answer.[15] This puts pressure on the idea that, in his use of reference classes, Boorse captures *natural* rather than social or evaluative norms.

BST faces a similar problem when it comes to environments. Physiological functions are very varied, specific, and fine-grained—often operating within a very narrow and situation-specific norm: clotting factors and blood platelets must clot blood when there is a bleed or damage to the blood vessel. But they must *not* clot when there

---

14 E.g., Garson and Piccinini, "Functions Must Be Performed at Appropriate Rates in Appropriate Situations," 1–20; Hausman, "Is an Overdose of Paracetamol Bad for One's Health?" 657–68; Hausman, "Health, Naturalism and Functional Efficiency," *Philosophy of Science* 79 (2012), 519–41; P. Schwartz, "Defining Dysfunction: Natural Selection, Design, and Drawing a Line," *Philosophy of Science* 74 (2007), 364–85.

15 E. Kingma, "What Is It to Be Healthy?" *Analysis* 67 (2007), 128–33.

isn't such damage. Statistically normal function must therefore be defined as what is normal relative to a particular environment or situational demand. But that raises a challenge similar to that of reference classes: Why are sunshine, sleeping in the sand, sprinting, and sex normal situations or environments, in which the statistically typical function is healthy, but Paracetamol overdose, poliovirus exposure, and a pneumococcal infection *not* situations or environments in which the statistically normal function is healthy? Again, the answer can't, on pain of circularity, be that the former set of environments is normal or healthy, and the latter unhealthy.[16]

Naturalism and Value

Whether or not BST can survive such criticisms, questions about values cannot be ignored. Even the staunchest naturalists acknowledge that a naturalist account of health is at best a theoretical notion, which will have to be supplemented with value judgments insofar as it is employed in an actual social or policy context.[17] It is *because* certain functions are also disvaluable that they become appropriate for medical treatment and/or social security. If naturalist accounts of health were never brought in contact with values, they could not be relevant to policy at all. So whether values form *part of* our best naturalist accounts or are merely added to it when theoretical accounts of health and disease meet practice, they matter. It is therefore to questions of value that we shall now turn.

16 E. Kingma, "Paracetamol, Poison and Polio: Why Boorse's Account of Function Fails to Distinguish Health and Disease," *British Journal for the Philosophy of Science* 61 (2010), 241–64. See Hausman, "Is an Overdose of Paracetamol Bad for One's Health?" 657–68; and Garson and Piccinini, "Functions Must Be Performed at Appropriate Rates in Appropriate Situations," 1–20, for a response; E. Kingma, "Situational Disease and Dispositional Function," *British Journal for the Philosophy of Science* 67/2 (2015), 391–404, responds to Hausman.

17 See, e.g., Boorse, "On the Distinction between Disease and Illness," 54–55 and 60; "Health as a Theoretical Concept," 544; "A Rebuttal on Health," 11, 12–13, 55, and 95–99.

Two Questions for Normativism

Normativists believe that health and disease are value-laden concepts. But although the vast majority of commentators seem to hold a version of this view, normativist accounts have not received anything like the amount of attention or detailed scrutiny that naturalist proposals have received. In this section I aim to partially rectify that, as well as consider *why* this is the case.

When we consider *normativism* about health and disease, a first thing to note is that this label unites an extraordinarily diverse group of views.[18] What these positions share is the view that health and disease are in some sense value-laden. But *how* values come into the disease concepts, *what* (those) values are, and *how* health and disease should be defined, are questions on which normativists are anything but unified.

We might depict this state of the literature as one where there is a small, dense, and homogenous core in the middle, which represents

18 Normativists include G. J. Agich, "Disease and Value: A Rejection of the Value-Neutrality Thesis," *Theoretical Medicine* 4 (1983), 27–41; K. D. Clouser, C. M. Culver, and B. Gert, "Malady a New Treatment of Disease," *The Hastings Center Report* 11 (1981), 29–37; K. D. Clouser, C. M. Culver, and B. Gert, "Malady," in J. M. Humber and R. F. Almeder (eds), *What Is Disease?* (Totowa, NJ: Humana Press, 1997); Cooper, "Disease," 263–82; R. Cooper, *Classifying Madness: A Philosophical Examination of the Diagnostic and Statistical Manual of Mental Disorders* (Dordrecht: Springer, 2005); H. T. Engelhardt, "Ideology and Etiology," *Journal of Medical Philosophy* 1 (1976), 256–68; K. W. M. Fulford, *Moral Theory and Medical Practice* (Cambridge: Cambridge University Press, 1989); W. Goosens, "Values, Health and Medicine," *Philosophy of Science* 47 (1980), 100–15; J. Margolis, "The Concept of Disease," *Journal of Medicine and Philosophy* 1 (1976), 238–55; L. Kopelman, "On Disease: Theories of Disease and the Ascription of Disease: Comments on 'The Concepts of Health and Disease,'" in H. T. Engelhardt and S. F. Spicker (eds), *Evaluation and Explanation in the Biomedical Sciences* (Dordrecht: Reidel, 1975), 143–50; L. Nordenfelt, *On the Nature of Health: An Action-Theoretic Approach* (Dordrecht: Reidel, 1987); L. Nordenfelt, *Quality of Life, Health and Happiness* (Aldershot: Ashgate, 1993); L. Nordenfelt, *Health, Science and Ordinary Language* (Amsterdam: Rodopi, 2001); L. Nordenfelt, "The Concepts of Health and Illness Revisited," *Medicine, Health Care, and Philosophy* 10 (2007), 5–10; L. Nordenfelt, "Establishing a Middle-Range Position in the Theory of Health: A Reply to my Critics," *Medicine, Health Care, and Philosophy* 10 (2007), 29–32; L. Reznek, *The Nature of Disease* (London: Routledge, 1987); and C. Whitbeck, "Four Basic Concepts of Medical Science," *PSA: Proceedings of the Biennial Meeting of the Philosophy of Science Association* 1 (1978), 210–22.

naturalism.[19] Around this densely compact naturalist core float many normativist positions, which are widely dispersed in all directions and appear to share or agree on almost nothing—except that they all reject naturalism.

In this kind of situation one would expect a sizeable literature on the relative merits of different normativist proposals: we should see normativists jockeying with each other to establish which normativist account is the best, as well as competing with naturalists to establish whether naturalism or normativism is right. But perhaps one of the most surprising features of the literature on health and disease is that former type of engagement is almost entirely lacking; normativists don't engage with one another. Consider, for example, Cooper who structures her paper, first, by discussing and rejecting naturalism, and, second, by developing her own alternative normativist account.[20] What she does *not* do is consider other main normativist accounts: she does not point out why they are unsatisfactory or defend why her account is superior or even needed. It is almost as if other normativists don't exist.[21] This is not a particular criticism of Cooper; it is typical of the normativist literature that nearly always proceeds by first criticizing Boorse—who has become the referential point of departure—before positing a normativist alternative.

The upshot of this is twofold. First, the relationship of different normativist accounts to each other is, at best, unclear. In the picture I just sketched, the "points" that represent normativist accounts are not in any way related or structured; it is not clear whether they are similar or dissimilar and whether they could be placed on—say—axes that would structure the option space. Second, normativist accounts have not had the benefit of scrutiny and critical engagement.

19 For—although, as we saw, naturalists can subtly differ on how they interpret *dysfunction*—they are otherwise in very close agreement both on what it means to give a value-free account of disease and on what that value-free account consists in: biological dysfunction.

20 Cooper, "Disease."

21 Though Cooper criticizes Aristotelian views in a separate paper: R. Cooper, "Aristotelian Accounts of Disease—What Are They Good For?" *Philosophical Papers* 36 (2007), 427–42.

For whatever one may think of naturalists' success in responding to their many challengers, it is fair to say that the resulting dialogues have improved the debate, forcing a clarification and sometimes improvement of positions, making sure that there is now a relative consensus on the strengths and weaknesses of naturalism; how naturalist claims are to be interpreted; and the implications of the different accounts. Normativist accounts, in contrast, are virtually untested territory.

Why do normativists not engage with one another? One reason may be that people simply have not realized that normativist accounts are so heterogeneous. Or perhaps they simply don't feel confident that normativism is attractive enough to warrant debates within normativism, rather than in defense of normativism. A third possibility is that because naturalism is widely seen as desirable, though probably false, the burden of proof seems (subtly) to fall on the naturalist, with normativism appearing as the default view. The outcome is that naturalism is defended by positing accounts that are then defended and criticized, resulting in their being well described and well tested, whereas normativism is defended by *criticizing naturalism*, rather than by a positive defense—which requires a clear articulation—of any particular refined version of a normativist account.

Whichever explanation is correct—and I think all have some merit—they do clearly indicate that normativism is not very easily, and perhaps should not at all be, discussed as one single position. But I have no space in this chapter to compare and contrast normativist accounts or consider them one by one in great detail. What I will do instead is posit two general questions that I think any normativist account has to face, which I hope will stimulate a more mature debate between, and resulting refinement of, different normativist positions. These questions concern the circumscription problem, and the relation between values, health, and disease.

The Circumscription Problem

The circumscription problem is the problem of giving good criteria that delineate, amongst *all* bad conditions—domestic disagreements, bad days at work, economic downturns, miserable weather, and flat tires—only those bad conditions that are also *diseases* (which none of the above are). This is an important question for any normativist account because value-laden definitions of disease nearly always overgeneralize. For any normativist account's central evaluative criterion there will always be many more conditions than *diseases* alone that will meet it. To illustrate this, and also by way of introduction to the different substantial normativist accounts on offer, I shall briefly discuss how each fares with respect to the circumscription problem. I contend that, at first sight, no single one performs very well.

First, consider Clouser, Culver, and Gert, who define disease as a condition, not caused by a rational belief or desire, that incurs or significantly increases the risk of incurring a harm or evil.[22] A "harm or evil" is to be interpreted liberally, to include limits to freedom or pain. Thus a broken leg is a disease because it limits your freedom and causes pain, but running in a marathon is not a disease because, even though it is very painful, it is due to a rational desire. This account, however, is far too inclusive. Bodily conditions that limit freedom or cause pain but that are not due to a rational belief or desire include many that we would not want to consider diseases. The need to sleep and go to the bathroom, for example, limit your freedom, but they are not diseases.[23] In a similar vein, the possession of ovaries and a womb puts you at significant risk of pain, discomfort, and limitations on your freedom in the form of menstruation—but nonetheless it is not a

22 Clouser, Culver, and Gert, "Malady a New Treatment of Disease," 29–37; Clouser, Culver, and Gert, "Malady," in Humber and Almeder.

23 See also M. Martin, "Malady and Menopause," *Journal of Medicine and Philosophy* 10 (1985), 329–37 and Boorse, "A Rebuttal on Health," 43–4. Note that although sleep may be rationally desired, the need for sleep is not due to a rational desire (in the way that, for example, blisters in the pursuit of a new garden hedge are).

disease. The same holds for being male, which limits one's freedom to produce milk for one's own children.

A second account by Whitbeck suffers from a similar problem. She defines diseases as conditions that people want to prevent because they interfere with the bearer's capacity to do things people commonly wish and expect to be able to do.[24] This is also too inclusive. Normal hair growth on legs, faces, and armpits, for example, is a physical process that many treat or prevent because it interferes with a culturally common desire to appear hairless in certain places. It is therefore a disease on Whitbeck's account.

Nordenfelt defines diseases as second-order inabilities to reach vital goals.[25] Vital goals, in his view, are the goals that are jointly necessary and sufficient to achieve minimal happiness. Second-order abilities are abilities to gain first-order abilities. Thus, a first-order ability is, say, playing the violin (which I cannot do). A second-order ability is the capacity to learn to play the violin if appropriate training is provided (my possession of which, given my age, is starting to look implausible). This account, too, seems hopelessly over-inclusive. First, consider that someone may have vital goals that constitute "expensive preferences" or that are *highly ambitious*. For such a person, conditions that we do not ordinarily consider diseases—such as lacking the ability to achieve amazing athletic or artistic prowess, or financial success, would become diseases.[26] Second, there are many conditions that would significantly impact most people's second-order abilities to reach vital goals, yet we still do not ordinarily think of them as *diseases*. These include, for example, the lack of social or financial resources.

---

24 Whitbeck, "Four Basic Concepts of Medical Science," 210–22.

25 Nordenfelt, *On the Nature of Health*; Nordenfelt, *Quality of Life, Health and Happiness*; Nordenfelt, *Health, Science and Ordinary Language*; Nordenfelt, "The Concepts of Health and Illness Revisited," 5–10; Nordenfelt, "Establishing a Middle-Range Position in the Theory of Health," 29–32.

26 Schramme, "A Qualified Defence of a Naturalist Theory of Health," 11–17.

It therefore seems that neither Whitbeck nor Clouser, Culver, and Gert, nor Nordenfelt have a convincing solution to the circumscription problem. Or, if they bite the bullet, then their accounts of disease are extremely revisionary—to the point of being thoroughly counterintuitive. And that requires substantial defense in its own right.

A fourth group of "normative" accounts is neo-Aristotelian.[27] I put "normative" in scare quotes, because the thinking that supports these accounts denies a distinction between facts and values and so might reject their characterization as normative. Simply put, Aristotelian accounts suppose there is a natural norm for all biological entities that governs how humans and other biological entities ought to be. These norms are grounded in the kind of thing an entity is. Thus, bees are colony-living pollen collectors. A "good" bee is therefore one that is good at living in the colony and good at collecting pollen. These natural norms apply to both what we might think of as our somatic realm—that we should have two arms, ten fingers, and a well-functioning liver, for example—as well as our mental lives and the way we should live: they provide an account of our flourishing. According to neo-Aristotelian accounts of health, health is Aristotelian normal function.

In response to these accounts one might once again argue that these accounts are over-inclusive. Cooper, for example, argues that neo-Aristotelians face a problem because they can't distinguish between vices and diseases: both impair flourishing. [28] As we saw in previous chapters of this volume, that is exactly in line with what earlier Aristotelians wanted to say: vices are diseases of the soul.[29] Nonetheless it does not map onto present discussions about disease, which are

---

27 P. Foot, *Natural Goodness* (Oxford: Clarendon Press, 2001); C. Megone, "Aristotle's Function Argument and the Concept of Mental Illness," *Philosophy, Psychiatry and Psychology* 5 (1998), 187–201; Megone, "Mental Illness, Human Function and Values," *Philosophy, Psychiatry and Psychology* 7 (2000), 45–65.

28 Cooper, "Aristotelian Accounts of Disease?," 427–42.

29 See the contributions of James Allen and Peter Adamson in the present volume.

precisely concerned with separating diseases from vices, bad choices, and bad personal traits—think, for example, of the insanity defense in court or the discussion of homosexuality and psychiatry that this chapter started with.[30] This account, too, is highly revisionary.

A fifth account of disease is offered by Cooper, who defines diseases as conditions that are (1) bad for the sufferer; (2) unlucky/abnormal (which serves to exclude ordinary conditions such as hair growth); and (3) are deemed within the remit of the medical profession.[31] Although this account also suffers from inclusivity—unwanted pregnancies, for example, are a disease on her account[32]—I think it is the best performer amongst contenders on the circumscription problem. Nonetheless I think the *method* by which Cooper achieves this success undermines it. After limiting conditions to those that are (1) bad and (2) unlucky/abnormal, the main job of solving the circumscription problem is performed by the third criterion: whether society considers these conditions within the remit of the medical profession. But that seems unsatisfactory. First, if we appeal to an account of disease in the hope that it would *help us decide* what conditions fall within the medical profession—as many commentators, including Cooper,[33] do—then this is hopelessly circular and unhelpful. It cannot *explicate*; it can merely track. Second, it makes diseases prone to *vary* with a societies' perception of the condition and of what medicine might do.[34]

In short, in the present overview we have seen not just quite *how* diverse normativist accounts are, but also that a serious problem for pretty

30 I don't mean to suggest that homosexuality is bad (I don't think it is); merely that, historically, *when* people agree on their disvaluation of homosexuality, they must still tackle the question of whether it is a disease or some other kind of trait.

31 Cooper's account is very similar to that offered by Reznek, *The Nature of Disease*. The difference between Cooper and Reznek is that Cooper gives an anthropological account of medical treatment, whereas Reznek defines it in terms of surgical and pharmacological interventions. The former is more convincing; think of recreational opiate use or doping in sport (Cooper, "Disease," 278).

32 Cooper, "Disease," 278–79, acknowledges and defends this commitment.

33 See, e.g., R. Cooper and C. Megone, "Introduction," *Philosophical Papers* 36 (2007), 339–41.

34 I provide more detail in E. Kingma, "Cooper on Disease" (unpublished).

much all normativist accounts is piecing together the *non-evaluative* components of their account in such a way as to convincingly solve the circumscription problem—for any evaluative component, it seems, is over-inclusive by nature.

How Do Health, Disease, and Value Relate?

The second important challenge for normativism is that they tell us something substantive about values beyond the general, not-particularly-informative and highly-open-to-different-interpretations claim that health and disease are *in some sense* value-laden. This challenge is particularly interesting if, like me, you are convinced that even supposedly naturalist accounts include evaluative components. This challenge comprises at least the following two questions: First, *how* are health and disease value-laden, or *how* do evaluative considerations enter into accounts of health and disease? Second, *what* are these values, and how do they need to be understood?

As the accounts discussed above illustrate, there are different ways in which values appear in accounts of definitions of health and disease, and this source of variation is not something that has received a lot of attention. Without claiming to be exhaustive, let me review some of the options. First, an evaluative criterion such as "bad" or "harmful" can appear as one of the criteria that a condition must meet to be a disease. This is how normativism is most widely understood.[35] But here is a different way in which concepts might be evaluative: health and disease could be so-called thick concepts, where the negative evaluation is not separable as one of the criteria but inextricably intertwined with a descriptive content.[36] I think this is an interesting possibility, but not one that seems to have been explored.[37] A third option is to make health

---

35 See, e.g., Cooper, "Disease," 263–82; and Reznek, *The Nature of Disease*.

36 See S. Kirchin (ed.), *Thick Concepts* (Oxford: Oxford University Press, 2013).

37 But see J. L. Nelson, "Health and Disease as 'Thick' Concepts in Ecosystemic Contexts," *Environmental Values* 4 (1995), 311–22, for an exception.

and disease derivative of some evaluative notion—as Nordenfelt does in this case of vital goals.[38] Fourth, one might fully equate (or reduce) health and disease to an evaluative criterion—as neo-Aristotelians do in the case of flourishing. Fifth—as we saw in the form of objections to Boorse—accounts may be value-laden if the reasons for selecting and employing particular descriptive concepts in particular roles is or was motivated by evaluative considerations.[39] Whether this latter version is still in opposition to naturalism is, I think, not obvious.

It appears, then, that there are many ways for concepts to be value-laden. Surely one important task for normativists—and one way in which the literature as a whole could make considerable progress—is to get very clear on *how* values enter health and disease concepts. Some normativists are clear about this, as we saw in the above accounts, but many (as will be clear in a moment) aren't.

A second question that normativists should answer takes us into ethics proper. It is what the relevant values involved in health and disease concepts are, and how they should be understood. Once again there is a wide scope for variety here; we can distinguish several different proposals in the accounts already discussed. Contrast, for example, the neo-Aristotelians' proposal that refers to an impairment of flourishing with Whitbeck's idea that the relevant evaluative criterion is an impairment of common wishes and expectations. These are very different. A much wider range of variation emerges if we also consider the things that normativists who fall short of offering a full account of health and disease say about values. Compare, for example: (1) a condition is a disease if and only if it is bad for you;[40] (2) to label something

38 See also Clouser, Culver and Gert, "Malady a New Treatment of Disease," 29–37; and Clouser, Culver, and Gert, "Malady," in Humber and Almeder; Whitbeck, "Four Basic Concepts of Medical Science," 210–22.

39 Kingma, "What Is It to Be Healthy?," 128–33; Kingma, "Health and Disease."

40 Clouser, Culver and Gert, "Malady a New Treatment of Disease," 29–37; and "Malady"; Cooper, "Disease," 263–82; and Reznek, *The Nature of Disease*; Wakefield, "Disorder as Harmful Dysfunction," 232–47; J. C. Wakefield, "The Concept of Medical Disorder," 373–88; and to some extent Goosens, "Values, Health and Medicine," 100–15.

a disease is to express disapproval of the condition;[41] (3) our concepts of health and disease are grounded in nonnaturalistic concerns, such as pain, discomfort, and disability;[42] (4) health and disease are relative to subjective goals, values, and desires;[43] (5) to label something a disease is to commit to an obligation to treat it;[44] and (6) health and disease are relative to social expectations.[45] All these claims are professed by normativists—but they are quite different. They do not just vary on the first question—how values appear in definitions of health and disease—but also on the second one: what those values are and how they are to be interpreted. Such variation is not, of course, a problem in itself; ethicists also hold very different views on what values are. Thus this might be taken to indicate that normativists simply hold different and quite sophisticated ethical views. For example, the claim that to label something a disease is to express disapproval of the condition might sound like a non-cognivist approach, whereas the claim that to label something a disease is to commit to treatment might sound like prescriptivism, while the claim that disease is relative to subjective

41 For example H. T. Engelhardt, "The Concepts of Health and Disease," in H. T. Engelhardt and S. F. Spicker (eds), *Evaluation and Explanation in the Biomedical Sciences* (Dordrecht: Reidel, 1975), 127 and 137; Margolis, "The Concept of Disease," 242; and Nordenfelt, "The Concepts of Health and Illness Revisited"; Whitbeck, "Four Basic Concepts of Medical Science," 210–22, disagrees.

42 Cooper, "Disease," 263–82; Engelhardt, "The Concepts of Health and Disease," 127; Engelhardt, "Ideology and Etiology," 262, Goosens, "Values, Health and Medicine"; Margolis, "The Concept of Disease," 242; and Nordenfelt, "The Concepts of Health and Illness Revisited," 5–10.

43 Nordenfelt, *On the Nature of Health*; Nordenfelt, *Health, Science and Ordinary Language*, and "The Concepts of Health and Illness Revisited"; Engelhardt, "Ideology and Etiology," 256–68; Whitbeck, "Four Basic Concepts of Medical Science," 210–22; and, depending on the account of badness that she adopts, Cooper, "Disease," 263–82. Goosens, "Values, Health and Medicine," denies this claim.

44 Engelhardt, "The Concepts of Health and Disease," 127, 137; R. L. Spitzer and J. Endicott, "Medical and Mental Disorder: Proposed Definition and Criteria," in R. L. Spitzer and D. F. Klein (eds), *Critical Issues in Psychiatric Diagnosis* (New York: Raven Press, 1978), 18; and possibly Cooper, "Disease." But Goosens, "Values, Health and Medicine"; Kopelman, "On Disease"; and Whitbeck, "Four Basic Concepts of Medical Science," disagree.

45 Engelhardt, "Ideology and Etiology," 265; L. S. King, "What Is Disease?" *Philosophy of Science* 21 (1954), 193–203; Margolis, "The Concept of Disease," 247; and Whitbeck, "Four Basic Concepts of Medical Science"; Kopelman, "On Disease," disagrees.

goals might be a version of moral naturalism. Nonetheless, I caution against reading normativists in this way. Most normativists profess the above claims indiscriminately and in often inconsistent ways;[46] sophisticated ethical positions cannot be readily applied to most forms of normativism.[47] It would be helpful, however, if normativists did develop such clearer and more sophisticated views. For the unanalyzed claim "health is something that is (essentially) good, and disease is something that is (essentially) bad"—or, worse, "Boorse is wrong"—is simply not good enough as an analysis of the relations among health, disease, and value. Our understanding of health and disease would be better served if less time was spent arguing whether naturalism or normativism was right, and more time was spent on *how* normativists get things right. This is not just a point that is relevant to normativism; it is also important if naturalists are right. For, as I argued, values are at the very least crucially implicated in the translation of theoretical, naturalist accounts to applied medical, social, and political domains; but they are probably also implicated in the nature of dysfunction, even according to naturalism.

Here is an additional reason why it is important to be clear about the exact role and understanding of values in any particular normativist position. An often-heard worry about normativism is that it would lead to social and historical relativism.[48] To illustrate this point people often refer to past "diagnoses" that we now consider not only to be mistaken but also to indicate—or so it is argued—an illegitimate intrusion of problematic social and evaluative judgment on disease. Examples include masturbation (sexual morale); hysteria (sexism and sexual morality); "drapetomania," which supposedly was the disease suffered by slaves who ran away from their master (racism); and

46 Though there are exceptions, e.g., Cooper, "Disease," 263–82; Foot, *Natural Goodness*, and Fulford, *Moral Theory and Medical Practice*.

47 See J. Simons, "Beyond Naturalism and Normativism: Reconceiving the 'Disease' Debate," *Philosophical Papers* 36 (2007), 343–70, for a similar conclusion.

48 See, e.g., Boorse, "Concepts of Health," 13–64.

the locking away of political dissidents in psychiatric institutions.[49] Normativists, it is thought, can't avoid such worries because they condone an account of health and disease that is determined by contingent and contemporary value judgments.

But whether any specific normativist account must be subject to this worry is by no means a foregone conclusion. It depends on the specific account of values it adopts. To contrast two examples, it seems that Whitbeck, who defines disease by reference to what people commonly wish and expect to be able to do, is committed to at least some forms of cultural and historical relativism. For common wishes and expectations vary substantially with time and place. But neo-Aristotelian accounts, for example, seem explicitly committed to a view of health and human flourishing that is grounded in what humans *are*. Such an account, depending on how precisely it is to be cashed out, may not be subject to social and historical relativism; surely a successful neo-Aristotelian account would not condone that drapetomania and political dissidence are diseases.

Health, Well-Being, and the Good Life

The original debate between naturalism and normativism was centrally concerned with the question whether health and disease were—on the one hand, and as naturalists argued—objective, value-free, and empirical concepts that can then feed into value-driven personal, medical, and political decision-making processes; or—on the other hand—whether they are already, and thoroughly, value-driven. I have cast doubt on this supposed opposition: these options aren't necessarily mutually exclusive, and there aren't single, homogeneous options to be contrasted on both sides. Yet it is not surprising that this "opposition"

49 See, e.g., H. T. Engelhardt, "The Disease of Masturbation: Values and the Concept of Disease," *Bulletin of the History of Medicine* 48 (1974), 234–48; Boorse, "Concepts of Health," gives more examples.

has received much attention and has seemed difficult to resolve. For, I shall now suggest, the two positions map onto two prominent social criticisms of medicine that pull in different directions. These criticisms focus on the relationships among medicine, health policy, and what we may broadly call "the pursuit of the good life."

The Cessation of Paternalism and the Rise of Informed Consent

The first of these criticisms has stressed the need to deemphasize the medical focus on the "good life." This criticism has at least two components. One, which we have discussed, relates to the anti-psychiatrists' worry that medicine should not *impose* or *take an implicit stance on* what the good life is. For this is far too likely to result in psychiatry's reinforcing a set of questionable social norms. Medicine does and should stick to "the facts."

But there is a similar drive that has nothing to do with psychiatry. In very general terms, the second half of the twentieth century has seen a well-documented change in the health-care professional's role *away* from paternalism—and thus away from judging what is best for the patient—toward an (idealized) model of medical decision-making in which the doctor merely provides information and leaves it up to the patient to judge what is best. This move is not so much motivated by worries about the implicit enforcing of social norms, but by a more general—and perhaps liberalist—resistance to and distrust of authority: a recognition of legitimate value-pluralism on the one hand, and an emphasis on personal autonomy on the other. The doctor, we think, may know what is "best" in terms of medical knowledge, but she does not know what is, all else being equal, best for you. Given that people do and should be able to develop very different conception of the good, only you are the judge of that. And even if the doctor does know what is best for you, we now think that it is *still* your right to make your own decisions—and decide otherwise. For that falls under the legitimate exercise of your autonomy.

This approach, however, *requires* that the doctor is able to refrain from value-judgment and restricts her pronouncements—much more narrowly—to the "facts at hand." These facts can then feed into an evaluation, a supported or shared decision-making process, and eventually a normative judgment by the patient. But this type of expectation of our health-care system practically *demands* naturalism: it requires a medical language, and medical concepts, that are "fact-focused" and distinct, if not divorced, from questions about value. On this kind of view, medicine is there for the former, not the latter. We have at least two kind of social demands, then, that broadly speaking push in a "naturalist" direction: in a direction that divorces medical, fact-based judgments from subjective, evaluative non-medical judgments. It is hardly surprising, then, that naturalism has received such prominence and seemed so appealing in the forty years since Boorse's first publication.

Focus on Well-Being, Not Function!

Yet in the same period, we have also seen a movement that pushes in the opposite direction. That is, the persistent complaint that modern medicine is too focused on reductive, biomedical considerations and is not sufficiently attuned to the good; it is not focused on what is important to us. Again, this criticism appears in many forms. A prominent one is the disability critique that has complained, loudly and justly, that a medical model aiming to restore them to the biomedically *normal* functioning, rather than to a position in which they can best do the things important to them, is not good health care.[50] In response we have, thankfully, generally abandoned attempts to biomedically *normalize* people with disabilities. We no longer attempt to teach the paralyzed to walk painfully and for short distances on crutches,

50 See, e.g., R. Amundson, "Against Normal Function," *Studies in History and Philosophy of Biological and Biomedical Sciences* 31 (2000), 33–53.

teach the deaf to speak and lip-read, or teach left-handed people to write with their right hands. Instead we favor wheelchairs, ramps, sign language, and left-handed tools that are much *better* ways of helping people with these disabilities/different abilities to do the important things able-bodied people do with ease: move, communicate, and open cans.

But we can find versions of this criticism in many places. One example is the criticism that medicine, especially medical research, is too bent on output parameters that may be "objective" or easily measurable but are not relevant to patients. Examples are physiological measures of pain or lung capacity rather than subjective pain experience or the ability to do the things that need doing. We can probably view the attraction of "holistic" medicine, "personalized" medicine, and perhaps even certain complementary and alternative treatments in a similar vein. Finally there has been a general move, especially at the end of life, to considering *quality* rather than *quantity* of life of utmost importance. This, too, embodies the idea that health-care professionals should not focus on (curing) diseases, prolonging life, or promoting some objective notion of health—but that they should try to better things for *people* and strive for them to have a good life.

This second kind of push, then, seems to pull away from naturalist accounts of disease. Or, at the least, it would say that the naturalist account of disease or health is not very relevant to health care; it is not what medicine should strive for.

I suggest, then, that the debate between naturalism and normativism exists in the context of two important social movements and critiques of medicine that pull in opposite directions. The one wishes to *divorce* medicine from any value-judgments, making doctors the providers of informative facts only and leaving patients as the evaluative interpreters and action guiders. This is the exact and naive model of naturalism about health and disease: an objective, value-free judgment of health or disease that then feeds into an evaluative decision or application. The other movement criticizes medicine for focusing on notions of health and disease that are *too divorced* from what really matters to us; medicine has lost sight of, and should return to, the promotion of values and the good life.

This social context not only mirrors the supposed naturalist/normativist opposition, but it also encounters the same problems that I previously argued are encountered by these two positions. For just as normativism suffers from expanding the category of diseases beyond all recognition unless it satisfactorily solves the circumscription problem, so the push for a health-care system more focused on the good life runs the risk of not being recognizable as health care anymore—for example, deciding to fund family cruises rather than chemotherapy for the terminally ill because that is what, in fact, will promote well-being. On the other hand, a health-care system focused on a narrow, naturalist account of disease is perhaps too much at risk of forgetting what really matters—well-being. Just like naturalism, it risks no longer being *relevant* to us. The more naturalized an account becomes, the weaker its connection to well-being will be.

It is all the more important, then, to ask what the right kind of connection between health and well-being should be, in light of the difficulties that face both normativists and naturalist accounts and in light of these two social criticisms.

The Relation between Health and Well-Being

One way one might answer that question is by taking the second type of criticism very seriously, and, as a result, start to equate "health" with "well-being." This is exactly what the World Health Organization (WHO) proposed when it famously and aspirationally defined health as "a state of complete physical, mental and social well-being."[51] But that strikes me as the wrong way to respond.

---

51 Preamble to the constitution of the WHO as adopted by the International Health Conference, New York, June 19–22, 1946; signed on July 22, 1946 by the representatives of sixty-one states (Official Records of the WHO, no. 2, p. 100) and entered into force on April 7, 1948.

There are many ways in which one might find this definition problematic, but I want to focus on two related ones. First, if health simply *is* well-being, then what is the point of having the concept of health? Surely merely having well-being (or health, but not both!) is enough? Second, and more substantially, equating health and well-being makes it impossible for us to rationally sacrifice our health in the pursuit of other goods or forms of well-being. So, for example, whilst I am still able to sacrifice my health to selflessly benefit others—by throwing myself in front of a car to save a child, or by donating a kidney to a stranger, for example—it would be logically impossible self-interestedly to sacrifice my health to pursue *other* aspects of my well-being. This means that were I to decide—say—to become unfit, overweight, sedentary, and overtired in the pursuit of my art, my philosophy, or some of my other worldly or spiritual projects that I believe constitute the good life for me, then either I am mistaken about what contributes to my well-being, or I don't sacrifice my health. Similarly if I regularly consume too much alcohol or binge on less-than-healthy food in the pursuit of a fun evening out, or decide to move to a city with worse air quality—which will worsen my asthma—in the pursuit of a better paid and more fulfilling job, I either don't sacrifice my health, or I am mistaken about these decisions' enhancing my well-being. But at least on the face of it, it seems plausible that we can, and do, often sacrifice our health in the pursuit of other things that are important to us. It also seems patronizing and distinctly implausible to think that all we should strive for in life is health.

Now the defender of the "health as well-being" approach might respond in two ways. First, they might say I am stuck in too much of an un-aspirational idea of health; if health genuinely is well-being then I do *not* sacrifice my health in the pursuit of a fulfilling job or fun evening out. Instead I sacrifice something on the physical dimension of health in the pursuit of its other dimensions: social and mental well-being. A different line of response concedes the other horn of the dilemma: I am mistaken in my assessment of well-being. Thus, the defender might say, in setting up these examples I confuse well-being with things that I value or desire; it is perfectly possible for me to sacrifice my well-being in pursuit of something I value, like world peace, the safety of my children, or a political goal. And this would create a similar dilemma. Thus, I might *value* my job in the city, but this simply leads to a conflict between my desires and my well-being/health, which are best served by staying in the country and would be compromised if I pursued city life.

Such responses are possible but hardly convincing. The first just serves to highlight quite *how* revisionary this concept of health as well-being is. Imagine responding to one's doctor, who counsels to cut down on smoking and take more exercise, as follows: "No, doc, I thought about it, and according to the WHO smoking is *good* for my health." That makes little sense. What does make sense is to say: "No, doc, taking everything else into consideration, that just isn't best for me." This connects to the first point: there is a good reason we have two concepts, one for health and one for well-being.

Of course, this boils down to the now familiar point that the attempted equation between health and well-being runs into serious circumscription problems, even if we take a sophisticated and objective stance of what well-being is. It also shows that at least our *ordinary* understanding of health is one that very clearly does *not* map onto well-being but leaves a clear gap between the two. This leads me to think that we can at least take one firm step forward in an investigation of the relationship between health, disease and value: health is *not* well-being. It is only of instrumental value—or, if one has an expansive view of intrinsically valuable things, it is only one out of very many things that are intrinsically valuable.

Is such a firm separation between health and well-being compatible with the kind of social criticisms that push in the direction of well-being? Yes. Consider the "disability" critique. One does not need to equate health with well-being to recognize that ramps and wheelchairs are the best way to help someone who can't walk. Indeed, such an argument may even be better supported by an instrumental view of health. Precisely because health is only of instrumental value, what matters to people is not their health, but what health (ordinarily) allows people to do: the ability to get around. And so the best way to help them is to improve that ability. One might, in fact, consider it a particular advantage of the instrumental view of health that it retains the ability simultaneously to recognize that a person who suffers paralysis does not have perfect, or even normal, health—which puts them within the scope of our health-care system—*and* that moving them in the direction of restoring their *health*—or some other kind of physical or functional "normality"—is not what matters. What matters is the restoration of what they see as important.

Furthermore, one need not define health as well-being if one is to criticize an overly reductive approach to medicine. The problem with focusing on measurable physiological parameters of pain or lung function—rather than experienced pain or breathlessness—is related to the problem of measuring what can be measured, rather than measuring what needs to be measured. It is not a problem of defining disease as dysfunction: pain and breathlessness still indicate what can be deemed a biological dysfunction, but at a higher or less reductive level.

Finally, an emphasis on quality over quantity of life is certainly compatible with viewing health as instrumental, perhaps even more so than an equation of health and well-being. Again, it allows us to recognize someone's illness, and that what really matters is not health or length of life but well-being.

What the instrumental view does particularly well, because it avoids the circumscription problem, is to provide a notion of disease that might still serve as some guide as to whom a health-care system should concern itself with, and what kind of conditions it should seek to relieve.[52] *How* it best relieves those conditions should—of course—be

---

52 Which is a main reason why people seek an account of health and disease in the first place. See, e.g., Cooper and Megone, "Introduction," 339–41.

guided by what matters: well-being, not health (which is only instrumental to it).

The view that health and disease are only instrumental to well-being does not, of course, commit to a "retreat" into some supposedly value-free concept of health and disease. As I have indicated, I am highly suspicious of both the desirability and likelihood of arriving at such a concept. Health is not well-being, but neither is health value-free. Health sits at an intermediate position: it has enough descriptive content to avoid the circumscription problem and play some guiding role in apportioning health care, but it also has enough evaluative content to be normatively relevant. I have indicated that the important and interesting work is in uncovering the right combination of the two, and I have indicated multiple ways in which values can be combined with other, more descriptive factors.

This chapter started by discussing the controversy surrounding psychiatry in the 1960s and 1970s. Those controversies have waxed and waned but have never really gone away. They do not, however, form the main contemporary focus for an interest in health and disease. Controversies in psychiatry have been surpassed by a contemporary focus on global health inequalities, the just distribution of health-care resources, and the rising costs of health care in the western world, as well as the (related) expansion of an interest in life-style diseases and responsibility in health and—in philosophy—a focus on enhancement debates. In these contexts, somatic medicine is, squarely, the focus. Difficult questions about psychiatry, concerning the relation between body and mind and the merging of the social and the biological, may increasingly be avoided by concentrating on the somatic level. But questions about naturalism, normativism, and social norms were never specific to psychiatry. They affect all of medicine and, as I have argued, mirror the social framework within which medicine operates. To make progress in these debates, we need to ask not *whether* values play a role but *how* they do so, and *which* values are important. As a small step in that investigation, I have argued that health cannot be equated with well-being but is instrumental to it. Investigating these questions further is important because in the context of allocating scarce resources between many needy people on a global scale, questions about the relation between health and value are unlikely to diminish in either importance or controversy.

Bibliography

Chinese Primary Sources

Baopuzi neipian 抱朴子內篇, by Ge Hong 葛洪. in *Baopuzi neipian jiaoshi* 抱朴子內篇校釋, ed. Wang Ming 王明. Taibei: Liren shuju, 1981.

Beiji qianjin yaofang 備急千金要方, by Sun Simiao 孫思邈. Taibei: Guoli zhongguo yiyao yanjiusuo, 1990.

Daode zhen jing zhu 道德真經註 DZ 682, by Heshang gong 河上公.

Guanzi Jin zhu jin yi 管子今註今譯, by Guan Zhong 管仲. ed. Li Mian 李勉. Taibei: Taiwan shangwu yinshuguan, 2013.

Han shu 漢書, by Ban Gu 班固, Yan Shigu 顏師古, and Ban Zhao 班昭. Beijing: Zhonghua shuju, 1962.

Huangdi neijing lingshu yijie 黃帝內經靈樞譯解. ed. Yang Weijie 楊維傑. *Zhongguo yiyao congshu* 中國醫藥叢書. Taibei: Tailian guofeng chubanshe, 1984.

Huangdi neijing suwen yijie 黃帝內經素問譯解. ed. Yang Weijie 楊維傑. *Zhongguo yiyao congshu* 中國醫藥叢書. Taibei: Tailian guofeng chubanshe, 1984.

Lie xian zhuan 列仙傳 DZ 294, att. Liu Xiang 劉向.

Lüshi chunqiu jiao shi 呂氏春秋校釋 by Lü Buwei 呂不韋. ed. Chen Qiyou 陳奇猷. Shanghai: Xuelin chubanshe, 1984.

Mawangdui Hanmu boshu 馬王堆漢墓帛書. ed. MWD zhengli xiaozu. Beijing: Wenwu chubanshe, 1985.

Suishu 隋書, by Wei Zheng 魏徵. in *Xin jiao ben Suishu fu suoyin* 新校本隋書附索引, ed. 楊家駱 Yang Jialuo. Taibei: Ding wen shuju, 1980.

Yangxing yanming lu 養性延命錄 DZ 838.

Yuanyou 遠遊, by Qu Yuan 屈原. in *Chuci buzhu* 楚辭補注, ed. Bai Huawen 白化文 and Wang Yi 王逸. *Zhongguo gudian wenxue jiben congshu* 中國古典文學基本叢書. Beijing: Zhonghua shuju, 1983.
Zhuangzi jishi 莊子集釋, by Zhuang Zhou 莊周. ed. Guo Qingfan 郭慶藩 and Wang Xiaoyu 王孝魚. Beijing: Zhonghua shuju, 1995.

Primary Sources

Adamson, P. and P. E. Pormann, "More than Heat and Light: Miskawayh's Epistle on Soul and Intellect," in *Muslim World* 102 (2012), 478–524.
Adamson, P. and P. E. Pormann (trans.), *The Philosophical Works of al-Kindī*. Karachi: Oxford University Press, 2012.
Alsted, J. H. *Panacea philosophica; id est, facilis, nova, et accurata methodus docendi & discendi universam encyclopædiam septem sectionibus distincta . . .* Herborn: N.p., 1610.
Anastassiou, A. and D. Irmer. *Testimonien zum Corpus Hippocraticum*. Götingen: Vandenhoeck, 1997–2012.
Andreae, T. *Disputatio philosophica inauguralis explicans naturam & phaenomena cometarum*. Duisburg: University of Duisburg, 1659.
Annas, J. and J. Barnes (trans.), *Sextus Empiricus: Outlines of Scepticism*. Cambridge: Cambridge University Press, 2000.
Arberry, A. J. (trans.), *The Spiritual Physick of Rhazes*. London: John Murray, 1950.
Arnoldus de Villa Nova, *Hec sunt opera Arnaldi de Villanova que in hoc volumine continentur*. Lyon: Fradin, 1504.
Biesterfeldt, H. H. *Galens Traktat "Dass die Kräfte der Seele den Mischungen des Körpers Folgen"—in arabischer Übersetzung*. Weisbaden: Franz Steiner, 1972.
Bivens, L. and J. Panksepp, *The Archaeology of Mind: Neuroevolutionary Origins of Human Emotions*. New York: W. W. Norton, 2013.
Boerhaave, H. *Institutiones Medicae*. Leiden: Severinus, 1730.
Boerhaave, H. *Dr. Boerhaave's Academical Lectures on the Theory of Physick*. 6 vols. London: W. Innys, 1742–1746.
Brague, R. (trans.), *Al-Razi: La Médecine spirituelle*. Paris: Flammarion, 2003.
Burnett, C. S. F. and D. Jacquart. *Constantine the African and ʿAlī ibn al-ʿAbbās al-Maǧūsī: The Pantegni and Related Texts*. Leiden: Brill, 1994.
Cabrol, B. *Alphabet anatomic*. Tournon: C. Michel and G. Linocier, 1594.
Cabrol, B. *Ontleeding des menschelycken lichaems. Eertijts in't Latijn beschreven door Bartholmaeus Cabrolius*. Amsterdam: Cornelis van Breugel voor Hendrick Laurentsz, 1633.
Calvi, M. F. *Hippocratis Coi medicorum omnium longe principis, Octoginta volumina . . .* Rome: Franciscus Minitius, 1525.

Carrara, E. (ed.), *Petrarch: Secretum*, in G. Martellotti et al. (eds), *F. Petrarca: Prose*. Milan: Ricciardi, 1955.

Chambers, E. *Cyclopoedia: or, an Universal Dictionary of Arts and Sciences*. 2 vols. London: W. Strahan et al., 1728.

Clarke, J. (trans.), *Rohault's System of Natural Philosophy, Illustrated with Dr S. Clarke's Notes, Taken Mostly out of Sir Isaac Newton's Philosophy, With Additions*. 2 vols. London: Printed for James, John, and Paul Knapton at the Crown in Ludgate Street, 1723.

Clauberg, J. *Notae in Cartesii Principia*, in J. T. Schalbruch (ed.), *J. Clauberg: Opera Omnia Philosophica*. 2 vols. Amstelodami: Ianssonio-Waesbergii, Boom, à Someren, & Goethals, 1691.

Cohen, R. S. and Fawcett, C. R. (trans.), *George Canguilhem: The Normal and the Pathological*. New York: Zone Books, 1991.

Coiter, V. *Externarum et internarum principalium humani corporis partium tabulae*. Nuremberg: Gerlatzenus, 1572.

Cottingham, J. et al., *Descartes' Philosophical Writings*. 3 vols. Cambridge: Cambridge University Press, 1985–1991.

De Franco, L. (ed.), *B. Telesio: De rerum natura*. 3 vols. Cosenza: Casa del Libro; Florence: La Nuova Italia, 1965–1976.

De Franco, L. (ed.), *B. Telesio: De somno libellus*, in L. De Franco (ed.), *B. Telesio: Varii de naturalibus rebus libelli*. Florence: La Nuova Italia, 1981, 379–414.

De Franco, L. (ed.), *B. Telesio: Quod animal universum ab unica animae substantia gubernatur*, in L. De Franco (ed.), *B. Telesio: Varii de naturalibus rebus libelli*. Florence: La Nuova Italia, 1981, 188–288.

Deichgräber, K. *Die griechische Empirikerschule: Sammlung der Fragmente und Darstellung der Lehre*. Berlin: Weidmann, 1965.

De Lacy, P. H. (ed. and trans.), *Galen: On the Doctrines of Hippocrates and Plato*. 3 vols. Berlin: Akademie Verlag, 1978–1984.

De Raey, J. *Cogitata de interpretatione*. Amstelaedami: Henricum Wetstenium, 1692.

Di Benedetto, A. (ed.), *Prose di Giovanni della Casa e altri trattatisti cinquecenteschi del comportamento*. Turin: UTET, 1970.

Diderot, D. *Oeuvres de Denis Diderot*. Paris: J. L. J. Brière, 1821.

Diels, H. and W. Kranz (eds. and trans.), *Die Fragmente der Vorsokratiker, Grieschisch und Deutsch*. 3 vols. Berlin: Weidmann, 1951–1952.

Dionis, P. *Cours de chirurgie*. Paris: Chez la Veuve d'Houry, 1708.

Dubois, J. *Iacobi Sylvii . . . Commentarius in Claudij Galeni de Ossibus ad Tyrones libellum, erroribus quamplurimis tam Graecis quam Latinis ab eodem purgatum*. Paris: Petrum Drouart, 1556.

Elkhadem, E. *Le Taqwīm al-ṣiḥḥa (Tacuini sanitatis) d'Ibn Buṭlān: un traité médical du XIe siècle*. Leuven: Peeters, 1990.

Frame, D. M. (trans.), *Montaigne: The Complete Works*. New York: Knopf, 2003.

Garbers, K. (ed.), *Ishāq ibn ʿImrān: Maqāla fī l-mālīḫuliyā, Abhandlung über die Melancholie, und Constantini Africani libri duo de melancholia: vergleichende kritische arabisch-lateinische Parallelausg*. Hamburg: Buske, 1977.

Giglioni, G. (ed.), *Bernardino Telesio: De rerum natura iuxta propria principia libri IX*. Rome: Carocci, 2013.

Green, R. M. (trans.), *Galen: Hygiene*. Springfield, IL: Charles C. Thomas, 1951.

Guyer, P. and Wood, H. (eds), *The Cambridge Edition of the Works of Immanuel Kant in Translation: Anthropology, History, and Education*. Cambridge: Cambridge University Press, 2007.

Hicks, R.D. (trans.), *Diogenes Laertius: Lives of Eminent Philosopher*. 2 vols. Cambridge: Harvard University Press, 1925.

Ibn Sīnā, *Kitāb al-Qānūn fī l-ṭibb*. 3 vols. Būlāq, 1877.

Inwood B. (trans.), *Seneca: Selected Philosophical Letters*. Oxford: Oxford University Press, 2007.

Jardine, L. and M. Silberthorne (eds), *Francis Bacon: The New Organon*. Cambridge: Cambridge University Press, 2000.

Johnston, I. (ed. and trans.), *Galen: On the Constitution of the Art of Medicine. The Art of Medicine. A Method of Medicine to Glaucon*. Cambridge, MA: Harvard University Press, 2016.

Jouanna, J. (ed.), *Hippocratis De natura hominis, Corpus Medicorum Graecorum*. Berlin: de Gruyter, 2002.

Kaske, C. V. and J. R. C. Binghamton (ed. and trans.), *M. Ficino: Three Books on Life*. New York: Center for Medieval and Early Renaissance Studies, 1989.

Kiernan, M. (ed.), *Bacon: The Essayes or Counsels, Civill and Morall*. Oxford: Clarendon Press, 1985.

Kiernan, M. (ed.), *The Oxford Francis Bacon*. Vol. 4. Oxford: Oxford University Press, 2000.

King, J. E. (trans.), *Cicero: Tusculan Disputations*. Cambridge: Harvard University Press, 1927.

Kraus, P. "Kitāb al-aḫlāq li-Jālīnūs [*On Character Traits*, by Galen]," *Bulletin of the Faculty of Arts of the University of Egypt* 5 (1937), 1–51.

Kraus, P. (ed.), *al-Rāzī: Rasā'il falsafiyya* (*Opera philosophica*). Cairo: Barbey, 1939, 15–96.

Kühn, K. G. (ed.), *Galen: Claudii Galeni Opera omnia*. 20 vols. Leipzig: Knobloch, 1821–33. Repr. Hildesheim, Olms, 1964–1965.

Laks, A. and G. W. Most (ed. and trans.), *Early Greek Philosophy*. 9 vols. Cambridge, MA: Harvard University Press, 2016.

Langholf, V. *Medical Theories in Hippocrates. Early Texts and the "Epidemics."* Berlin: de Gruyter, 1990.

Littré, É. (ed. and trans.), *Hippocrate: Oeuvres complètes: Traduction nouvelle avec le texte grec en regard, collationné sur les manuscrits et toutes les éditions*. 10 vols. Amsterdam: Adolf M. Hakkert, 1961–1962 [repr. 1973–1982].

Mahdi, M. (ed.) *Al-Fārābī: Kitāb al-Milla*. Beirut: Dār al-Mashriq, 2001.

Mattock, J. N. (trans.), "A Translation of the Arabic Epitome of Galen's Book *Peri Ethon*," in S. M. Stern et al. (eds), *Islamic Philosophy and the Classical Tradition*. Oxford: Cassirer, 1972, 235–60.

More, H. *An Account of Virtue, or, Dr. Henry More's Abridgment of Morals Put into English*. London: Printed for Benj. Tooke, 1690.

Morra, G. (ed.), *P. Pomponazzi: Tractatus de immortalitate animae*. Bologna: Nanni and Fiammenghi, 1954.

Omrani, A. [ʿĀdil ʿUmrānī], *Maqāla fī l-mālīḫūliyā* (*Traité de la mélancolie*). Carthage: Académie tunisienne des Sciences, des Lettres et des Arts Beït al-Hikma, 2009.

Pauw, P. *A. du Laurens: Historia anatomia humani corporis*. Paris: Excudebat Iametus Mettayer and Marcus Ourry, 1600.

Pauw, P. *Primitiae anatomicae de humani corporis ossibus*. Leiden, 1615.

Peck, A. L. (trans.), *Aristotle: History of Animals*. Vol. 1. Cambridge, MA: Harvard University Press, 1965.

Pheasant, F. (trans.), *Aline Rousselle: Porneia. On Desire and the Body in Antiquity*. Oxford: Blackwell, 1988. First published as *Rousselle, A. Porneia*. Paris: Presses Universitaires de France, 1983.

Pormann, P. E. (ed.), *Rufus of Ephesus On Melancholy*. Tübingen: Mohr Siebeck, 2008.

Potter, P. (trans.), *Hippocrates*. Vol. 5: *Affections. Diseases 1. Diseases 2*. Cambridge, MA: Harvard University Press, 2014.

Rees, G. and M. Wakely (eds), *Francis Bacon: The Instauratio Magna*. Part 3: *Historia naturalis et experimentalis: historia ventorum and historia vitae et mortis*. Oxford: Clarendon Press, 2007.

Regius, H. *Physiologia, sive Cognitio sanitatis: tribus disputationibus in Academiâ Ultrajectinâ publicê proposita*. Vltraiecti: Ex Officinâ AEgidii Roman, Academiae Typographi, 1641.

Regius, H. *Fundamenta physices*. Amstelodami: Ludovicum Elzevirium, 1646.

Regius, H. *Fundamenta medica*. Ultrajecti: Theodorum Ackersdycium, 1647.

Regius, H. *Philosophia naturalis*. Amstelodami: Ludovicum Elzevirium, 1654.

Ricci, P. G. (ed.), *F. Petrarch: Invective contra medicum*, in G. Martellotti et al. (eds), *F. Petrarca: Prose*. Naples: Ricciardi, 1955, 648–93.

Ross, D. (trans.), *Aristotle: the Nicomachean Ethics*. Oxford: Oxford University Press, 2009.

Ruysch, F. *Thesaurus anatomicus*, in *Opera Omnia*. 4 vols. Amsterdam: Jansson-Waesberg, 1720–1733.

Santo, M. *Commentaria in Avicennae textum*, in *Ad communem medicorum chirurgicorum usum commentaria in Avicennae textum*. Venice: Lucantonio Giunta, 1543.

Santo, M. *Oratio de laudibus medicinae*, in *Ad communem medicorum chirurgicorum usum commentaria in Avicennae textum*. Venice: Lucantonio Giunta, 1543.

Serés, G. (ed.), *J. Huarte de San Juan: Examen de ingenios para las ciencias*. Madrid: Cátedra, 1989.

Sezgin, F. (ed.) *Abū Zayd al-Balḫī: Sustenance for Body and Soul, Benefits for Souls and Bodies* (*Maṣāliḥ al-Abdan wa-l-Anfus*). Frankfurt: Institute for the History of Arabic Islamic Science, 1984 and 1998.

Singer, C. (trans.), *Galen: On Anatomical Procedures*. Oxford: Oxford University Press, 1956.

Singer, P. (trans.), *Galen: Selected Works*. Oxford World Classics. Oxford: Oxford University Press, 1997.

Singer, P. N. (ed.), *Galen: Psychological Writings*. Cambridge: Cambridge University Press, 2013.

Shapiro, L. (ed. and trans.), *The Correspondence between Princess Elisabeth of Bohemia and René Descartes*. Chicago: University of Chicago Press, 2007.

Schiefsky, M. J. (trans. and comm.), *Hippocrates: On Ancient Medicine*. Leiden: Brill, 2005.

al-Ṣiddīqī al-Bakrī, Ḥ. (ed.), *Abū Bakr al-Rāzī: Al-Kitāb al-Manṣūrī fī l-ṭibb* [*Book for al-Manṣūr on Medicine*]. Kuwait: Maʿhad al-Maḫṭūṭāt al-ʿarabīya, 1987.

Speeding, J., R. Ellis and D. Heath (eds), *The Works of Francis Bacon*. 15 vols. London: Longman, 1860.

Spon, C. (ed.) *Cardano: Opera omnia*. 10 vols. Lyon: Jean-Antoine Huguetan and Marc-Antoine Ravaud 1663 [repr. Stuttgart-Bad Cannstatt: Frommann, 1966].

Strachey, J. (ed.) *The Standard Edition of the Complete Psychological Works of Sigmund Freud*. 24 vols. London: Hogarth Press, 1957.

Tissot, S. A. *Oeuvres de Monsieur Tissot*. Lausanne: François Grasset, 1784.

Totelin, L. M. V. *Hippocratic Recipes: Oral and Written Transmission of Pharmacological Knowledge in Fifth- and Fourth-Century Greece.* Leiden: Brill, 2009.

Usener, H. (ed.), *Epicurea*. Leipzig: Teubner, 1887.

de Vauzelles, J. *Les simulachres et historiées faces de la mort, autant élégamment pourtraictes, que artificiellement imaginées*. Lyon: soubz l'eseu de Coloigne, 1538.

Walzer, R. and M. Frede (trans.), *Galen: Three Treatises on the Nature of Science.* Indianapolis: Hackett, 1985.

Zedler, J. H. *Grosses vollständiges Universal-lexicon der Wissenschafften und Künste*. 64 vols. Halle: J. H. Zedler, 1732–1750.

Zurayk, C. (ed.) *Miskawayh, Tahḏīb al-aḫlaq*. Beirut: al-Nadin al-Lubnāniyya, 1966.

Zurayk, C. (trans.), *Miskawayh: The Refinement of Character.* Beirut: American University of Beirut, 1968.

Secondary Sources

Adamson, P. "Miskawayh's Psychology," in P. Adamson (ed.), *Classical Arabic Philosophy: Sources and Reception*. London: Warburg Institute, 2007, 39–54.

Adamson, P. "Platonic Pleasures in Epicurus and al-Rāzī," in P. Adamson (ed.), *In the Age of al-Fārābī: Arabic Philosophy in the Fourth/Tenth Century.* London: Warburg Institute, 2008, 71–94.

Adamson, P. (ed.), *Interpreting Avicenna*. Cambridge: Cambridge University Press, 2013.

Adamson, P. and H. H. Biesterfeldt, "The Consolations of Philosophy: Abū Zayd al-Balḫī and Abū Bakr al-Rāzī on Sorrow and Anger," in P. Adamson and P. E. Pormann (ed.), *Philosophy and Medicine in the Formative Period of Islam.* London: Warburg Institute, 2018, 190–205.

Adler, J. "The Education of Ehrenfried Walther von Tschirnhaus (1651–1708)," *Journal of Medical Biography* 23 (2015), 27–35.

Agich, G. J. "Disease and Value: A Rejection of the Value-Neutrality Thesis," *Theoretical Medicine* 4 (1983), 27–41.

Albala, K. *Eating Right in the Renaissance*. Berkeley: University of California Press, 2002.

Allan, S. and C. Williams, *The Guodian Laozi: Proceedings of the International Conference, Dartmouth College, May 1998*. Society for the Study of Early China and Institute of East Asian Studies. Berkeley: University of California, 2000.

Allen, J. "Pyrrhonism and Medicine," in R. Bett (ed.), *The Cambridge Companion to Ancient Scepticism.* Cambridge: Cambridge University Press, 2010, 232–34.

Allen, M. J. B. "Life as a Dead Platonist," in M. J. B. Allen, V. Rees, and M. Davies (eds), *Marsilio Ficino: His Theology, His Philosophy, His Legacy*. Leiden: Brill, 2002, 159–78.

Amundson, R. "Against Normal Function," *Studies in History and Philosophy of Biological and Biomedical Sciences* 31 (2000), 33–53.

Ananth, M. *In Defense of an Evolutionary Concept of Health: Nature, Norms and Human Biology*. Aldershot: Ashgate, 2008.

Annas, J. (ed.) and R. Woolf (trans.), *Cicero: On Moral Ends*. Cambridge: Cambridge University Press, 2001.

Ariew, R. "Descartes and the Tree of Knowledge," *Synthèse* 92 (1992), 101–16.

Aron, W. "Baruch Spinoza and Medicine," *The Hebrew Medical Journal* 2 (1963), 255–82.

Arrizabalaga, J., J. Henderson, and R. French. *The Great Pox: The French Disease in Renaissance Europe*. New Haven, CT: Yale University Press, 1997.

Badaloni, N. "Sulla costruzione e la conservazione della vita in Bernardino Telesio (1509–1588)," *Studi Storici* 30/1 (1989), 25–42.

Barker-Benfield, G. J. *The Culture of Sensibility: Sex and Society in Eighteenth-Century Britain*. Chicago: University of Chicago Press, 1992.

Bayer, R. *Homosexuality and American Psychiatry: The Politics of Diagnosis*. Princeton, NJ: Princeton University Press, 1987.

Beebe, B., and F. Lachman. *The Origins of Attachment: Infant Research and Adult Treatment*. New York: Routledge, 2014.

Bell, D. "Who Is Killing What or Whom? Some Notes on the Internal Phenomenology of Suicide," in S. Briggs, A. Lemma, and W. Crouch (eds), *Relating to Self Harm and Suicide: Psychoanalytic Perspectives on Practice, Theory and Prevention*. London: Routledge, 2008, 38–45.

Biesterfeldt, H. H. "Abū Zayd," in U. Rudolph (ed.), *Grundriss der Geschichte der Philosophie: 8.–10. Jahrhundert*. Basel: Schwabe, 2012, 156–67.

Blatt, S. *Experiences of Depression*. Washington, DC: APA Press, 2004.

Bokenkamp, S. R. "Simple Twists of Fate: The Daoist Body and Its Ming," in C. Lupke (ed.), *The Magnitude of Ming: Command, Allotment, and Fate in Chinese Culture*. Honolulu: University of Hawai'i Press, 2005, 151–65.

Boltz, W. G. "Textual Criticism and the Ma Wang Tui Lao Tzu. Review of *D. C. Lau: Chinese Classics: Tao Te Ching*," *Harvard Journal of Asiatic Studies* 44/1 (1984), 185–224.

Boltz, W. G. "Lao Tzu Tao Te Ching 老子道德經," in M. Loewe (ed.), *Early Chinese Texts: A Bibliographical Guide*. Early China Special Monograph Series. Berkeley: Society for the Study of Early China, 1993, 269–92.

Boorse, C. "On the Distinction between Disease and Illness," *Philosophy and Public Affairs* 5 (1975), 49–68.
Boorse, C. "What a Theory of Mental Health Should Be," *Journal for the Theory of Social Behaviour* 6 (1976), 61–84.
Boorse, C. "Wright on Functions," *Philosophical Review* 85 (1976), 70–86.
Boorse, C. "Health as a Theoretical Concept," *Philosophy of Science* 44 (1977), 542–73.
Boorse, C. "Concepts of Health," in D. van de Veer and T. Regan (eds), *Health Care Ethics: An Introduction*. Philadelphia: Temple University Press, 1987, 359–93.
Boorse, C. "A Rebuttal on Health," in J. M. Humber and R. F. Almeder (eds), *What Is Disease?* Totowa, NJ: Humana Press, 1997, 1–134.
Boorse, C. "A Rebuttal on Functions," in A. Ariew, R. Cummins, and M. Perlman (eds), *Functions: New Essays in the Philosophy of Psychology and Biology* (Oxford: Oxford University Press, 2002), 63–112.
Boorse, C. "Concepts of Health and Disease," in F. Gifford (ed.), *Philosophy of Medicine*. Amsterdam: Elsevier, 2010, 1–52.
Boorse, C. "Replies to My Critics," *Journal of Medicine and Philosophy* 39 (2014), 648–82.
Boorse, C. "A Second Rebuttal on Health," *Journal of Medicine and Philosophy* 39 (2014), 683–724.
Bos, E.-J. "The Correspondence between Descartes and Henricus Regius," PhD diss., Utrecht University, 2002.
Brann, N. L. *The Debate over the Origin of Genius during the Italian Renaissance*. Leiden: Brill, 2001.
Brennessel, B., M. Drout, and R. Gravel, "A Reassessment of the Efficacy of Anglo-Saxon Medicine," *Anglo-Saxon England* 34 (2005), 183–95.
Brittain, C. *Philo of Larissa: The Last of the Academic Sceptics*. Oxford: Oxford University Press, 2001.
Brockliss, L. and C. Jones, *The Medical World of Early Modern France*. Oxford: Oxford University Press, 1997.
Broman, T. H. *The Transformation of German Academic Medicine, 1750–1820*. Cambridge: Cambridge University Press, 1996, 42–72.
Brown, D. J. "Cartesian Functional Analysis," *Australiasian Journal of Philosophy* 90 (2012), 75–92.
Brown, M. *The Art of Medicine in Early China: The Ancient and Medieval Origins of a Modern Archive*. Cambridge: Cambridge University Press, 2015.
Brown, T. M. "Descartes, Dualism, and Psychosomatic Medicine," in W. F. Bynum et al. (eds), *The Anatomy of Madness: Essays in the History of Psychiatry*. Vol. 1. New York: Tavistock, 1985, 40–62.

Byl, S. and W. Szafran. "La phrénitis dans le Corpus Hippocratique: Etude philologique et médicale," *Vesalius* 2/2 (1996), 98–105.

Bylebyl, J. J. "The School of Padua: Humanistic Medicine in the Sixteenth Century," in C. Webster (ed.), *Health, Medicine and Mortality in the Sixteenth Century*. Cambridge: Cambridge University Press, 1979, 335–70.

Carmichael, G. *Plague and the Poor in Renaissance Florence*. Cambridge: Cambridge University Press, 1986.

Cavallo, S. and T. Storey. *Healthy Living in Late Renaissance Italy*. Oxford: Oxford University Press, 2013.

Cedzich, U.-A. "Corpse Deliverance, Substitute Bodies, Name Change, and Feigned Death: Aspects of Metamorphosis and Immortality in Early Medieval China," *Journal of Chinese Religons* 29 (2001), 1–68.

Chan, A. Kam-leung. *Two Visions of the Way: A Study of the Wang Pi and the Ho-Shang Kung Commentaries on the Lao-Tzu*. Albany: State University of New York Press, 1991.

Chan, A. "Laozi," in E. N. Zalta (ed.), *The Stanford Encyclopedia of Philosophy*, 2014.

Chen Banghuai 陳邦懷. "Zhanguo xingqi ming kaoshi 战国行气铭考释," *Guwenzi yanjiu* 古文字研究 7 (1982).

Cipolla, C. *Public Health and the Medical Profession in the Renaissance*. Cambridge: Cambridge University Press, 1976.

Clark, A. *Surfing Uncertainty: Prediction, Action, and the Embodied Mind*. Oxford: Oxford University Press, 2016.

Cohen, K. *Metamorphosis of a Death Symbol: The Transi Tomb in the Late Middle Ages and the Renaissance*. Berkeley: University of California Press, 1973.

Cohen, C. et al., "The Lasting Impact of Early-Life Adversity on Individuals and Their Descendants: Potential Mechanisms and Hope for Intervention," *Genes, Brains, and Behavior* 15/1 (2016), 155–68.

Coleman, W. "Health and Hygiene in the *Encyclopédie*: A Medical Doctrine for the Bourgeoisie," *Journal of the History of Medicine and Allied Sciences* 29 (1974), 399–421.

Campany, R. F. *To Live as Long as Heaven and Earth: A Translation and Study of Ge Hong's Traditions of Divine Transcendents*. Berkeley: University of California Press, 2002.

Campany, R. F. "Living Off the Books: Fifty Ways to Dodge *Ming* 命 in Early Medieval China," in C. Lupke (ed.), *The Magnitude of Ming: Command, Allotment, and Fate in Chinese Culture*. Honolulu: University of Hawai'i Press, 2005, 129–50.

Campany, R. F. *Making Transcendents: Ascetics and Social Memory in Early Medieval China*. Honolulu: University of Hawai'i Press, 2009.

Clouser, K. D., C. M. Culver, and B. Gert. "Malady a New Treatment of Disease," *The Hastings Center Report* 11 (1981), 29–37.

Clouser, K. D., C. M. Culver, and B. Gert. "Malady," in J. M. Humber and R. F. Almeder (eds), *What Is Disease?* Totowa, NJ: Humana Press, 1997, 173–217.

Condren, C. et al. (eds), *The Philosopher in Early Modern Europe: The Nature of a Contested Identity*. Cambridge: Cambridge University Press, 2006.

Cooper, J. M. (ed.), *Plato: Complete Works*. Indianapolis: Hackett, 1997.

Cooper, R. "Disease," *Studies in History and Philosophy of Biological and Biomedical Sciences* 33 (2002), 263–82.

Cooper, R. *Classifying Madness: A Philosophical Examination of the Diagnostic and Statistical Manual of Mental Disorders*. Dordrecht: Springer, 2005.

Cooper, R. "Aristotelian Accounts of Disease—What Are They Good For?" *Philosophical Papers* 36 (2007), 427–42.

Cooper, R. and C. Megone, "Introduction," *Philosophical Papers* 36 (2007), 339–41.

Cooter, R. and C. Stein. "Cracking Biopower," *History of the Human Sciences* 23/2 (2010), 109–28.

Corneanu, S. *Regimens of the Mind: Boyle, Locke, and the Early Modern Cultura Animi Tradition*. Chicago: University of Chicago Press, 2011.

Cottingham, J. *Philosophy and the Good Life*. Cambridge: Cambridge University Press, 1998.

Csikszentmihalyi, M. *Material Virtue*. Leiden: Brill, 2004.

Cummins, R. "Functional Analysis," *Journal of Philosophy* 72 (1975), 741–65.

Damasio, A. *Descartes' Error: Emotion, Reason and the Human Brain*. New York: Avon Books, 1995.

Damasio, A. and Carvalho, G. B. "The Nature of Feelings: Evolutionary and Neurobiological Origins," *Nature Reviews Neuroscience* 14 (2013), 143–52.

Dancy, J. "Moral Particularism," in E. N. Zalta (ed.), *The Stanford Encyclopedia of Philosophy* (online; Fall 2013 Edition).

Daston, L. "Introduction: The Coming into Being of Scientific Objects," in L. Daston (ed.), *Biographies of Scientific Objects*. Chicago: University of Chicago Press, 2000, 1–14.

Davies, P. S. *Norms of Nature*. Cambridge: MIT Press, 2001.

Dayan, P. et al. "The Helmholtz Machine," Neural Computation 7/5 (1995), 899–904.

Dean-Jones, L. "Autopsia, Historia and What Women Know: The Authority of Women in Hippocratic Gynaecology," in D. Bates (ed.), *Knowledge and the*

Scholarly Medical Traditions: A Comparative Study. Cambridge: Cambridge University Press, 1995, 41–58.

DeLacy, M. *The Germ of an Idea: Contagionism, Religion, and Society in Britain, 1660–1730*. Houndmills, UK: Palgrave Macmillan, 2016.

Demiéville, P. "Philosophy and Religion from Han to Sui," in D. C. Twitchett and M. Loewe (eds), *The Cambridge History of China*. Vol. 1: *The Ch'in and Han Empires, 221 B.C–A.D. 220*. Cambridge: Cambridge University Press, 1986, 808–72.

Despeux, C. "Gymnastics: The Ancient Tradition," in L. Kohn and Y. Sakade (eds), *Taoist Meditation and Longevity Techniques*. Ann Arbor: Center for Chinese Studies University of Michigan, 1989, 223–61.

Despeux, C. *Immortelles De La Chine Ancienne: Taoïsme Et Alchimie Féminine*. Puiseaux: Pardès, 1990.

Despeux, C. "Taixi," in F. Pregadio (ed.), *The Encyclopedia of Taoism*. London: Routledge, 2008, 953—54.

Detlefson, K. "Descartes on the Theory of Life and Methodology in the Life Sciences," in P. Distelzweig et al. (eds), *Early Modern Medicine and Natural Philosophy*. Dordrecht: Springer, 2015, 141–72.

Des Chene, D. *Spirits and Clocks: Machine and Organism in Descartes*. Cornell: Cornell University Press, 2001.

Des Chene, D. "Life and Health in Descartes and After," in S. Gaukroger et al. (eds), *Descartes' Natural Philosophy*. New York: Routledge, 2002, 723–35.

DeWoskin, K. J. *Doctors, Diviners, and Magicians of Ancient China: Biographies of Fang-Shih*. New York: Columbia University Press, 1983.

DiMeo, M. "'Such a Sister Became Such a Brother': Lady Ranelagh's Influence on Robert Boyle," *Intellectual History Review* 25 (2015), 21–36.

Dodds, E. R. *The Greeks and the Irrational*. Berkeley: University of California Press, 1951.

Edelstein, L. "The Dietetics of Antiquity," in O. Temkin and C. L. Temkin (eds), *Ancient Medicine: Selected Papers of Ludwig Edelstein*. Baltimore, MD: Johns Hopkins University Press, 1967, 303–16.

van der Eijk, P.J. *Diocles of Carystus*. Leiden: Brill, 2000.

van der Eijk, P. "Therapeutics," in R. J. Hankinson (ed.), *The Cambridge Companion to Galen*. Cambridge: Cambridge University Press, 2008, 283–303.

Engelhardt, H. T. "The Disease of Masturbation: Values and the Concept of Disease," *Bulletin of the History of Medicine* 48 (1974), 234–48.

Engelhardt, H. T. "The Concepts of Health and Disease," in H. T. Engelhardt and S. F. Spicker (eds), *Evaluation and Explanation in the Biomedical Sciences*. Dordrecht: Reidel, 1975.

Engelhardt, H. T. "Ideology and Etiology," *Journal of Medical Philosophy* 1 (1976), 256–68.

Engelhardt, U. "Qi for Life: Longevity in the Tang," in L. Kohn and Y. Sakade (eds), *Taoist Meditation and Longevity Techniques*. Ann Arbor: Center for Chinese Studies University of Michigan, 1989, 263–96.

Enlow, M., et al. "Mother–Infant Attachment and the Intergenerational Transmission of Posttraumatic Stress Disorder," *Development and Psychopathology* 26/1 (2014), 41–65.

Farquhar, J. "Problems of Knowledge in Contemporary Chinese Medical Discourse," *Social Science and Medicine* 24/12 (1987), 1013–21.

Farquhar, J. *Knowing Practice: The Clinical Encounter of Chinese Medicine*. Studies in the Ethnographic Imagination. Boulder, CO: Westview Press, 1994.

Farquhar, J. and Wang Jun, "Knowing the Why but Not the How: A Dilemma in Contemporary Chinese Medicine," *Asian Medicine* 5/1 (2011), 57–79.

Fauconnier, G., and M. Turner, *The Way We Think: Conceptual Blending and the Mind's Hidden Capacities*. New York: Basic Books, 2002.

El-Fekkak, B. "Cosmic Justice in al-Fārābī's Virtuous City: Healing the Medieval Body Politic," PhD diss., King's College London, 2011.

Foot, P. *Natural Goodness*. Oxford: Clarendon Press, 2001.

Forschner, M. *Die stoische Ethik*. Darmstadt: Wissenschaftliche Buchgesellschaft, 1995.

Frede, M. "An Anti-Aristotelian Point of Method in Three Rationalist Doctors," in K. Ierodiokonou and B. Morison (eds), *Episteme etc. Essays in Honour of Jonathan Barnes*. Oxford: Oxford University Press, 2011, 115–37.

French, R. and J. Arrizabalaga, "Coping with the French Disease: University Practitioners' Strategies and Tactics in the Transition from the Fifteenth to the Sixteenth Century," in R. French et al. (eds), *Medicine from the Black Death to the French Disease*. Aldershot: Ashgate, 1998, 248–87.

French, R. K. *Robert Whytt, the Soul, and Medicine*. London: Wellcome Institute for the History of Medicine, 1969.

French, R. K. *Canonical Medicine: Gentile da Foligno and Scholasticism*. Leiden: Brill, 2001.

Friston, K. "Prediction, Perception and Agency," *International Journal of Psychophysiology* 83 (2012), 248–52.

Fulford, K. W. M. *Moral Theory and Medical Practice*. Cambridge: Cambridge University Press, 1989.

Gabbey, A. "'A Disease Incurable': Scepticism and the Cambridge Platonists," in R. H. Popkin and A. J. Vanderjagt (eds), *Scepticism and Irreligion in the Seventeenth and Eighteenth Centuries*. Leiden: Brill, 1993, 71–91.

Gabbey, A. "Spinoza's Natural Science and Methodology," in D. Garrett (ed.), *The Cambridge Companion to Spinoza*. Cambridge: Cambridge University Press, 1996, 142–91.

Garber, D. "*Semel in vita*: The Scientific Background to Descartes' *Meditations*," in A. O. Rorty (ed.), *Essays on Descartes' Meditations*. Berkeley: University of California Press, 1986, 81–116.

García-Ballester, L. "On the Origins of the 'Six Non-Natural Things,' in Galen," in J. Kollesch and D. Nickel (eds), *Galen und das Hellenistiche Erbe*. Stuttgart: Steiner, 1993, 105–15.

Gariepy, T. P. "Mechanism without Metaphysics: Henricus Regius and the Establishment of Cartesian Medicine," PhD diss., Yale University, 1990.

Garson, J. and G. Piccinini, "Functions Must Be Performed at Appropriate Rates in Appropriate Situations," *British Journal for the Philosophy of Science* 65 (2014), 1–20.

Gaukroger, S. *The Collapse of Mechanism and the Rise of Sensibility*. Oxford: Clarendon Press, 2010.

Giglioni, G. "The Hidden Life of Matter: Techniques for Prolonging Life in the Writings of Francis Bacon," in J. R. Solomon and C. Gimelli Martin (ed.), *Francis Bacon and the Refiguring of Early Modern Thought: Essays to Commemorate the Advancement of Learning (1605–2005)*. Aldershot: Ashgate, 2005, 129–44.

Giglioni, G. "Coping with Inner and Outer Demons: Marsilio Ficino's Theory of the Imagination," in Y. Haskell (ed.), *Diseases of the Imagination and Imaginary Disease in the Early Modern Period*. Turnhout: Brepols, 2011, 19–51.

Giglioni, G. "Spirito e coscienza nella medicina di Bernardino Telesio," in G. Ernst and R. M. Calcaterra (eds), "*Virtù ascosta e negletta*": *La Calabria nella modernità*. Milan: Angeli, 2011, 154–68.

Giglioni, G. "The Many Rhetorical Personae of an Early Modern Physician: Girolamo Cardano on Truth and Persuasion," in S. Pender and N. S. Struever (eds), *Rhetoric and Medicine in Early Modern Europe*. Farnham: Ashgate, 2012, 173–93.

Giglioni, G. "Medicine," in P. Ford, J. Bloemendal and C. Fantazzi (eds), *Brill's Encyclopaedia of the Neo-Latin World* (*Macropaedia*). Leiden: Brill, 2014, 679–90.

Giglioni, G. "Medicine for the Mind in Early Modern Philosophy," in J. Sellars (ed.), *The Routledge Handbook of the Stoic Tradition*. London: Routledge, 2016, 189–203.

Godfrey-Smith, P. "A Modern History Theory of Functions," *Noûs* 28 (1994), 344–62.

Golinski, J. *British Weather and the Climate of Enlightenment.* Chicago: University of Chicago Press, 2010, 140–50.

Goosens, W. "Values, Health and Medicine," *Philosophy of Science* 47 (1980), 100–15.

Gopnik, A. "Van Gogh's Ear," *The New Yorker*, January 4, 2010.

Graham, A. C. "How Much of *Chuang Tzu* Did Chuang Tzu Write?" *in Studies in Chinese Philosophy and Philosophical Literature*. New York: State University of New York Press, 1986, 283–321.

Graham, A. C. *Disputers of the Tao: Philosophical Argument in Ancient China*. La Salle, IL: Open Court, 1989.

Grant, M. *Galen on Food and Diet*. London: Routledge, 2000.

Grattan, J. G. H. and C. Singer, *Anglo-Saxon Magic and Medicine: Illustrated Specially from the Semi-Pagan Text Lacnunga*. London: Oxford University Press, 1952.

Grensemann, H. *Hippokratische Gynäkologie*. Wiesbaden: Franz Steiner, 1982.

Guan Feng 關鋒. "'Zhuangzi: Wai, za pian' chu tan《莊子：外雜篇》初探," in ed. Zhuangzi zhexue yanjiu bianji bu 莊子哲學研究編輯部, Zhuangzi zhexue taolun ji 莊子哲學討論集. Beijing: Zhonghua, 1962.

Guo Moruo 郭沫若, "Gudai wenzi zhi bianzheng de fazhan 古代文字辯證的發展," *Kaogu xuebao* 考古學報 29 (1972), 2–13.

Gutas, D. "Medical Theory and Scientific Method in the Age of Avicenna," in D. C. Reisman (ed.), *Before and After Avicenna: Proceedings of the First Conference of the Avicenna Study Group*. Leiden: Brill, 2003, 145–62. Reprinted in P. E. Pormann (ed.), *Islamic Medical and Scientific Tradition*. 4 vols. Routledge: London, 2011, 1: 33–47.

Hankinson, R. J. (ed.), *The Cambridge Companion to Galen*. Cambridge: Cambridge University Press, 2008.

Hanson, A. E. "The Medical Writers' Woman," in D. M. Halperin et al. (eds), *Before Sexuality: The Construction of Erotic Experience in the Ancient Greek World*. Princeton, NJ: Princeton University Press, 1990, 309–38.

Han Wei 韓巍 (ed.), *Laozi—Beijing Daxue cang Xi Han zhushu* 老子—北京大學藏西漢竹書. Vol 2. Shanghai: Shanghai guji chubanshe, 2012.

Harper, D. *Early Chinese Medical Literature: The Mawangdui Medical Manuscripts*. London: Kegan Paul Intl., 1998.

Harper, D. "Warring States: Natural Philosophy and Occult Thought," in M. Loewe and E. L. Shaughnessy (eds), *The Cambridge History of Ancient China: From the Origins of Civilization to 221 B.C.* Cambridge: Cambridge University Press, 1999, 813–84.

Harper, D. "The Textual Form of Knowledge: Occult Miscellanies in Ancient and Medieval Chinese Manuscripts, Fourth Century B.C. to Tenth Century A.D," in F. Bretelle-Establet (ed.), *Looking at It from Asia: The Processes That Shaped the Sources of History of Science*. Boston Studies in the Philosophy of Science. Dordrecht: Springer Netherlands, 2010, 37–80.
Harrison, P. *The Fall of Man and the Foundations of Science*. Cambridge: Cambridge University Press, 2007.
Hatfield, G. "*Descartes' Metaphysical Physics* by Daniel Garber; *Kant and the Exact Sciences* by Michael Friedman," *Synthese* 106 (1996), 113–38.
Hatfield, G. *Routledge Philosophy Guidebook to Descartes and the Meditations*. London: Routledge, 2002.
Hatfield, G. "Animals," in J. Broughton and J. Carriero (eds), *Companion to Descartes*. Oxford: Blackwell, 2008, 404–25.
Hausman, D. "Is an Overdose of Paracetamol Bad for One's Health?" *British Journal for the Philosophy of Science* 62 (2011), 657–68.
Hausman, D. "Health, Naturalism and Functional Efficiency," *Philosophy of Science* 79 (2012), 519–41.
Hawkes, D. *Ch'u Tz'u, the Songs of the South: An Ancient Chinese Anthology*. New York: Beacon Press, 1962.
Heilbron, J. "Was there a Scientific Revolution?" in J. Z. Buchwald and R. Fox (eds), *The Oxford Handbook of the History of Physics*. Oxford: Oxford University Press, 2013, 7–24.
Henderson, J. "The Black Death in Florence: Medical and Communal Responses," in S. Bassett (ed.), *Death in Towns: Urban Responses to the Dying and the Dead, 100–1600*. Leicester: Leicester University Press, 1992, 136–50.
Henderson, J. *The Renaissance Hospital: Healing the Body and Healing the Soul*. New Haven, CT: Yale University Press, 2006.
Hendrischke, B. "Religious Ethics in the *Taiping Jing*: The Seeking of Life," *Daoism: Religion, History and Society* 4 (2012), 53–94.
Henricks, R. G. "On the Chapter Divisions in the *Lao-Tzu*," *Bulletin of the School of Oriental and African Studies* 45 (1982), 501–24.
Henricks, R. G. *Philosophy and Argumentation in Third-Century China: The Essays of Hsi K'ang*. Princeton, NJ: Princeton University Press, 1983.
Hobson, A. *Ego Damage and Repair: Towards a Psychodynamic Neurology*. London: Karnac, 2014.
Hobson, A. *Psychodynamic Neurology: Dreams, Consciousness, and Virtual Reality*. New York: Taylor and Francis, 2015.
Hobson, A. Hong, C. and K. Friston, "Virtual Reality and Consciousness Inference in Dreaming," *Frontiers in Psychology* 5 (2014), art. 1133.

Hopkins, J. "Psychoanalysis, Metaphor, and the Concept of Mind," in M. P. Levine (ed.), *The Analytic Freud: Philosophy and Psychoanalysis*. London: Routledge, 2000, 11–35.

Hopkins, J. "Evolution, Emotion, and Conflict," in M. Chung and C. Feltham (eds), *Psychoanalytic Knowledge*. Houndmills, UK: Palgrave Macmillan, 2003, 132–56.

Hopkins, J. "Conscience and Conflict: Darwin, Freud, and the Origins of Human Aggression," in D. Evans and P. Cruse (eds), *Emotion, Evolution, and Rationality*. Oxford: Oxford University Press, 2004, 225–48.

Hopkins, J. "Psychoanalysis Representation and Neuroscience: The Freudian Unconscious and the Bayesian Brain," in A. Fotopolu, D. Pfaff, and M. A. Conway (eds), *From the Couch to the Lab: Psychoanalysis, Neuroscience and Cognitive Psychology in Dialogue*. Oxford: Oxford University Press, 2012, 230–65.

Hopkins, J. "Understanding and Healing: Psychiatry and Psychoanalysis in the Era of Neuroscience," in K. W. M. Fulford et al. (eds), *The Oxford Handbook of the Philosophy of Psychiatry*. Oxford: Oxford University Press, 2013, 1264–92.

Hopkins, J. "The Significance of Consilience: Psychoanalysis, Attachment, Neuroscience, and Evolution," in S. Boag et al. (eds), *Psychoanalysis and Philosophy of Mind: Unconscious Mentality in the 21st Century*. London: Karnac, 2015, 47–136.

Hopkins, J. "Free Energy and Virtual Reality in Neuroscience and Psychoanalysis," *Frontiers in Psychology* 7 (2016), art. 922.

Hotson, H. *Johann Heinrich Alsted 1588–1638: Between Renaissance, Reformation, and Universal Reform*. Oxford: Clarendon Press, 2000.

Howard, R. (trans.) *Michel Foucault: Madness and Civilisation: A History of Insanity in the Age of Reason*. London: Tavistock, 1961.

Howhy, J. *The Predictive Mind*. Oxford: Oxford University Press, 2013.

Hsu, E. 許小麗 "Outward Form (*Xing* 形) and Inward *Qi* 氣: The 'Sentimental Body' in Early Chinese Medicine," *Early China* 32 (2009), 103–24.

Hulsewé, A. F. P. "*Han Shu* 漢書," in M. Loewe (ed.), *Early Chinese Texts: A Bibliographical Guide*. Berkeley: Society for the Study of Early China, 1993, 269–92.

Hutton, S. "Of Physic and Philosophy: Anne Conway, Francis Mercury van Helmont and Seventeenth-Century Medicine," in A. Cunningham and O. Grell (eds), *Religio Medici: Medicine and Religion in Seventeenth-Century England*. Surrey: Aldershot, 1996, 218–46.

Hutton, S. *Anne Conway: A Woman Philosopher*. Cambridge: Cambridge University Press, 2004.

Hutton, S. "Making Sense of Pain: Valentine Greatrakes, Henry Stubbe and Anne Conway," in L. Jardine and G. Manning (eds), *Testimonies: States of Mind and States of the Body in the Early Modern Period*. Dordrecht: Springer (forthcoming).

Iliouchine, A. "A Study of the Central Scripture of Laozi (Laozi Zhongjing)." MA thesis, McGill University, 2011.

Iskandar, A. Z. "Ar-Rāzī, the Clinical Physician (*Ar-Rāzī aṭ-Ṭabīb al-Iklīnī*)," in P. E. Pormann, (ed.), *Islamic Medical and Scientific Tradition*. 4 vols. Routledge: London, 2011, 1: 207–53.

Jalobeanu, D. "Francis Bacon and Justus Lipsius: Natural Philosophy, Natural Theology and the Stoic Discipline of the Mind," in J. Papy and H. Hirai (eds), *Justus Lipsius and Natural Philosophy*. Bruxelles: Wetteren Universal Press, 2007, 107–21.

Jolly, K. L. *Popular Religion in Late Saxon England: Elf Charms in Context*. Chapel Hill: University of North Carolina Press, 1996.

Jones, M. L. *The Good Life in the Scientific Revolution: Descartes, Pascal, Leibniz and the Cultivation of Virtue*. Chicago: University of Chicago Press, 2006.

Joosse, N. P. and P. E. Pormann, "Archery, Mathematics, and Conceptualising Inaccuracies in Medicine in 13th Century Iraq and Syria," *Journal of the Royal Society of Medicine* 101 (2008), 425–27.

Joutsivuo, T. *Scholastic Tradition and Humanist Innovation: The Concept of Neutrum in Renaissance Medicine*. Helsinki: Finnish Academy of Science and Letters, 1999.

Kaltenmark, M. *Le Lie-Sien Tchouan: Biographies légendaires des immortels Taoïstes de l'antiquité*. Pékin: Centre d'études sinologiques de Pékin, 1953.

Kass, L. R. "Regarding the End of Medicine and the Pursuit of Health," *The Public Interest* 40 (1975), 11–42.

Keegan, D. J. "The Huang-Ti Nei-Ching: The Structure of the Compilation, the Significance of the Structure." PhD diss., University of California, 1988.

Kendell, R. "The Concept of Disease and Its Implications for Psychiatry," *British Journal of Psychiatry* 127 (1975), 305–15.

King, H. "Medical Texts as a Source for Women's History," in A. Powell (ed.), *The Greek World*. London: Routledge, 1995, 199–218.

King, H. *Hippocrates' Woman: Reading the Female Body in Ancient Greece*. London: Routledge, 1998.

King, H. *The Disease of Virgins: Green Sickness, Chlorosis and the Problems of Puberty*. London: Routledge, 2004.

King, H. *Health in Antiquity*. London: Routledge, 2005.

King, L. S. "What Is Disease?" *Philosophy of Science* 21 (1954), 193–203.

Kingma, E. "What Is It to Be Healthy?" *Analysis* 67 (2007), 128–33.

Kingma, E. "Paracetamol, Poison and Polio: Why Boorse's Account of Function Fails to Distinguish Health and Disease," *British Journal for the Philosophy of Science* 61 (2010), 241–64.

Kingma, E. "Health and Disease: Social Constructivism as a Combination of Naturalism and Normativism," in H. Carel and R. Cooper (eds), *Health, Illness and Disease: Philosophical Essays*. Durham, NC: Acumen, 2012, 37–56.

Kingma, E. "Naturalist Accounts of Disorder," in K. W. M. Fulford et al. (eds), *Oxford Handbook of Philosophy and Psychiatry*. Oxford: Oxford University Press, 2013.

Kingma, E. "Situational Disease and Dispositional Function," *British Journal for the Philosophy of Science* 67/2 (2015), 391–404.

Kirchin, S. (ed.) *Thick Concepts*. Oxford: Oxford University Press, 2013.

Kitcher, P. *The Lives to Come: The Genetic Revolution and Human Possibilities*. New York: Touchstone, 1996.

Kleeman, T. F. *Great Perfection: Religion and Ethnicity in a Chinese Millennial Kingdom*. Honolulu: University of Hawaii Press, 1998.

Kleeman, T. F. *Celestial Masters: History and Ritual in Early Daoist Communities*. Harvard-Yenching Institute Monograph Series. Cambridge, MA: Harvard University Asia Center, 2016.

Klibansky, R., E. Panofsky, and F. Saxl, *Saturn and Melancholy: Studies in the History of Natural Philosophy, Religion, and Art*. London: Nelson, 1964.

Knoblock, J. and J. Riegel. *The Annals of Lü Buwei*. Stanford: Stanford University Press, 2000.

Kohn, L. "Daoyin among the Daoists: Physical Practice and Immortal Transformation in Highest Clarity," in V. Lo (ed.), *Perfect Bodies: Sports, Medicine and Immortality: Ancient and Modern*. London: British Museum, 2012, 111–20.

Kopelman, L. "On Disease: Theories of Disease and the Ascription of Disease: Comments on 'The Concepts of Health and Disease,'" in H. T. Engelhardt and S. F. Spicker (eds), *Evaluation and Explanation in the Biomedical Sciences*. Dordrecht: Reidel, 1975, 143–50.

Koudounaris, P. *The Empire of Death*. New York: Thames and Hudson, 2011.

Kristeller, P. O. *The Philosophy of Marsilio Ficino*. New York: Columbia University Press, 1943.

KUBO Teruyuki 久保 輝幸, "Retsu sen den' no bōshitsu shita senden ni soku nitsuite 『列仙伝』の亡失した仙伝2則について," *Jinbun gaku ronshū* 人文学論集 29 (2011), 109—28.

Kurtz, J. *The Discovery of Chinese Logic*. Leiden: Brill, 2011.

Laing, R. D. *The Divided Self.* London: Tavistock, 1959.
Lagerwey, J. "Deux Écrits Taoïstes Anciens," *Cahiers d'Extrême-Asie* 14 (2004), 139–71.
Latour, B. "A Textbook Case Revisited. Knowledge as Mode of Existence," in E. J. Hackett, O. Amsterdamska, M. Lynch and J. Wacjman (eds), *The Handbook of Science and Technology Studies*. Cambridge, MA: MIT Press, 2007, 83–112.
Lau, D. C. *Laozi: Dao De Jing*. London: Penguin Books, 1963.
Lave, J. and E. Wenger, *Situated Learning: Legitimate Peripheral Participation*. Cambridge: Cambridge University Press, 1991.
Leitao, D. *The Pregnant Male as Myth and Metaphor in Classical Greek Literature*. Cambridge: Cambridge University Press, 2012.
Lennon, T. M. "Bayle and Late Seventeenth Century Thought," in J. P. Wright and P. Potter (eds), *Psyche and Soma: Physicians and Metaphysicians on the Mind-Body Problem from Antiquity to Enlightenment*. Oxford: Oxford University Press, 2000, 197–216.
Leong, E. "Making Medicine in the Early Modern Household," *Bulletin of the History of Medicine* 82 (2008), 145–68.
Levin, R., and T. Nielsen, "Nightmares, Bad Dreams, and Emotion Dysregulation," *Current Directions in Psychological Science* 18/2 (2009), 84–88.
Lewis, M. E. "Warring States Political History," in M. Loewe and E. L. Shaughnessy (eds), *The Cambridge History of Ancient China: From the Origins of Civilization to 221 B.C.* Cambridge: Cambridge University Press, 1999, 587–649.
Lewis, M. E. *Writing and Authority in Early China*. SUNY Series in Chinese Philosophy and Culture. Albany: State University of New York Press, 1999.
Li Jianmin 李建民. "They Shall Expel Demons: Etiology, the Medical Canon and the Transformation of Medical Techniques before the Tang," in J. Lagerwey and M. Kalinowski (eds), *Early Chinese Religion*. Leiden; Boston: Brill, 2009, 1103–50.
Li Ling 李零, *Zhongguo fangshu xu kao* 中国方术续考. Beijing: Zhonghua shuju, 2000.
Li Ling 李零, *Zhongguo fang shu kao (xiuding ben)* 中国方术正考 (修訂本). Beijing: Dongfang chubanshe, 2001.
Lin Fushi 林富士, "Zhongguo zaoqi daoshi de 'yizhe' xingxiang: yi Shenxian zhuan wei zhu de chubu tantao 中國早期道士的「醫者」形象: 以《神仙傳》為主的初步探討," in *Zhongguo zhonggu shiqi de zongjiao yu yiliao* 中國中古時期的宗教與醫療. Taibei: Lianjing, 2008.

Lin Fushi 林富士, "Zhongguo zaoqi daoshi de yiliao huodong ji qi yiliao kaoshi: yi Han Wei Jin Nanbeichao shi de 'zhuanji' ziliao weizhu de chubu tantao 中國早期道士的醫療活動及其醫術考釋：以漢魏晉南北朝時期的「傳記」資料為主的初步探討," in *Zhongguo zhonggu shiqi de zongjiao yu yiliao* 中國中古時期的宗教與醫療. Taibei: Lianjing, 2008.

Lin Fushi 林富士, "Donghan shiqi de jiyi yu zongjiao 東漢時期的疾疫與宗教," in Zhongguo zhonggu shiqi de zongjiao yu yiliao 中國中古時期的宗教與醫療. Taibei: Lianjing chubanshe, 2008.

Little, B. W. and M. Little, *Moral Particularism*. Oxford: Oxford University Press, 2000.

Liu Shufen 劉淑芬. "Tang, Song shiqi sengren, guojia, he yiliao de guanxi—Cong Yaofangdong dao huiminju 唐、宋時期僧人、國家和醫療的關係—從藥方洞到惠民局" in Li Jianmin (ed.), 李建民, *Cong yiliao kan zhongguo shi* 從醫療看中國史. Taipei: Lianjing, 2008.

Lloyd, G. and N. Sivin. *The Way and the Word: Science and Medicine in Early China and Greece*. New Haven, CT: Yale University Press, 2002.

Lo, V. "The Influence of 'Yangsheng' Culture on Early Chinese Medicine," PhD diss., School of Oriental and African Studies, 1998.

Lo, V. "Tracking the Pain. *Jue* and the Formation of a Theory of Circulating *Qi* through the Channels," *Sudhoffs Archiv* 83/2 (1999), 191–211.

Lo, V. "Crossing the *Neiguan* 內關 'Inner Pass': A *Nei/Wai* 內外 'Inner/Outer' Distinction in Early Chinese Medicine," *East Asian Science Technology and Medicine EASTM* 17 (2000), 15–65.

Lo, V. "The Influence of Nurturing Life Culture," in E. Hsu (ed.), *Innovation in Chinese Medicine*. Needham Research Institute Studies. Cambridge: Cambridge University Press, 2001, 13–50.

Lo, V. "Spirit of Stone: Technical Considerations in the Treatment of the Jade Body," *Bulletin of the School of Oriental and African Studies* 65/1 (2002), 99–128.

Lo, V. "The Han Period," in T. J. Hinrichs and L. L. Barnes (eds), *Chinese Medicine and Healing: An Illustrated History*. Cambridge, MA: Belknap Press of Harvard University Press, 2013, 31–64.

Lo 羅維前, V. and Li Jianmin 李建民. "Manuscripts, Received Texts, and the Healing Arts," in M. Nylan and M. Loewe (eds), *China's Early Empires: A Re-Appraisal*. Cambridge: Cambridge University Press, 2010, 367–97.

Loewe, M. "The Religious and Intellectual Background," in D. C. Twitchett and M. Loewe (eds), *The Cambridge History of China*. Vol. 1: *The Ch'in and Han Empires, 221 B.C.–A.D.* 220. Cambridge: Cambridge University Press, 1986, 649–725.

Lindeman Nelson, J. "Health and Disease as 'Thick' Concepts in Ecosystemic Contexts," *Environmental Values* 4 (1995), 311–22.

Lloyd, H. M. *The Discourse of Sensibility: The Knowing Body in the Enlightenment*. New York: Springer, 2013.

Loemker, L. E. (ed. and trans.), *Philosophical Papers and Letters: A Selection*. Dordrecht: D. Reidel, 1969.

Ma Jixing 馬繼興, *Mawangdui gu yishu kaoshi* 馬王堆古醫書考釋. Changsha: Hunan kexue jishu chubanshe, 1992.

Mair, V. H., ed. *Experimental Essays on Chuang-Tzu*. Honolulu: University of Hawaii Press, 1983.

Malinowski, J. and C. Horton. "Metaphor and Hyperassociativity: The Imagination Mechanisms behind Emotion Assimilation in Sleep and Dreaming," *Frontiers in Psychology* 6 (2015), art. 1132.

Maltsberger, J. T. "Self Break-up and the Descent into Suicide," in S. Briggs, A. Lemma and W. Crouch (eds), *Relating to Self Harm and Suicide: Psychoanalytic Perspectives on Practice, Theory and Prevention*. London: Routledge 2008, 38–44.

Manning, G. "Descartes' Health Machines and the Human Exception," in D. Garber and S. Roux (eds), *The Mechanization of Natural Philosophy*. Dordrecht: Springer, 2013, 237–62.

Manning, G. "Descartes and the Bologna Affair," *British Journal for the History of Science* 47 (2014), 1–13.

Mansfeld, J. "Plato and the Method of Hippocrates," *Greek, Roman and Byzantine Studies* 2 (1980), 341–62.

Margolis, J. "The Concept of Disease," *Journal of Medicine and Philosophy* 1 (1976), 238–55.

Martin, M. "Malady and Menopause," *Journal of Medicine and Philosophy* 10 (1985), 329–37.

Maspero, H., Kierman, F. A. (trans.): *Taoism and Chinese Religion*. Amherst: University of Massachusetts Press, 1981.

McDonald, G. C. "Concepts and Treatment of Phrenitis in Ancient Medicine," PhD diss., Newcastle University, 2009.

McNally, R. J. *Panic Disorder: A Conceptual Analysis*. New York: Guilford, 1994.

Megone, C. "Aristotle's Function Argument and the Concept of Mental Illness," *Philosophy, Psychiatry and Psychology* 5 (1998), 187–201.

Megone, C. "Mental Illness, Human Function and Values," *Philosophy, Psychiatry and Psychology* 7 (2000), 45–65.

Menn, S. *Descartes and Augustine*. Cambridge: Cambridge University Press, 2002.

Mercer, C. *Leibniz's Metaphysics: Its Origins and Development*. Cambridge: Cambridge University Press, 2001.

Mercer, C. "Platonism in Early Modern Natural Philosophy: The Case of Leibniz and Conway," in C. Horn and J. Wilberding (eds), *Neoplatonic Natural Philosophy*. Oxford: Oxford University Press, 2012, 103–26.

Mewaldt, J. *Galeni In Hippocratis De natura hominis commentaria*. Leipzig: Teubner, 1914.

Miller, G. *The Adoption of Inoculation for Smallpox in England and France*. Philadelphia: University of Pennsylvania Press, 1957.

Miller, J. and B. Inwood (eds), *Hellenistic and Early Modern Philosophy*. Cambridge: Cambridge University Press, 2003.

Millikan, R. G. *Language, Truth and Other Biological Categories*. Cambridge, MA: MIT Press, 1984.

Millikan, R. G. "In Defense of Proper Functions," *Philosophy of Science* 56 (1989), 288–302.

Mills, S. "The Challenging Patient: Descartes and Princess Elisabeth on the Preservation of Health," *Journal of Early Modern Studies* 2 (2013), 101–22.

Morris, C. J. *Judith Scott: Bound and Unbound*. Brooklyn: Brooklyn Museum of Art, 2014.

Mosquera, D., A. Gonzalez, and A. Leeds, "Early Experience, Structural Dissociation, and Emotional Dysregulation in Borderline Personality Disorder: The Role of Insecure and Disorganized Attachment," *Borderline Personality Disorder and Emotion Dysregulation* 1/1 (2014), art. 15, 1–8.

Moss, J. "Appearances and Calculations: Plato's Division of the Soul," *Oxford Studies in Ancient Philosophy* 34 (2008), 35–68.

Mouy, P. *Le Développement de la physique Cartésienne*. Paris: Vrin, 1934, 73–85.

Mullan, J. *Sentiment and Sociability: The Language of Feeling in the Eighteenth Century*. Oxford: Clarendon Press, 1988.

Mulsow, M. *Frühneuzeitliche Selbsterhaltung: Telesio und die Naturphilosophie der Renaissance*. Tübingen: Niemeyer, 1998.

Munt, A. H. "The Impact of Dutch Cartesian Medical Reformers in Early Enlightenment German Culture (1680–1720)," PhD diss. University College London, 2005.

Murphy, D. "A Comparison of the Guodian and Mawangdui Laozi Texts," MA thesis, University of Massachusetts, 2006.

Murphy, D. and R. L. Woolfolk, "Conceptual Analysis versus Scientific Understanding: An Assessment of Wakefield's Folk Psychiatry," *Philosophy, Psychiatry and Psychology* 7 (2000), 271–93.

Murphy, D. and R. L. Woolfolk, "The Harmful Dysfunction Analysis of Mental Disorder," *Philosophy, Psychiatry and Psychology* 7 (2000), 241–52.

Nagel, T. *Mortal Questions*. Cambridge: Cambridge University Press, 1979.

Neander, K. "Functions as Selected Effects: The Conceptual Analyst's Defense," *Philosophy of Science* 58 (1991), 168–84.

Neander, K. "The Teleological Notion of 'Function,'" *Australasian Journal of Philosophy* 69 (1991), 454–68.

Needham, J. and Wang Ling, *Science and Civilisation in China*. Vol. 2. Cambridge: Cambridge University Press, 1956.

Ngo Van Xuyet and Fan Ye, *Divination, magie et politique dans la Chine ancienne: Essai*. Paris: Presses universitaires de France, 1976.

Nicholson-Smith, D. (trans.), *J. Laplanche and J.-B. Pontalis: The Language of Psychoanalysis*. London: W. W. Norton, 1973.

Nickerson, P. S. "The Great Petition for Sepulchral Plaints," in S. R. Bokenkamp (ed.), *Early Daoist Scriptures*. Berkeley: University of California Press, 1997, 230–60.

Nicoud, M. *Les régimes de santé au Moyen Âge: Naissance et diffusion d'une écriture médicale en Italie et en France (XIIIe–XVe siècle)*. 2 vols. Rome: École française de Rome, 2007.

Nordenfelt, L. *On the Nature of Health: An Action-Theoretic Approach*. Dordrecht: Reidel, 1987.

Nordenfelt, L. *Quality of Life, Health and Happiness*. Aldershot: Ashgate, 1993.

Nordenfelt, L. *Health, Science and Ordinary Language*. Amsterdam: Rodopi, 2001.

Nordenfelt, L. "The Concepts of Health and Illness Revisited," *Medicine, Health Care, and Philosophy* 10 (2007). 5–10.

Nordenfelt, L. "Establishing a Middle-Range Position in the Theory of Health: A Reply to My Critics," *Medicine, Health Care, and Philosophy* 10 (2007), 29–32.

Nussbaum, M. *The Therapy of Desire: Theory and Practice in Hellenistic Ethics*. Princeton, NJ: Princeton University Press, 1996.

Nutton, V. "The Rise of Medical Humanism: Ferrara, 1464–1555," *Renaissance Studies* 11 (1997), 2–19.

Nutton, V. "Avoiding Distress," in P. N. Singer, *Galen: Psychological Writings*. Cambridge: Cambridge University Press, 2014.

Nylan, M. "The 'Chin Wen/Ku Wen' Controversy in Han Times," *T'oung Pao* 80/1.3 (1994), 83–145.

Nylan, M. "*Yin-Yang*, Five Phases, and *Qi*," in M. Nylan and M. Loewe (eds), *China's Early Empires: A Re-Appraisal*. Cambridge: Cambridge University Press, 2010, 398–414.

O'Keefe, T. "The Cyrenaics on Pleasure, Happiness, and Future-Concern," *Phronesis* 47 (2002), 395–416.

Ogata Toru 大形 徹, "Rensen den ni miru doutokuteki sennin no houga『列仙傳』にみる道徳的仙人の萌芽," *Jinbungaku ronshou* 人文学論集 33 (2015), 29–38.

Onozawa Seiichi 小野沢精一, Fukunaga Mitsuji 福永光司, and Yamanoi Yū 山井湧 (eds), *Ki no shisō: Chūgoku ni okeru shizenkan to ningenkan no tenkai* 気の思想: 中国における自然観と人間観の展開. Tokyo: Tōkyō daigaku shuppankai, 1978.

Özkan, Z. *Die Psychomatik bei Abū Zaid al-Balḫī (gest. 934 AD)*. Frankfurt a.M.: Institute for the History of Arabic Islamic Science, 1990.

Overwien, O. *Die Sprüche des Kynikers Diogenes in der griechischen und arabischen Überlieferung*. Stuttgart: Franz Steiner, 2005.

Overwien, O. (ed.), "Zur Funktion der *Summaria Alexandrinorum* und der *Tabulae Vindobonenses*," in U. Schmitzer (ed.), *Enzyklodädie der Philologie: Themen und Methoden der klassischen Philologie heute*. Göttingen: Ruprecht, 2013, 187–207.

Packham, C. *Eighteenth-Century Vitalism: Bodies, Culture, Politics*. Houndsmills, UK: Palgrave Macmillan, 2012.

Papineau, D. "Mental Disorder, Illness and Biological Dysfunction," in A. Griffiths (ed.), *Philosophy, Psychology and Psychiatry*. Cambridge: Cambridge University Press, 1994, 73–82.

Park, K. *Secrets of Woman: Gender, Generation, and the Origins of Human Dissection*. Brooklyn: Zone Books, 2006.

Penny, B. "Immortality and Transcendence," in L. Kohn (ed.), *Daoism Handbook*. 109–33. Leiden: Brill, 2000.

Penny, B. "*Lie Xian Zhuan*," in F. Pregadio (ed.), *The Encyclopedia of Taoism*. London: Routledge, 2008, 653—54.

Perkins, F. "Metaphysics in Chinese Philosophy," in E. N. Zalta (ed.), *The Stanford Encyclopedia of Philosophy*, 2015.

Perlman, M. "The Modern Philosophical Resurrection of Teleology," *The Monist* 87 (2004), 3–51.

Perring, C. D. "Indeterminacy and Resentment," *Philosophy, Psychiatry, and Psychology* 17/3 (2010), 263–64.

Petocz, A. Freud, *Psychoanalysis, and Symbolism*. Cambridge: Cambridge University Press, 1999.

Pigeaud, J. *La maladie de l'âme: Étude sur la relation de l'âme et du corps dans la tradition médico-philosophique antique*. Paris: Belles Lettres, 1981, 2nd edition 1989.

Pormann, P. E. *Mirror of Health: Medical Science during the Golden Age of Islam*. London: Royal College of Physicians, 2013.

Pormann, P. E. "The Dispute between the Philarabic and Philhellenic Physicians and the Forgotten Heritage of Arabic Medicine," in P. E. Pormann, (ed.), *Islamic Medical and Scientific Tradition*. 4 vols. Routledge: London, 2011, 2:283–316.

Pormann, P. E. "Medical Methodology and Hospital Practice: The Case of Tenth-century Baghdad," in: P. Adamson (ed.), *In the Age of al-Fārābī: Arabic Philosophy in the 4th/10th Century*. London: Warburg Institute, 95–118; reprinted in P. E. Pormann, *Islamic Medical and Scientific Tradition*. 4 vols. Routledge: London, 2011, 2:179–206.

Pormann, P. E. and N. P. Joosse, "Commentaries on the Hippocratic *Aphorisms* in the Arabic Tradition: The Example of Melancholy," in P. E. Pormann (ed.), *Epidemics in Context: Greek Commentaries on Hippocrates in the Arabic Tradition*. Berlin: de Gruyter, 2012, 211–49.

Pormann, P. E. "The Formation of the Arabic Pharmacology: Between Tradition and Innovation," *Annals of Science* 68 (2011), 493–515.

Pormann, P. E. "Medical Education in Late Antiquity: From Alexandria to Montpellier," in H. F. J. Horstmanshoff and C. R. van Tilburg (eds), *Hippocrates and Medical Education: Selected Papers Read at the XIIth International Hippocrates Colloquium, Universiteit Leiden, 24–26 August 2005*. Leiden: Brill, 2010, 419–41.

Pormann, P. E. "Qualifying and Quantifying Medical Uncertainty in 10th-century Baghdad: Abu Bakr al-Razi," *Journal of the Royal Society of Medicine* 106 (2013), 370–2.

Pormann, P. E. and K. Karimullah, "The Arabic Commentaries on the Hippocratic *Aphorisms*: Introduction," *Oriens* 45/1–2 (2017), 1–52.

Pormann, P. E. and E. Selove, "Two New Texts on Medicine and Natural Philosophy by Abū Bakr Muḥammad ibn Zakarīyā' al-Rāzī," *Journal of the American Oriental Society* 137 (2017), 279–99.

Porter, R. "Cleaning Up the Great Wen: Public Health in Eighteenth-Century London," *Medical History*, suppl. 11 (1991), 61–75.

Porter, R. *Disease, Medicine and Society in England, 1550–1860*. Cambridge: Cambridge University Press, 1992, 17–26.

Porter, D. *Health, Civilization, and the State: A History of Public Health from Ancient to Modern Times*. London: Routledge, 1999.

Po-tuan, Chang (ed.), *Introduction à L'alchimie intérieure Taoïste: De l'unité et de la multiplicité*, translated by I. Robinet. Paris: Editions du Cerf, 1995.

Pregadio, F. "Destiny, Vital Force, or Existence? On the Meanings of *Ming* in Daoist Internal Alchemy and Its Relation to *Xing* or Human Nature," *Daoism: Religion, History & Society* 6 (2014), 157–218.

Puzzlio, G. et al. "Active Inference, Homeosatis Regulation, and Adaptive Behavioural Control," *Progress in Neurobiology* 134 (2015), 17–35.

Rankin, A. *Panaceia's Daughters: Noblewomen as Healers in Early Modern Germany*. Chicago: University of Chicago Press, 2013.

Raphals, L. "Craft Analogies in Chinese and Greek Argumentation," in E. Ziolkowski (ed.), *Literature, Religion, and East-West Comparison: Essays in Honor of Anthony C. Yu*. Wilmington: University of Delaware, 2005, 181–201.

Raphals, L. "Chinese Philosophy and Chinese Medicine," in E. N. Zalta (ed.), *The Stanford Encyclopedia of Philosophy*, 2015.

Rather, L. J. "The 'Six Things Non-Natural': A Note on the Origins and Fate of a Doctrine and Phrase," *Clio Medica* 3 (1968), 337–47.

Reill, P. H. *Vitalizing Nature in the Enlightenment*. Berkeley: University of California Press, 2005.

Reissland, N. et al. "Do Facial Expressions Develop before Birth?" *PLoS ONE* 6/8 (2011), e24081.

Reissland, N., B. Francis, and J. Mason. "Can Healthy Fetuses Show Facial Expressions of 'Pain' or 'Distress'?" *PLoS ONE* 8/6 (2013), e65530.

Reznek, L. *The Nature of Disease*. London: Routledge, 1987.

Richardson, G. *Iconology: or, A Collection of Emblematical Figures, Containing Four Hunded and Twenty-Four Remarkable Subjects, Moral and Instructive; in Which Are Displayed the Beauty of Virtue and Deformity of Vice*. London: Printed for the author by G. Scott, 1779.

Rickett, W. A. "Kuan Tzu 管子," in M. Loewe (ed.), *Early Chinese Texts: A Bibliographical Guide*. Berkeley: Society for the Study of Early China, 1993, 244–51.

Rickett, W. A. *Guanzi: Political, Economic, and Philosophical Essays from Early China: A Study and Translation*. 2 vols. Princeton, NJ: Princeton University Press, 1998.

Riley, J. C. *Population Thought in the Age of the Demographic Revolution*. Durham: Carolina Academic Press, 1985.

Riskin, J. *Science in the Age of Sensibility: The Sentimental Empiricists of the French Enlightenment*. Chicago: University of Chicago Press, 2002.

Rodis-Lewis, G. "Descartes and the Unity of the Human Being," in J. Cottingham (ed.), *Descartes*. Oxford: Oxford University Press, 1998, 197–210.

Rosenthal, F. *The Classical Heritage in Islam*. London: Routledge, 1994.

Roth, H. D. "Chuang Tzu 莊子," in M. Loewe (ed.), *Early Chinese Texts: A Bibliographical Guide*. Berkeley: Society for the Study of Early China, 1993, 56–66.

Roth, H. D. *Original Tao: Inward Training and the Foundations of Taoist Mysticism*. New York: Columbia University Press, 1999.

Rusnock, A. *Vital Accounts: Quantifying Health and Population in Eighteenth-Century England and France*. Cambridge: Cambridge University Press, 2002.

Saban, R. "Les premières représentations anatomiques des squelettes humain imprimées en Alsace au XVe siècle," *113e Congrès nationale des sociétés savantes 1988, Questions de l'histoire de la médecine* (1991), 27–46.

Sakade Y. *Taoism, Medicine and Qi in China and Japan*. Osaka: Kansai University Press, 2007.

Saks, E. *The Center Cannot Hold: My Journey Through Madness*. New York: Hyperion, 2008.

Salguero, C. P. "Fields of Merit, Harvests of Health: Some Notes on the Role of Medical Karma in the Popularization of Buddhism in Early Medieval China," *Asian Philosophy* 23/4 (2013), 341–49.

Salguero, C. P. *Translating Buddhist Medicine in Medieval China*. Philadelphia: University of Pennsylvania Press, 2014.

Scadding, J. G. "Health and Disease: What Can Medicine Do for Philosophy?" *Journal of Medical Ethics* 14 (1988), 118–24.

Scadding, J. G. "The Semantic Problem of Psychiatry," *Psychological Medicine* 20 (1990), 243–48.

Schäfer, D. *The Crafting of the 10,000 Things: Knowledge and Technology in Seventeenth-Century China*. Chicago: University of Chicago Press, 2011.

Schaps, D. "The Woman Least Mentioned: Etiquette and Women's Names," *Classical Quarterly* 27/2 (1977), 323–30.

Schipper, K. M. "The Inner World of the Lao-Tzu Chung-Ching," in Chun-chieh Huang and E. Zurcher (eds), *Time and Space in Chinese Culture*. Leiden: Brill, 1995, 114–31.

Schipper, K. M. and Franciscus V. (eds.), *The Taoist Canon: A Historical Companion to the Daozang*. Chicago: University of Chicago Press, 2004.

Schmaltz, T. *Early Modern Cartesianisms: Dutch and French Constructions*. Oxford: Oxford University Press, 2016.

Schmitter, A. "Responses to Vulnerability: Medicine, Politics and the Body in Descartes and Spinoza," in S. Pender and N. S. Struever (eds), *Rhetoric and Medicine in Early Modern Europe*. Surrey: Ashgate, 2012, 141–71.

Schoenfeldt, M. "Aesthetics and Anesthetics: The Art of Pain Management in Early Modern England," in J. Frans van Dijkhuizen and K. A. E. Enenkel (eds),

The Sense of Suffering: Constructions of Physical Pain in Early Modern Culture. Leiden: Brill, 2009, 19–38.

Schofield, M. "Academic Therapy: Philo of Larissa and Cicero's Project in the Tusculans," in G. Clark and T. Rajak (eds), *Philosophy and Power in the Graeco-Roman World: Essays in Honour of Miriam Griffin.* Oxford: Oxford University Press, 2002, 91–109.

Schöpf, V. et al. "The Relationship between Eye Movement and Vision Develops before Birth," *Frontiers in Human Neuroscience* 8 (2014), 775.

Schramme, T. "A Qualified Defence of a Naturalist Theory of Health," *Medicine, Health Care, and Philosophy* 10 (2007), 11–17.

Schwartz, P. "Defining Dysfunction: Natural Selection, Design, and Drawing a Line," *Philosophy of Science* 74 (2007), 364–85.

Skues, R. *Sigmund Freud and the History of Anna O.* Houndmills, UK: Palgrave Macmillan, 2006.

Shapiro, L. "The Health of the Body Machine? Or Seventeenth-Century Mechanism and the Concept of Health," *Perspectives on Science* 11 (2003), 421–42.

Shapiro, L. "Descartes on Human Nature and the Human Good," in C. Fraenkel, D. Perinetti, and J. E. H. Smith (eds), *The Rationalists: Between Tradition and Innovation.* Dordrecht: Springer, 2011, 13–26.

Sherrer, G. B. "Philalgia in Warwickshire: F. M. van Helmont's Anatomy of Pain Applied to Lady Anne Conway," *Studies in the Renaissance* 5 (1958), 196–206.

Simmons, A. "Sensible Ends: Latent Teleology in Descartes' Account of Sensation," *Journal of the History of Philosophy* 39 (2001), 49–75.

Simons, J. "Beyond Naturalism and Normativism: Reconceiving the 'Disease' Debate," *Philosophical Papers* 36 (2007), 343–70.

Singer, C. "Galen's Elementary Course on Bones," *Proceedings of the Royal Society of Medicine* 45 (1952), 767–76.

Siraisi, N. G. *Avicenna in Renaissance Italy: The Canon and Medical Teaching in Italian Universities* after 1500. Princeton, NJ: Princeton University Press, 1987.

Siraisi, N. G. *History, Medicine, and the Traditions of Renaissance Learning.* Ann Arbor: University of Michigan Press, 2007.

Siraisi, N. G. *Communities of Learned Experience: Epistolary Medicine in the Renaissance.* Baltimore, MD: Johns Hopkins University Press, 2013.

Sivin, N. "On the Word 'Taoist' as a Source of Perplexity: With Special Reference to the Relations of Science and Religion in Traditional China," *History of Religions* 17/3–4 (1978), 303–30.

Sivin, N. "Why the Scientific Revolution Did Not Take Place in China—or Didn't It?" *Chinese Science* 5 (1982), 45–66.

Sivin, N. "Science and Medicine in Chinese History," in P. S. Ropp and T. H. Barrett (eds), *Heritage of China: Contemporary Perspectives on Chinese Civilization*. Berkeley: University of California Press, 1990.

Sivin, N. "Huang Ti Nei Ching 黃帝內經," in M. Loewe (ed.), *Early Chinese Texts: A Bibliographical Guide*. Berkeley: Society for the Study of Early China, 1993, 196–215.

Sivin, N. "State, Cosmos, and Body in the Last Three Centuries B.C.," *Harvard Journal of Asiatic Studies* 55/1 (1995), 5–37.

Sivin, N. "Taoism and Science," in *Medicine, Philosophy and Religion in Ancient China*. Aldershot: Variorum, 1995, 303–30.

Skenazi, C. *Aging Gracefully in the Renaissance: Stories of Later Life from Petrarch to Montaigne*. Leiden: Brill, 2013.

Smith, J. E. H. *Divine Machines: Leibniz and the Life Sciences*. Princeton, NJ: Princeton University Press, 2011.

Smith, J. E. H. "Heat, Action, Perception: Models of Living Beings in German Medical Cartesianism," in M. Dobre and T. Nyden (eds), *Cartesian Empiricisms*. Dordrecht: Springer, 2013, 105–24.

Smith, J. E. H. "Early Modern Medical Eudaimonism," in P. Distelzweig et al. (eds), *Early Modern Medicine and Natural Philosophy*. Dordrecht: Springer, 2015, 325–41.

Smith, J. Z. "On Comparison," in *Drudgery Divine: On the Comparison of Early Christianities and the Religions of Late Antiquity*. Chicago: University of Chicago Press, 1990, 36–53.

Spitzer, R. L. and J. Endicott, "Medical and Mental Disorder: Proposed Definition and Criteria," in R. L. Spitzer and D. F. Klein (eds), *Critical Issues in Psychiatric Diagnosis*. New York: Raven Press, 1978, 15–40.

Stanley-Baker, M. "Cultivating Body, Cultivating Self: A Critical Translation and History of the Tang Dynasty *Yangxing Yanming Lu* (Records of Cultivating Nature and Extending Life)," MA thesis, Indiana University, Bloomington, 2006.

Stanley-Baker, M. "Daoists and Doctors: The Role of Medicine in Six Dynasties Shangqing Daoism," PhD diss., University College London, 2013.

Stanley-Baker, M. "Drugs, Destiny, and Disease in Medieval China: Situating Knowledge in Context," *Daoism: Religion, History and Society* 6 (2014), 113–56.

Steger, F. "Antike Diätetik—Lebensweise und Medizin," *NTM Zeitschrift für Geschichte der Wissenschaften, Technik und Medizin* 12 (2004), 146–60.

Steinke, H. and M. Stuber, "Medical Correspondence in Early Modern Europe. An Introduction," *Gesnerus* 61 (2004), 139–60.

Steinschneider, M. *Die hebraeischen Uebersetzungen des Mittelalters und die Juden als Dolmetscher.* Berlin: Kommissionsverl. des Bibliographischen Bureaus, 1893.

Storms, G. *Anglo-Saxon Magic.* Halle: Jijhoff, 1948. Repr. Folcroft: Folcroft Library Editions, 1975.

Strazzoni, A. "A Logic to End Controversies: The Genesis of Clauberg's *Logica Vetus et Nova,*" *Journal of Early Modern Studies* 2 (2013), 123–49.

Strickmann, M. *Chinese Magical Medicine.* Stanford: Stanford University Press, 2002.

Strohmaier, G. "Die Ethik Galens und ihre Rezeption in der Welt des Islams," in J. Barnes and J. Jouanna (eds), *Galien et la Philosophie.* Vandoeuvres: Fondation Hardt, 2003, 307–29.

Sullivan, M. "In What Sense Is Contemporary Medicine Dualistic?," *Culture, Medicine and Psychiatry* 10/4 (1986), 331–50.

Switankowsky, I. "Dualism and Its Importance for Medicine," *Theoretical Medicine* 21 (2000), 567–80.

Szabó, S. P. "The Term Shenming: Its Meaning in the Ancient Chinese Thought and in a Recently Discovered Manuscript," *Acta Orientalia Academiae Scientiarum Hungaricae* 56/2–4 (2003), 251–74.

Szasz, T. S. "The Myth of Mental Illness," *American Psychologist* 15 (1960), 113–18.

Szasz, T. S. *The Myth of Mental Illness.* London: Paladin, 1972.

Tavor, O. "Embodying the Way: Bio-Spiritual Practices and Ritual Theories in Early and Medieval China," PhD diss., University of Pennsylvania, 2012.

Tecusan, M. *The Fragments of the Methodists.* Leiden: Brill, 2004.

Temkin, O. *Galenism: Rise and Decline of a Medical Philosophy.* Ithaca, NY: Cornell University Press, 1973.

Thomas, K. *Ends of Life: Roads to Fulfillment in Early Modern England.* Oxford: Oxford University Press, 2009.

Toombs, S. K. "Illness and the Paradigm of Lived Body," *Theoretical Medicine* 9/2 (1988), 201–26.

Trevisani, F. *Descartes in Germania: La ricezione del Cartesianesimo nella Facolta filosofica e medica di Duisburg (1652–1703).* Milano: F. Angeli, 1992.

Trevisani, F. *Descartes in Deutschland: Die Rezeption des Cartesianismus in den Hochschulen Nordwestdeutschlands.* Wien: LIT, 2011.

Unschuld, P. U. *Huangdi Neijing Suwen: Nature, Knowledge, Imagery in an Ancient Chinese Medical Text.* Berkeley: University of California Press, 2003.

Unschuld, P. U., and H. Tessenow. *Huangdi Neijing Suwen: An Annotated Translation of Huang Di's Inner Classic—Basic Questions.* 2 vols. Berkeley: University of California Press, 2011.

Van Helmont, F. M. (ed.), *J. B. van Helmont: Sextuplex digestio alimenti humani*, in F. M. van Helmont (ed.), *Ortus medicinae, id est, initia physicae unaudita*. Amsterdam: Lodewijk Elzevier, 1648, 208–25.

Verbeek, T. *Descartes and the Dutch: Early Receptions to Cartesian Philosophy, 1637–1650*. Carbondale: Southern Illinois University Press, 1992.

Verbeek, T. "Tradition and Novelty: Descartes and Some Cartesians," in T. Sorell (ed.), *The Rise of Modern Philosophy: The Tension between the New and Traditional Philosophies from Machiavelli to Leibniz*. Oxford: Clarendon, 1993, 167–75.

Vila, A. *Enlightenment and Pathology: Sensibility in the Literature and Medicine of Eighteenth-Century France*. Baltimore, MD: Johns Hopkins University Press, 1998.

Voelke, A.-J. "Soigner par le Logos: La therapeutique de Sextus Empiricus," in A.-J. Voelke (ed.), *Le Scepticisme antique: Perspectives historiques et systématiques*. Geneva: Cahiers de la revue de Théologie et Philosophie, 1990.

Volkan, V. *A Nazi Legacy: Depositing, Transgenerational Transmission, Dissociation, and Remembering Through Action*. London: Karnac Books, 2015.

Von Staden, H. *Herophilus: The Art of Medicine in Early Alexandria*. Cambridge: Cambridge University Press, 1989.

Von Staden, H. "'In a Pure and Holy Way': Personal and Professional Conduct in the Hippocratic Oath?" *Journal of the History of Medicine and Allied Sciences* 51 (1995), 404–37.

Voss, S. "Descartes: Heart and Soul," in J. P. Wright and P. Potter (eds), *Psyche and Soma: Physicians and Metaphysicians on the Mind-Body Problem from Antiquity to Enlightenment*. Oxford: Oxford University Press, 2000, 173–96.

Wakefield, J. C. "The Concept of Medical Disorder: On the Boundary between Biological Facts and Social Values," *American Psychologist* 47 (1992), 373–88.

Wakefield, J. C. "Disorder as Harmful Dysfunction: A Conceptual Critique of DSM-III-R's Definition of Mental Disorder," *Psychological Review* 99 (1992), 232–47.

Wakefield, J. C. "Dysfunction as a Value-Free Concept: A Reply to Sadler and Agich," *Philosophy, Psychiatry and Psychology* 2 (1995), 233–46.

Wakefield, J. C. "Evolutionary versus Prototype Analyses of the Concept of Disorder," *Journal of Abnormal Psychology* 108 (1999), 374–99.

Wakefield, J. C. "Mental Disorder as a Black Box Essentialist Concept," *Journal of Abnormal Psychology* 108 (1999), 465–72.

Wakefield, J. C. "Spandrels, Vestigial Organs, and Such: Reply to Murphy and Woolfolk's 'The Harmful Dysfunction Analysis of Mental Disorder,'" *Philosophy, Psychiatry and Psychology* 7 (2000), 253–69.

Wakefield, J. C. "Taking Disorder Seriously: A Critique of Psychiatric Criteria for Mental Disorders from the Harmful-Dysfunction Perspective," in T. Millon, R. F. Krueger, and E. Simonsen (eds), *Contemporary Directions in Psychopathology: Scientific Foundations of the DSM-V and ICD-11*. New York: Guilford Press, 2010, 275–302.

Wakefield, J. C. "The Biostatistical Theory versus the Harmful Dysfunction Analysis, Part 1: Is Part-Dysfunction a Sufficient Condition for Medical Disorder?" *Journal of Medicine and Philosophy* 39 (2014), 648–82.

Wakefield, J. C. and J. C. Baer, "The Cognitivization of Psychoanalysis: Toward an Integration of Psychodynamic and Cognitive Theories," in W. Bordon (ed.), *Reshaping Theory in Contemporary Social Work: Toward a Critical Pluralism in Clinical Practice*. New York: Columbia University Press, 2010, 51–80.

Wallis, F. *Medieval Medicine: A Reader*. Toronto: University of Toronto Press, 2010.

Walzer, R. "New Light on Galen's Moral Philosophy," *The Classical Quarterly* 43 (1949), 82–96.

Wang Aihe. *Cosmology and Political Culture in Early China*. Cambridge: Cambridge University Press, 2006.

Watt, D. F. and J. Panksepp, "Depression: An Evolutionarily Conserved Mechanism to Terminate Separation Distress? A Review of Aminergic, Peptidergic, and Neural Network Perspectives," *Neuropsychoanalysis* 11/1 (2009), 7–51.

Wear, A. "Medicine in Early Modern Europe," in L. I. Conrad, et al. (eds), *The Western Medical Tradition: 800 BC to AD 1800*. Cambridge: Cambridge University Press, 1995, 215–362.

Wear, A. "Place, Health, and Disease: The Airs, Waters, Places Tradition in Early Modern England and North America," *Journal of Medieval and Early Modern Studies* 38 (2008), 443–65.

Weaver, H. (trans.), *Philippe Ariès: The Hour of our Death*. New York: Alfred A. Knopf, 1981.

Weisser, O. *Ill Composed: Sickness, Gender, and Belief in Early Modern England*. New Haven, CT: Yale University Press, 2016.

Whitbeck, C. "Four Basic Concepts of Medical Science," *PSA: Proceedings of the Biennial Meeting of the Philosophy of Science Association* 1 (1978), 210–22.

Wilhelm, H. "Eine Chou-Inschrift über Atemtechnik," *Monumenta Serica* 13 (1948), 385.

Wilkinson, E. P. *Chinese History: A New Manual*. Cambridge, MA: Harvard University Asia Center, 2012.

Williams, B. *Moral Luck*. Cambridge: Cambridge University Press, 1981.

Williams, E. A. *A Cultural History of Medical Vitalism in Enlightenment Montpellier*. Burlington, VT: Ashgate, 2003.

Williston, B. and A. Gombay (eds), *Passion and Virtue in Descartes*. Amherst, MA: Humanity Books, 2003.

Wilms, S. "Nurturing Life in Classical Chinese Medicine: Sun Simiao on Healing without Drugs, Transforming Bodies and Cultivating Life," *Journal of Chinese Medicine* 93 (2010), 5–13.

Wilson, A. *The Making of Man-Midwifery: Childbirth in England, 1660–1770*. Cambridge, MA: Harvard University Press, 1995.

Zhang Ruixian 張瑞賢, Wang Jiakui 王家葵, and M. 徐源 Stanley-Baker, "The Earliest Stone Medical Inscription," in V. Lo (ed.), *Imagining Chinese Medicine*. Leiden: Brill, 2017, 373–88.

Zhou Yimou 周一謀, and Xiao Zuotao 蕭佐桃, *Mawangdui yi shu kao zhu* 馬王堆醫書考注. Tianjin: Tianjin kexue jishu chubanshe, 1988.

Zhou Yimou 周一谋, *Mawangdui yixue wenhua* 马王堆医学文化. Shanghai: Wenhui chubanshe, 1994.

Index of Terms

Index of Authors and Works